SUSTAINABLE GROUNDWATER RESOURCES IN AFRICA

Sustainable Groundwater Resources in Africa

Water supply and sanitation environment

Editors

Yongxin Xu & Eberhard Braune

UNESCO Chair in Hydrogeology
University of the Western Cape, South Africa

CRC Press
Taylor & Francis Group
Boca Raton   London   New York   Leiden

CRC Press is an imprint of the
Taylor & Francis Group, an **informa** business

A BALKEMA BOOK

Copyright © 2010 UNESCO

Published jointly by:

United Nations
Educational, Scientific and
Cultural Organization

International
Hydrological
Programme

UNESCO
UNESCO International Hydrological Programme (IHP)
1 rue Miollis
75732 Paris Cedex 15
France
Tel: 33 (0) 1 4568 4004
Fax: 33 (0) 1 4568 5811
Website: http://www.unesco.org/water

and

CRC Press/Balkema
P.O.Box 447, 2300 AK Leiden, The Netherlands
e-mail: Pub.NL@taylorandfrancis.com
www.crcpress.com – www.balkema.nl – www.taylorandfrancis.com – www.routledge.com

First issued in paperback 2017

CRC Press/Balkema is an imprint of the Taylor & Francis Group, an informa business, London, UK

Typeset by Vikatan Publishing Solutions (P) Ltd., Chennai, India

Library of Congress Cataloging-in-Publication Data

Sustainable groundwater resources in Africa : water supply and sanitation environment /
editors, Yongxin Xu & Eberhard Braune.
    p. cm.
  "Published jointly by UNESCO."
  Includes bibliographical references.
  ISBN 978-0-415-87603-2 (hardcover : alk. paper) 1. Water resources development --
Africa. 2. Groundwater -- Africa. 3. Water-supply -- Africa. 4. Sanitation -- Africa. 5. Human
ecology -- Africa. I. Xu, Y. (Yongxin) II. Braune, E. III. Unesco.

  HD1699.A1S87 2010
  333.91'04096--dc22

                                2009036650

ISBN 13: 978-1-138-11185-1 (pbk)
ISBN 13: 978-0-415-87603-2 (hbk)

Table of contents

List of figures

I – *Best practice guidelines*

II – *Case studies*

List of tables

Preface

Water for Life!

What depth and breadth of meaning lies in these few words—from the religious to the scientific, to the developmental. No wonder that the UN family chose these words when they declared

'Water for Life 2005–2015 an International Decade for Action'.

Water is crucial for sustainable development, including the preservation of our natural environment and the alleviation of poverty and hunger. Water is indispensable for human health and well-being. The global concern is that poor sanitation and water supplies are the engines that drive cycles of disease, poverty and powerlessness in developing nations. Action to improve this situation must be seen as an essential step to enable the poorest people to escape poverty.

Therefore a decade of action! The primary goal of the 'Water for Life' Decade is to promote efforts to fulfill international commitments, in particular the Millennium Development Goals (MDGs), made on water and water-related issues by 2015. In addition, to get the seriously lagging sanitation back on track towards achieving its 2015 goal, 2008 was declared as the International Year of Sanitation.

In this situation, it pleases me greatly to see UNESCOs involvement in this book, 'Sustainable groundwater resources—within the water supply and sanitation environment in Africa' that tries to link the important contributing topics of water and sanitation services and water resources management. Also important for me is that the spotlight is put on Africa, which lags behind other regions in achieving the MDGs, and on groundwater, continuing to be the much neglected water resource.

The aim of the book is to illustrate the issues in achieving sustainable groundwater supplies in the challenging local community environment and it does this through a variety of case studies from across the continent, contrasted with groundwater best implementation practice. In trying to find a suitable framework for integrated water, sanitation and hygiene delivery in which these best practices could be rolled out effectively and efficiently, the authors make some interesting observations.

"Logically, the IWRM approach should achieve this integration, but both sanitation and sustainable groundwater resource utilization do not yet fit comfortably under this approach. ... Historically, an emphasis on technical aspects within this extremely complex system has been at the expense of the hydro-social and public health components."

Empowering people, reducing poverty, improving livelihoods and promoting economic growth, while at the same time ensuring sustainable ecosystems, requires local, appropriate and widely replicable solutions. The big challenge is that these are not just groundwater- as well as sanitation-technical solutions, but major societal and institutional ones.

What a challenge for the UNESCO IHP theme 'Water and society' unfolding, what a challenge for UNESCOs Science and Education arms working together and what a challenge for water and health sectors as a whole and the respective UN family members working together for 'Water for Life'.

The book is timely and, while not yet offering comprehensive solutions, it provides many pointers, which I encourage readers, in particular decision-makers, to consider and develop further. It will need the coordinated effort of many to address the 'hydro-social' challenge.

Professor Kader Asmal
University of the Western Cape
Bellville, South Africa

Strategic Advisory Group on Water and Sanitation
UN Children's Fund and World Health Organisation

Acknowledgements

The authors of this book place emphasis on two aspects: the first aspect deals with roles of groundwater resources in the water supply and sanitation provision environment in Africa and advocates a range of best practice measures aimed at sustainable utilization of groundwater for community growth and development under the framework of IWRM. The second aspect makes use of case studies to paint a picture of current status and some pitfall experiences the hydrogeologists went through and to highlight some important issues and challenges still facing Africa, especially in west, central, eastern and southern Africa. All the authors and their affiliated organisations are gratefully acknowledged for their valuable, scientific and technological contributions.

Especially, our gratitude goes to Prof. Dr. Ken R. Bradbury, Mr. Phil Hobbs, Mr. Thokozani Kanyerere, Dr. Nick Robins who were invited by the editor to provide expert inputs on the specific topics, i.e., Borehole Construction (Hobbs), Protection Zoning in Fractured Rock Aquifers (Bradbury), Drought Proofing and Spring Protection (Robins), and Ecosan Approach (Kanyerere).

The book has benefitted greatly from a peer-review process. The following reviewers are thanked for their critical comments and constructive inputs to the improvement of the quality of the manuscripts initially submitted to the editor of the book:

- the reviewers from Africa: Shafick Adams (South Africa), Pete Ashton (South Africa), Zayed Brown (South Africa), Josue B Chishugi (D.R.Congo), Phil Hobbs (South Africa), Thokazani Kanyerere (Malawi), Roger Parsons (South Africa), Yasmin Rajkumar (South Africa);
- the reviewers from overseas: Ken R Bradbury (USA), Kerstin Danert (Switzerland), Jonathan Levy (USA), Walt Kelly (USA), Alex K Makarigakis (UNESCO Addis Ababa), Nick Robins (UK), Yong Wu (China).

Publishing of this book is made possible through financial support from the UNESCO IHP and the Flemish Government. Dr. Alex Makarigakis, Mr. Ernesto Fernandez Polcuch and Dr. Alice Aureli, all from the UNESCO, are especially thanked for their advice and assistance. The need for such a book was identified and initiated through the Framework for Education and Training in Water (FET Water) programme in South Africa by Dr. Alex Makarigakis, whose encouragement and support during the conceptualisation, drafting, compilation and editing of this book is greatly appreciated.

We would like to acknowledge the University of the Western Cape for its support and the constructive discussions with our postgraduates and colleagues through the network of the UNESCO Chair in Hydrogeology on a wide range of issues regarding groundwater sustainable utilization in Africa. Messrs Josue Chishugi and Jaco Nel of the UNESCO Chair at the University of the Western Cape provided proof reading and technical support to the production of this book.

Professor Dr. Yongxin Xu
and
Professor Dr. Eberhard Braune

UNESCO Chair in Hydrogeology
University of the Western Cape
Bellville, South Africa

List of contributors

T.A. Abiye	*School of Geosciences, University of the Witwatersrand, Johannesburg, South Africa*
D. Adekile	*Water Surveys and Resources Development Limited, Kaduna, Nigeria*
A. Alassane	*Department of Earth Sciences, Faculty of Sciences and Technology, University of Abomey-Calavi, Cotonou, Benin*
B.F. Alemaw	*Department of Geology, University of Botswana, Gaborone, Botswana*
J. Anscombe	*GITEC Consult, Zambia*
M. Boukari	*Department of Earth Sciences, Faculty of Sciences and Technology, University of Abomey-Calavi, Cotonou, Benin*
K.R. Bradbury	*Wisconsin Geological and Natural History Survey, Madison, Wisconsin, USA*
E. Braune	*Department of Earth Sciences, University of the Western Cape, South Africa*
R.C. Carter	*Water Aid, London, UK*
T.R. Chaoka	*Department of Geology, University of Botswana, Gaborone, Botswana*
J.B. Chishugi	*Department of Earth Sciences, University of the Western Cape, South Africa and Department of Geology and Mineralogy, Official University of Bukavu, South Kivu, D.R. Congo*
P.E. Crane	*Department of Civil Engineering and Geological Sciences, University of Notre Dame, Notre Dame, Indiana, USA*
K. Danert	*SKAT, St Gallen, Switzerland*
A. Faye	*Department of Geology, Faculty of Sciences and Technology, Cheikh Anta Diop University – Campus of Dakar, Senegal*
S.C. Faye	*Department of Geology, Faculty of Sciences and Technology, Cheikh Anta Diop University – Campus of Dakar, Senegal*
L. Feenstra	*TNO Built Environment and Geosciences, The Netherlands*
J. Griffioen	*DELTARES, The Netherlands*
P. Hobbs	*Council for Scientific and Industrial Research, Pretoria, South Africa*
M. Holland	*Department of Geology, University of Pretoria, Pretoria, South Africa*
L. Kanowa	*Ministry of Local Government and Housing, Zambia*
T.O.B. Kanyerere	*Department of Earth Sciences, University of the Western Cape, South Africa and Department of Geography and Earth Sciences, University of Malawi, Malawi*
M. Karen	*Earth Science Systems, Botswana*
R.C. Leyland	*Department of Geology, University of Pretoria, Pretoria, South Africa*
A. MacDonald	*British Geological survey, Edinburg, United Kingdom*
G. Mahed	*Council for Geoscience, Pretoria, South Africa*
A.K. Makarigakis	*Science Programme Specialist, UNESCO Addis Ababa Office, Addis Ababa, Ethiopia*

T. Mkandawire	*Department of Water Development, Ministry of Irrigation and Water Development, Malawi*
J.M. Nel	*Department of Earth Sciences, University of the Western Cape, South Africa*
M.G.M. Nkhata	*Department of Water Development, Ministry of Irrigation and Water Development, Malawi*
Y. Rajkumar	*Department of Water Affairs and Forestry, Durban, South Africa and Department of Earth Sciences, University of the Western Cape, South Africa*
N.S. Robins	*British Geological Survey, Maclean Building, Wallingford, Oxfordshire, UK*
S.E. Silliman	*Department of Civil Engineering and Geological Sciences, University of Notre Dame, Notre Dame, Indiana, USA*
S. Vasak	*International Groundwater Resources Assessment Centre (IGRAC), Utrecht, The Netherlands*
M.A. Wienecke	*Habitat Research & Development Centre, Windhoek, Namibia*
K.T. Witthueser	*Department of Geology, University of Pretoria, Pretoria, South Africa*
Y. Xu	*Department of Earth Sciences at the University of the Western Cape, South Africa*

I
Best practice guidelines

1

Water supply and sanitation issues in Africa

Eberhard Braune
Department of Earth Sciences at the University of the Western Cape, South Africa

Alexandros K. Makarigakis
Science Programme Specialist, UNESCO, Addis Abeba

Governments in Africa face a twin challenge in terms of water supply and sanitation. Firstly to close the gap in rural areas where only two in five people have access to water supply and fewer than one in five have access to sanitation; and the urban population explosion at a rate unique in history, which is characterized by growing slums, unemployment, poor access to water, sanitation and health services. In both these environments, groundwater resources have a major role to play and their sustainable utilization has become of strategic importance.

The approaches to achieve sustainable resource utilization are not only groundwater-technical solutions, but are part of the overall challenges faced and solutions considered for sustainable community water supply and sanitation itself. The purpose of this chapter is therefore to understand these challenges, its drivers, its impacts and its strategies towards sustainability.

Some recent statistics on water supply and sanitation will be provided together with a discussion of the impact of adequate water services on development in Africa. Challenges, strategies and programmes to achieve the Millennium Development Goals in terms of water supply and sanitation will be discussed together with an outlook for a more sustainable service provision.

1.1 CONTRIBUTION OF WATER SUPPLY AND SANITATION TO AFRICAN DEVELOPMENT

Water plays a pivotal role in society; it is critical for economic development; for human health and social welfare, especially for the poor; and for environmental sustainability. A good summary of the water and development situation in Africa is provided in the Africa Regional Document to the 4th World Water Forum (World Water Council, 2006). Water is a key factor in Africa's development. None-the-less, the African continent has to date only used a small proportion (5%) of its available water resources. The water crisis referred to in the Africa Water Vision (Economic Commission for Africa et al., 2000) is therefore much more complex than a simple lack of water availability. Elements of the crisis include (Braune and Xu, 2008): (1) large spatial and temporal variability of resource availability, along with the more arid climate prevalent in about 60 percent of the African continent; (2) a wide-spread lack of coping capability to manage the irregular availability of water; (3) inadequate access to the most basic water and sanitation services, leading to living conditions which are not conducive to social and economic development; (4) rapid uncontrolled urbanization, leading to megacities with very poor water services and squalid living conditions; and (5) the poor utilization of water in the agricultural sector, still the most important sector in the African economy.

Alleviation of poverty is the key development challenge, with over 300 million people or about 40% of the total population still living in extreme poverty in 2001 (less than US$1 a day) (World Water Council, 2006). Africa also bears the brunt of the world's HIV/AIDs pandemic. To date, 13 million people have died of HIV/AIDS and 26 million are living with the virus, more than 60% of those infected world-wide. This is seriously hindering socio-economic growth and development (World Water Council, 2006).

The Poverty-Environment Partnership (2006) has come up with four key dimensions of poverty reduction:

- Enhanced livelihoods security: the ability of poor people to use their assets and capabilities to make a living in conditions of greater security and sustainability;
- Reduced health risk: the mitigation of environmental and social determinants that put the poor and most vulnerable (especially women and children) at risk from different diseases, disabilities, poor nutrition and premature death;
- Reduced vulnerability: the reduction of threats from environmental, economic and political hazards, including sudden impact shocks and long-term trends;
- Pro-poor economic growth: enhanced economic growth is essential for poverty reduction in most parts of the world, but the quality of growth, and in particular the extent to which it creates new opportunities for the poor, also matter.

Different aspects of water management contribute significantly to each of these. With this understanding the African Water Vision was formulated (Economic Commission for Africa et al., 2000):

> "An Africa where there is an equitable and sustainable use and management of water resources for poverty alleviation, socio-economic development, regional cooperation, and the environment."

As part of this vision, Africa has embraced the Millennium Development Goals, the centrepiece of the global development agenda, intended to reduce significantly by 2015 vicious cycles of global poverty, hunger, disease, illiteracy, environmental degradation and gender inequality.

1.2   MILLENNIUM DEVELOPMENT GOALS AND WATER

While water is a key to each of the development goals, the initial water focus has been on Target 10: Halve by 2015, the proportiation of people without sustainable access to safe drinking water and sanitation (UN Millennium Project Task Force on Water and Sanitation, 2005). This introduced a strong shift all over Africa from a water resource management and sustainability focus to service delivery. With this shift went a decentalization of service delivery, and often completely new institutions at national level, separate from the traditional water resource management institutions, to regulate and support this process.

A summary of water's diverse contribution to the different MDGs is provided below as further context to the water and sanitation issues in Africa.

Contribution of water supply and sanitation to the Millennium Development Goals.

Goal 1. **Eradicate** **extreme** **poverty** **and hunger**	Access to water supply and sanitation, often used in broad definitions of 'poverty', is a Millennium Target in itself (as part of Goal 7). It is invariably the poor who must spend much of their resources (money and time) carrying water to their homes; it is the poor who carry the greatest burden of productivity-sapping disease as a result of not having access to safe water and sanitation. But water is essential to economic development, which can create productive livelihoods for the poor. Water can also offer important direct opportunities for the poor to address their food and income needs. In many rural communities, the availability of food on which to subsist is dependent on the uncertainties of nature's cycles—on whether the rains come and the rivers flow. Creating conditions in which the poor can benefit from opportunities offered by access to water is one of the more important contributions that IWRM can make to poverty reduction.

(Continued)

Contribution of water supply and sanitation to the Millennium Development Goals. (*Continued*)

Goal 2. **Achieve** **universal** **primary** **education**	The challenge of primary education may seem removed from that of water until it is recognised that in many communities, children's time is a valuable commodity and school attendance competes with work such as carrying water. Water-related disease also affects school attendance. And the availability of adequate sanitation is a key determinant, for girls in particular, of attendance at school—for example, a study in the Nokali district of Pakistan showed that installing water and separate sanitation facilities for girls increased their attendance by 15%.
Goal 3. **Promote** **gender equality** **and empower** **women**	The burden of reproducing families and sustaining households has always fallen disproportionately on women. The fetching and storing of water is a task which takes much of their time and that of their female children in many poor communities. Women are also often the primary users of water for productive activities such as agriculture. Properly applied, IWRM approaches can ensure that they have a voice in decisions about water that affect them and can gain access to water to help boost their incomes. Any intervention that makes safe water more easily available is a direct contribution to the promotion of gender equality, lightening the domestic burden on women, and enabling them to participate more actively and effectively in the affairs of their community.
Goal 4. **Reduce** **child** **mortality**	In most poor communities, the health of children is directly related to the quality of their immediate nurturing environment, in which water and sanitation services and their management play a key role. Children are at risk when they are without safe water to drink, without adequate water to stay clean, without some sanitation facility to remove human wastes safely and when their care-givers are without the knowledge or power to make decisions about these issues. In developing countries, water-related diseases are almost always amongst the most important causes of death of children under the age of five, using deaths from diarrhoea as a proxy. More than 1.5 million children under five die every year from diarrhoea (more than from malaria and HIV/AIDS combined).
Goal 5. **Improve** **maternal** **health**	The burden of fetching water and dealing with water-related disease in the family falls disproportionately on women and puts pressure on their own health. Measures that help women to reduce this burden and to improve family health, will contribute to improved maternal health specifically, as well as to gender equality more generally.
Goal 6. **Combat** **HIV/AIDS,** **malaria and** **other diseases**	Access to safe water and sanitation services can help to reduce poverty—which in turn is an important determinant of HIV/AIDS—and help to keep HIV-infected people healthy and productive. Effective water management at local level can also help to reduce malaria and other diseases endemic in poor communities such as dengue fever, which is now spreading more rapidly than malaria.
Goal 7. **Ensure** **environmental** **sustainability** **(including the** **target of halving** **the number of** **people without** **access to water** **and sanitation)**	Water is key to the sustainable utilization of land, plant and animal resources. In many countries the main environmental problems, whether it is pollution, erosion or the loss of biodiversity in wetlands and estuaries, relate to water. If the water resources environment is not managed and protected, it will not be able to sustain human communities. A direct contribution offered by IWRM to Goal 7 is to facilitate, in a structured way, the achievement of a balance between economic, social objectives and activities, and environmental sustainability. Similarly, IWRM can help to ensure that the provision of water supply and sanitation services (the other dimension of Goal 7) is reliable and sustainable. Certainly, the disposal of waste water from sanitation is a major environmental challenge in many countries, best addressed through IWRM. Similarly, the reliability of domestic water supplies in dry seasons often depends on influencing the behaviour of other water users.

(*Continued*)

Contribution of water supply and sanitation to the Millennium Development Goals. (*Continued*)

Goal 8. **Develop a global** **partnership for** **development**	Water is a resource that knows no political boundaries. Just as many communities depend on water shared with their neighbours, so too do many countries. What is also shared between countries is the common commitment to achieve the MDGs and, if water is key to meeting the MDGs, cooperation in its management is critical. There are many ways in which countries will need to cooperate if the MDGs are to be reached, by no means limited to financial and technical support for specific activities. Integrated water resource management is one mechanism through which such partnerships can be built, particularly where rivers and lakes are shared between more than one country.

UN Millennium Project Task Force on Water and Sanitation (2005).

1.3   WATER SUPPLY AND SANITATION COVERAGE

The most recent summary statistics on the water supply and sanitation situation in Africa are (Kaberuka, 2008):

- 340 million Africans lack access to safe water;
- 500 million Africans have no access to sanitation.

These statistics become more meaningful when viewed in an international context and in terms of their distribution across the continent and broken down into urban and rurual areas as shown in Figure 1.1.

The number of people globally who lack access to an improved drinking water source has fallen below one billion for the first time since data were first compiled in 1990. At present 87 percent of the world population has access to improved drinking water sources, with current trends suggesting that more than 90 percent will do so by 2015 (WHO and UNICEF, 2004). In comparison only some 62 percent of Africans had access to an improved water supply in 2000 (UNEP, 2007). This represents the lowest total water supply coverage of any region of the world. The seriousness of this backlog is illustrated when viewed in the world map produced out of the World Bank water data & statistics (The World Bank, 2008). To meet MDGs, numbers served should double, to 700 million by 2015. Even then, 200 million would remain unserved (The World Bank, 2008).

Urban areas are better supplied, with 85 percent of the population having access to improved water supplies. In rural areas, the average is 47 percent. The total African population with access to improved sanitation was 60 percent in 2000. Again, urban populations fared better, with an average 84 percent having improved sanitation compared to an average 45 percent in rural areas (The World Bank, 2008).

The implications of this backlog in water supply and sanitation in Africa are major and can be illustrated with one impact statistic. Poor water supply and sanitation lead to high rates of water-related diseases such as ascariasis, cholera, diarrhoea, dracunculiasis, dysentery, eye infections, hookworm, scabies, schistosomiasis and trachoma. About 3 million people in Africa die annually as a result of water-related diseases. In 1998, 72 percent of all reported cholera cases in the world were in Africa (UNEP, 2007).

There are big country to country variations in these statistics. A distributed water supply and sanitation coverage in Africa is shown in Figure 1.1 below, for the urban, rural and overall situation (UNEP, 2002). The situation is particulat severe in sub-Saharan Africa. In 16 of the 54 African countries, sanitation coverage is still below 25 percent (Brocklehurst, 2008). In some countries, coverage is actually declining and many others need to double or treble the rate of increase to be able to catch up.

A fuller understanding of these statistics and the implication of the task ahead towards achieving the MDGs for water supply and sanitation can be obtained when considering

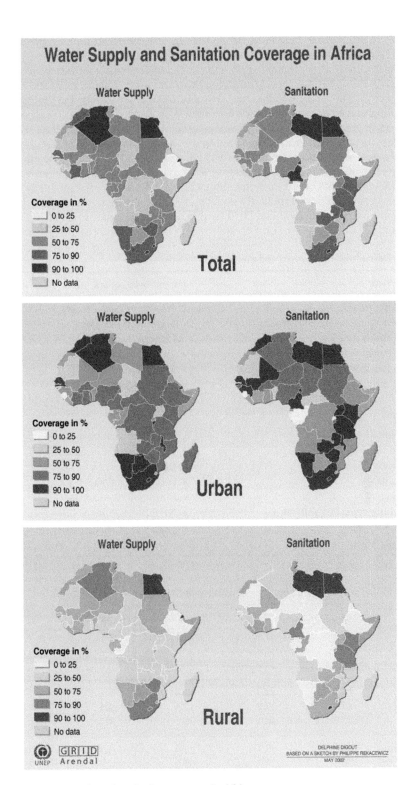

Figure 1.1.   Water supply and sanitation coverage in Africa.
Source: Global Water Supply and Sanitation Assessment 2000 Report.

the goals as they had originally been defined of an improved and an unimproved service (Brocklehurst, 2008) (Table 1.1).

The official global data on water supply and sanitation coverage are collected and disseminated by the WHO/UNICEF Joint Monitoring Programme for Water Supply and Sanitation. The most recent JMP report (WHO and UNICEF, 2008) assesses—for the first time ever—global, regional and country progress using an innovative "ladder" concept. This shows sanitation practices in greater detail, enabling experts to highlight trends in using improved, shared and unimproved sanitation facilities and the trend in open defecation. Similarly, the 'drinking water ladder' shows the percentage of the world population that uses water piped into a dwelling, plot or yard; other improved water sources such as hand pumps, and unimproved sources.

The special challenges of urban water supply and sanitation need some further elaboration. Urban growth will be particularly notable in Africa and Asia where the urban population will double between 2000 and 2030 (within a single generation) by 2030 (Table 1.2).

Table 1.1.   Definitions of improved and unimproved services.

Improved service	Unimproved service
Improved **water supply**	Not improved **water supply**
• Piped water into dwelling, plot or yard	• Unprotected dug well
• Public tap/stand pipe	• Unprotected spring
• Tubewell/borehole	• Cart with small tank/drum
• Protected dug well	• Tanker-truck
• Protected spring	• Surface water
• Rain water collection	
• Bottle water	
Improved **sanitation**	Unimproved **sanitation**
• Flush/pour flush to:	• Pit latrine without slab/open pit
○ piped sewer system	• Bucket
○ septic tank	• Hanging toilet/hanging latrine
○ pit latrine	• Flush/pour flush to elsewhere
• Ventilated Improved Pit (VIP) latrine	• No facilities, bush or field (open defecation)
• Pit latrine with slab	• Shared or public facilities
• Composting toilet	

Table 1.2.   Proportion of total population in urban area (Alabaster 2008).

Proportion (%) of total population in urban areas (derived from UNDESA statistics, 2004)

Nations and regions	Rural areas	Urban areas <0.5 million	Urban area 0.5 million– 4.9 million	Urban area 5.0 million– 9.9 million	Mega–cities >10 million
Africa	**62.9**	**22.3**	**12.4**	**1.1**	**1.3**
Asia	62.9	18.4	12.4	2.5	3.9
Europe	27.3	46.1	20.5	4.7	1.4
Latin America & Caribbean	24.5	37.1	23.4	3.7	11.3
North America	20.9	29.8	35.6	4.3	9.4
Oceana	27.3	31.7	41	–	–
World	52.9	24.5	15.7	2.7	4.1

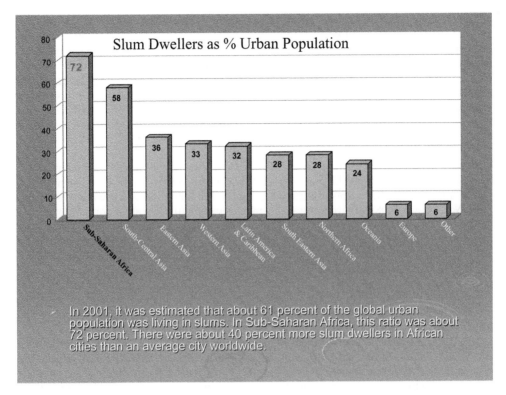

Figure 1.2.   Percentage of slum dwellers in urban population from: Alabaster (2008).

In urban centres (of all sizes) of poorer countries, poverty issues mostly concern health related impacts of inadequate water and sanitation, and limited or no waste disposal. Poor people, living predominantly in slums, will make up a large part of future urban growth in Africa (Figure 1.2). Ignoring this basic fact will make it impossible either to plan for the inevitable and massive growth in urban areas or to use urban dynamics to help provide basic water and sanitation services to relieve poverty (Alabaster, 2008).

Informal settlements have poor housing structures and poor facilities and services. Lack of secure land tenure and lack of effective 'voice' often prevent their residents from getting even minimally adequate services. In 2001, it was estimated that about 61 percent of the global urban population was living in slums. In Sub-Saharan Africa, this ratio was about 72 percent. There were about 40 percent more slum dwellers in African cities than an average city worldwide (Alabaster, 2008).

1.4   CHALLENGES AND STRATEGIES TO ACHIEVE THE MDGs FOR WSS

To make progress in the water sector as in other sectors, Sub-Saharan Africa (SSA) needs both institutional development and investment finance. To achieve the Millennium Development Goals (MDGs) in water supply and sanitation, the number of people served must more than double, from 350 million in 2000 to 720 million in 2015. The expected annual cost of meeting the MDG target for water is between US$1.7 and 2.1 billion, and just as much is likely to be needed for sanitation (The World Bank, 2008; Winpenny, 2003). Most countries are undertaking WSS sector reforms, and some have achieved good progress in expanding access to services and improving operating performance.

Governments face a twin challenge: to close the gap in rural areas—where only two in five people have access to water supply and fewer than one in five have access to sanitation—and to keep up with rapid population growth in urban areas. Utility performance in WSS at present is mostly very poor. The need is to simultaneously increase investments and to build implementation capacity by scaling up the reform agenda and strengthening institutions. Challenges and responses will be discussed below, mainly based on The World Bank (2008), with regard to water supply in both the rural and the urban situation and with regard to sanitation.

1.4.1   *Rural water supply*

Given the highest backlogs here, a strong focus on rural water supply, sanitation, and hygiene is needed if the Millennium Development Goals (MDGs) are to be met. Much has been learned about how to make investments in rural water supply and sanitation effective and sustainable (The World Bank, 2008). First, decentralizing management to the lowest appropriate level, coupled with close community involvement in planning, financing, implementation, and operations provides a solid foundation for sustainable services. Second, implementing rural WSS within a broad development context allows institutions to respond to and support a range of community needs in a cost-effective and holistic manner. Third, integrating sanitation and hygiene measures into rural WSS projects ensures that health benefits from increasing water supply coverage are realized. Fourth, addressing post-construction sustainability ensures that institutions, funds, and expertise are available to keep rural water supply systems viable and functional.

The challenge facing the sector today is how to scale up these experiences in order to meet the MDGs. Increased financing is clearly needed, but so is client capacity to implement and ensure the sustainability of investments. Progress can be made by moving to programmatic approaches, with donors harmonizing their use of support instruments and aligning them within government-designed development strategies. Such an approach can enable wide-scale sector reform, improve the predictability of financing, remove externally imposed bottlenecks, and optimize the impact of both government and external agency support.

1.4.2   *Urban water supply*

To reach the MDGs in urban areas will require reforms to increase financial viability and improve institutional performance in the WSS sector, in order to provide a basis for expanding access and improving the quality of service (The World Bank, 2008). Institutional models for service provision need to fit local circumstances—using the private sector, public entities, communities, or a mix to deliver services can all be effective ways to improve performance.

Reaching the new urban population is a complex task. Informal settlements have poor housing structures and poor facilities and services. Lack of secure land tenure and lack of effective 'voice' often prevent their residents from getting even minimally adequate services. Interventions need to be synchronized with policies towards informal settlements or slum upgrading, decentralization and fiscal policy.

WSS services are natural monopolies and have significant externalities, so need some form of regulation. But well-functioning utilities are allowed considerable autonomy, including substantial control to generate and retain revenues, and to use these revenues in day-to-day operations.

Even for a utility that functions efficiently, reaching the poor often requires targeted interventions. Partnerships between governments, utilities, and civil society can play an important role here. And, where traditional utilities cannot or will not provide service to the poor, standards and by-laws should be revised to allow small-scale providers to operate efficiently. Including poor people and other traditionally excluded groups in priority setting and decision making is critical to making good use of limited public resources, by permitting interventions to build on local knowledge and priorities.

Water supply should never be tackled without simultaneous attention to water conservation. Without this focus, huge wastage is almost inevitable (infrastructure break-down, lack of recycling and reuse). The latter is commonly found in urban areas and can represent 20% and more loss of the total volume. If alternatively, water was recycled and reused in combination with a near zero loss system, this could provide an untapped resource for the future. This solution still lacks general support, because people and institutions are not yet ready to invest in it. Nationally and internationally, however, there is ongoing pressure and a step by step approach to change this situation in developing countries, including Africa, The key approach is awareness-raising by governments, via their parastatals, and other tailor-made programmes through the UN family.

1.4.3 *Sanitation*

The sanitation situation is much worse than for water supply and has been referred to as a neglected and unspoken crisis. An almost a five-fold increase in people achieving access to sanitation every year would be required if the target is to be met (Brocklehurst, 2008). It differs from water supply in that it is mainly an on-site, domestic issue. Strategies to deal with the crisis will therefore have to include a large element of awareness raising and education. Sanitation promotion focuses on stimulating demand for ownership and use of a physical good. Access to basic sanitation refers to access to facilities that hygienically separate human excreta from human, animal, and insect contact. Hygiene promotion focuses on changing personal behavior. Both are essential to maximize the benefits of investments in clean drinking water.

Even where there is general agreement and evidence on the economic and health benefits of adequate sanitation services, there is not always an effective demand for action. Governments' limited spending on sanitation often reflects political, more than technical or economic, constraints in the context of competing demands for resources. The strategy package to address the large sanitation backlog should therefore include the following (The World Bank, 2008):

- Sound policies based on the development of demand-responsive and community-driven services matched to available resources;
- expansion of the menu of services and technologies; and
- inclusion and expansion of hygiene promotion.

Besides the key issue of capital investment, investment in human capital is equally essential. Sanitation remains a personal and household issue, rather than a shared community service. The appropriateness of sanitation technology (e.g. no flush toilets in water-scarce areas), the siting of the toilet (e.g. downstream of a groundwater source), the opportunity of waste recycling as a source of nutrients for local food production, and the wise consideration of cultural issues, all require informed decisions. To achieve participation and support this personal decision-making in efficient ways remains a major challenge.

1.4.4 *General*

Overall, it is important that governments take ownership of targets like the MDGs and localise them by breaking them down according to national and local realities: the state and distribution of existing infrastructure and water resources, available appropriate technologies and their supply chains, resource flows and gaps, the existence and capacities of development actors that will play a role in implementing plans.

A good example in this regard is Uganda, where the Poverty Eradication Action Plan made water and sanitation one of its priorities. As a result, total investments in the sector rose between 1997 and 2002 from over US$3 million to US$31 million (or from 0.5% to 2.5% of the national budget over the same period). This level of investments increased national access to water services from 39.4% in 1996 to 51% in 2003, or over 2 million newly served (CSD NGO Consortium, 2007).

An integrated sector approach is nowhere more necessary than in the water and sanitation sector, with its multiple users, providers, multiple funding streams and the varied impacts of the infrastructure services on the water resources and the environment. Donors need to support the efforts of governments to undertake sector co-ordination, starting with harmonising their support to developing countries. The plethora of donor procedures and reporting requirements diminish government capacity significantly, especially in those countries where the sector is heavily dependent on external assistance (CSD NGO Consortium, 2007).

Institutional reforms already underway to address these problems will have to deliver quickly and new strategies need to be developed in fragile environments where governments are not in a position to lead. By prioritizing WSS services in poverty reduction strategies (PRS) and applying business principles to their implementation, governments can strengthen sector governance and transparency (WSP Africa, 2008).

While access to basic water and sanitation is generally accepted as a human right, the question whether water should be a free good or an economic good is far from resolved and has major implications for a sustainable water supply and sanitation service. Water, in a number of countries around the world, can be free (at least for an initial volume set by the state as a human rights related volume) or it bears a tariff that in most cases is irrelevant of its availability and cost for delivery (which would include cost for operation and maintenance). As an example the tariff in countries that have little or no water scarcity can range from $0.00 (Dublin, Ireland) to $7.00 (Berlin, Germany), whereas in areas which approach physical water scarcity, it can range from $0.21 (Mexico City, Mexico) to $2.16 (Dubai, UAE; source: International Water Management Institute, Global Water Intelligence). There is no simple answer to whether or not water is a basic human right or an economic commodity and it is a topic that most often is highly politicized and sensitive.

A good summary of the challenges and implementation strategies for water and sanitation in Africa has been provided by Winpenny (2008) (Table 1.3).

Table 1.3.   Approaches for a more sustainable WSS delivery in Africa (Winpenny, 2008).

MDGs: reasons for lack of progress	• External agents the main drivers of change • NGO activities off-budget & weakly coordinated • Poor understanding of wider linkages of W&S, weak implementation capacity • Urban bias & neglect of rural areas • Sanitation & hygiene neglected & lagging • Uncoordinated & unclear Ministerial responsibilities • Water utilities unreformed
MDGs progress— success factors	• Governments "own" and drive reforms • Links between water, sanitation, poverty & growth understood & reflected in policy • Civil society bodies are effective advocates & involved in policymaking • Active reforms of water sector undertaken • Decentralisation of responsibility is "bedded in" and working • Ministerial responsibilities clear, coordinated & functional
Connect W&S with hygiene, health & poverty	• Watsan is central to anti-poverty strategies • Get W&S into PRSPs • Target medical profession & Ministries of Health for dialogue & cooperation
Engagement of other actors	• Public utilities: weak capacity to expand services: need "rules of engagement" & reform agenda • Local business (informal & small/medium scale) to be recognised & involved • Framework for foreign operators? Clarity of policy, regulation, pro-poor contract clauses • NGOs in policy & administrative vacuum: more involvement required

There is widening acceptance that hitting the water and sanitation targets of the Millennium Development Goals (MDGs) contributes significantly to progress in achieving the other targets. But we are currently lagging behind on these water and sanitation targets, especially in Africa. A more recent concern is that the targets for integrated water resource management (IWRM) and water efficiency (WEP) plans, which help to ensure long-term sustainability of water services, are also at risk of being missed (CSD NGO Consortium, 2007). This emphasizes the need for an integrated approach to water services delivery and water resources management.

1.5   VARIOUS PROGRAMMES TO ACHIEVE GOAL

Water supply and sanitation delivery in Africa is impemented through a variety of actors and programmes. Some of the key ones are listed in Table 1.4 together with their role and mode

Table 1.4.   Some water supply and sanitation delivery programmes in Africa.

AMCOW	The African Ministers Council on Water (AMCOW) was estalished in 2002	The mission of AMCOW is to provide political leadership, policy direction and advocacy in the provision, use and management of water resources for sustainable socio-economic development and maintenance of African ecosystems
NEPAD	The NEPAD Water & Sanitation Programme	NEPADs WSIP falls within the larger framework of the partnership initiative, the aims of which include developing regional infrastructure, harmonizing sectoral procedures, enhancing finacial flows towards investment in infrastructure, and developing skills and knowledge for the installation, operation and maintenance of infrastructure (World Water Council, 2006)
African Development Bank (AfDB)	The Rural Water & Sanitation Initiative	Launched in 2004 to address the problem of low access to water supply and sanitation in rural Africa, the RWWSI has grown to be the largest initiative in the Bank. Since then 17 programmes worth 1.8 billion US dollars (of which the Bank contributed 750 million dollars) have been financed
African Water Facility (AWF)	The AWF is led by AMCOW and is established as a Special Water Fund by the AfDB	The African Water Facility, only in its second year, has already provided 24 million Euros worth of financing to strengthen water governance, sustainable management of national trans-boundary water resources, feasibility studies and pilot investments
World Bank	The Water and Sanitation Program (WSP) is a multi-donor partnership of The World Bank	The World Bank is by far the largest provider of development assistance to Africa and attaches great importance to capacity-building efforts in Africa. WSP-AF (Africa) strives to make an impact in three critical entry points: • Promoting sector reform, improved governance, and the development of country-owned roadmaps • Assisting countries in developing sustainable financing strategies to implement large-scale programs • Providing capacity-building support to both regional and national policymakers and service providers WSP works directly with governments at the local and national level

(Continued)

Table 1.4. *(Continued)*

UNICEF	UNICEF is funded entirely by the voluntary contributions of individuals, businesses, foundations and governments	UNICEF supports child health and nutrition, good water and sanitation, quality basic education for all boys and girls, and the protection of children from violence, exploitation, and AIDS
World Health Organization (WHO)	WHO is the directing and coordinating authority for health within the United Nations system	It is responsible for providing leadership on global health matters, shaping the health research agenda, setting norms and standards, articulating evidence-based policy options, providing technical support to countries and monitoring and assessing health trends. WHO has recognized the importance of water and sanitation from its inception
Water Supply and Sanitation Collaborative Council (WSSCC)	WSSCC was created in 1990 as a follow-up to the International Drinking Water Supply and Sanitation Decade (1981–1990). Its institutional home is in WHO	The Collaborative Council emphasizes the need to view water, sanitation and hygiene (WASH) as an inseparable trinity for development. To achieve its mission, the Collaborative Council divides its activities into three programme areas, namely Advocacy & Communications; Networking & Knowledge Management; and the Global Sanitation Fund. At the core of its action programmes is the notion that the energy and initiative of local people with a stake in their local water supply and sanitation situations can generate and implement economical and sustainable solutions to the crisis of inadequate water, sanitation and hygiene. (www.wsscc[at]who.int)
JMP	WHO/UNICEF Joint Monitoring Programme for Water Supply and Sanitation	JMP is the official UN mechanism tasked with monitoring progress towards MDG Target 7c on drinking water supply and sanitation
UN-HABITAT	The UN-HABITAT Water and Sanitation Trust Fund (WSTF) was established in 2002	The key objectives for the fund are: • to create an enabling environment for pro-poor investment in water and sanitation in urban areas; and • to support capacity building at local level to manage these investments in a sustainable manner. A major thrust of the WSTF is the Water for African Cities Programme which was being demonstated in seventeen African countries by 2006 (World Water Council, 2006)
European Union (EU)	The EU Water Initiative (EUWI) was launched in 2002 at the WSSD in Johannesburg	The EUWI is a comprehensive partnership designed to help countries achieve water and sanitation targets. Immediate objectives in an Action Plan of its Water Supply and Sanitation Group are: • Increasing demand for investment in water supply and sanitation for the poor • Strengthening underlying institutions, building capacity and making better use of existing human and institutional resources • Enhancing funding for the supply, management and development of water resources, and sanitation • Improving coordination between the actors involved in water resources management

of operation to help identify potential linkages for mainstreaming groundwater resources management and protection as part of the thrusts of these overall drivers.

The Rural Water & Sanitation Initiative of the African Development Bank (AfDB) is the main continent-wide driver for improved water and sanitation delivery. It strongly encourages countries to raise their own funding and manage their own programmes.

An example in this regard is the Masibambane programme in South Africa. Masibambane—meaning let's work together, is a water sector support programme, now in its third 5 year phase, led by the Department: Water Affairs and Forestry (www.dwaf.gov.za). The programme is a partnership between the Department of Provincial and Local Government, the South African Local Government Association, the European Union and its member states; the Swiss Government and Ireland Aid. The Masibambane Sector Wide Support Approach works from the premise of coordinated strategies and joint implementation involving all players in the water sector: national and provincial government, municipalities, civil society, donors, water utilities and the private sector. It facilitates donor coordination, bringing resources together in a consolidated budget, allocated to the achievement of sector goals and objectives. Through this process Masibambane has created a platform for addressing "soft issues" related to the quality and sustainability of delivery, which are commonly noted, but frequently fall by the wayside. These also open the door for a more systematic inclusion of local groundwater sources, as highlighted in the conclusions to this chapter.

1.6   FROM WATER FOR DOMESTIC USE TO A LIVELIHOODS FOCUS

There has been a growing recognition in Africa that the water services focus has been too narrow and did not allow for the full developmental role of water. Institutions like the International Water Management Institute (IWMI) and the Programme for Land and Agrarian Studies (PLAAS) have been advocating a Livelihoods or Multi-Use System (MUS) approach to address this challenge (Tapela, 2008). Even the comprehensive IWRM approach has largely left poor neighbourhoods without adequate or secure access to water for their basic needs and livelihoods.

> *"Empowering people, reducing poverty, improving livelihoods and promoting economic growth ought to be the basic objectives of IWRM. But as currently understood and used, IWRM often tends to focus on second generation issues, discouraging attention to making water available to poor people for productive and domestic uses..."* Merrey et al., (2008).

While in development terms, basic needs has a very broad meaning, the water sector's interpretation and active implementation has been very narrow, as illustrated by the quotations taken from Tapela (2008):

> "The basic needs approach to development aims at meeting the basic needs of the poor in any country in the shortest possible time. There is a general consensus in defining these basic needs as food, education, health, housing, and sanitation. Basic needs also include such nonmaterial aspects as fundamental human rights and freedoms, self reliance, and participation".

A statement issued at the close of the 1977 Mar del Plata conference, which is the earliest comprehensive international water conference, became a landmark recognition of the right to water for basic needs:

> *"... all peoples, whatever their stage of development and their social and economic conditions, have the right to have access to drinking water in quantities and of a quality equal to their basic needs".*

The challenge of poverty and inequality in Africa requires that the current focus on human rights to water be broadened beyond access to drinking water and sanitation service provision, to embrace people's livelihoods as a whole. This also has major implication for sustainable water resource utilization, including groundwater and other locally available water resources.

The MUS approach recognizes that "people's water needs are integrated and are part and parcel of their multi-faceted livelihoods". The approach takes people's multiple uses of water as the starting point for planning and providing integrated services to enhance livelihoods of especially the poor (van Koppen et al., 2006: 2).

The beginnings of a new approach, at least at a strategic level, can be illustrated by recent developments in South Africa. In terms of the Accelerated and Shared Growth Initiative in South Africa, Government was mandated in 2004 to halve poverty and unemployment by 2014. This ambitious goal requires, first of all, a growth rate around 5% average, but also that the fruits of growth must be shared in such a way that the social objectives of the policy can be met.

The water sector responded with a Water for Growth and Development strategy with focus on multiple uses of water for poverty alleviation and improved livelihood, including some of the following suggested responses (DWAF, 2008):

- Strengthen sector collaboration between Water Services & Water Resources Management
- Engage with other sectors to ensure good governance
- Enhance integrated participatory planning for multiple use water services as part of Integrated Development Plans of Local Government
- Place development benefits of water at core of financial strategies and National Development Plans

All these measures indicate a more holistic approach to water services provision and to proper management of the available water resources.

1.7   CONCLUSIONS

The African continent has by far the lowest water supply and sanitation coverage in the world. This has a major impact on development, particularly given that about 40% of the total population are still living in extreme poverty. While African countries are committed to meeting the Millennium Development Goals for Water Supply and Sanitation, major obstacles remain. The situation is most serious in rural as well as in informal urban areas. Both these environments present unique services challenges which go way beyond the availability of secure water sources. These range from the general problems of lack of individual government ownership of the service delivery targets, appropriate policies, institutional capacity and cooperative governance approaches to the specific difficulties encountered in these environments, in particular the financing, planning and implementing of hundreds of small dispersed schemes, all lacking financial and operational accountability.

What is widely recommended is an integrated sector approach in the water and sanitation sectors, and within this an itegration of water services provision and water resources management. Capacity must urgently be built to scale up experiences of such integrated approaches. The trend to move away from a basic needs approach, catering for domestic water supply only, to a more holistic water for livelihoods approach augers well for a much more systematic and sustainable approach to water supply and water resources management, including the utilization of local groundwater resources.

Such major transformation of society has to be knowledge-driven to be effective and sustainable. It is amazing how little appreciation their still is in the need to invest in knowledge creation and its systematic utilization, not only to the average global citizen in his role as a consumer but also to the scientists, and policymakers who study water and create legislation for the resource.

Better understanding is achieved through better data and information on the interacting land/water/human systems. The old adadage "You can't manage what you can't measure" clearly applies. Thus until we have better knowledge of the subject matter, our management of freshwater systems will be inefficient and uncoordinated.

REFERENCES

African Development Bank Group (2005). The Rural Water Supply and Sanitation Initiative. United Nations, New York, 22 April 2005.

Alabaster, G. (2008). Meeting the WATSAN MDGs in Africa—The urban sector. Presentation at First African Water Week, Tunis, 26–28 March 2008.

AMCOW (2008). Roadmap for the Africa Groundwater Commission. UNEP/UNESCO/UWC: Nairobi.

Braune, E. and Xu, Y. (2008). Groundwater management issues in southern Africa—An IWRM perspective. Water SA 34 (6) (IWRM Special Edition).

Braune, E., Hollingworth, B., Xu, Y., Nel, M., Mahed, G. and Solomon, H. (2008). Protocol for the Assessment of the Status of Sustainable Utilization and Management of Groundwater Resources – With Special Reference to Southern Africa. Pretoria. WRC Report No. TT 318/08, Water Research Commission: Pretoria.

Brocklehurst, C. (2008). A Snapshot of Sanitation in Africa. AfricaSan: Second African Conference on Sanitation and Hygiene. Durban, South Africa, 18.2.08.

Burke, J.J. and Moench, M.H. (2000). Groundwater and Society: Resources, Tensions and Opportunities. New York: United Nations.

CSD NGO Consortium (2007). CSD 13: Water and Sanitation. www.watertreaty.org.

DWAF (2008). Department of Water Affairs and Forestry Seminar on Water for Growth and Development, 17.8.08. Stockholm Water Week 2008.

Economic Commission for Africa (2006). Southern Africa Water Development Report. ECA/SA/TPUB/2005/4.

Economic Commission for Africa, Organisation for African Unity and African Development Bank (2000). Safeguarding Life and Development in Africa. A Vision for Water Resources Management in the 21st Century. African Caucus Presentations—Second World Water Forum, The Hague, The Netherlands, 18 March 2000.

GWP (2008). The SADC Multi-Stakeholder Water Dialogue. Maseru, Lesotho, 4–15 May 2008. GWP Information Note. http://www.wssinfo.org/

Jallow, S. (2008). Rural Water Supply and Sanitation Initiative—Achievements & Challenges. First African Water Week, Tunis, 26–28 March 2008.

Kaberuka, D. (2008). Statement by African Development Group. At the Welcome Dinner in Honor of Ministers and Guests. First African Water Week, Tunis, 26–28 March 2008.

Koppen, van, B., Moriarty, P. and Boelee, E. (2006). Multiple-Use Water Services to Advance Millennium Development Goals. IWMI Research Report 98. Colombo.

Merrey, D.J., Penning de Vries, F.W.T. and Sally, H. (2008). Integrating Livelihoods' into Integrated Water Resources Management: Taking the Integration Paradigm to its Logical Next Step for Developing Countries. IWMI: Pretoria.

Poverty-Environment Partnership (2006). Linking Poverty Reduction and Water Management. UNDP and the Stockholm Environment Institute.

SADC (2009). Outcomes of SADC/AMCOW Workshop on Groundwater for Development in the SADC IWRM Initiative. November 18–19 2008, Gaborone, Botswana. www.sadc-groundwater.org.

Tapela, B. (2008). Water for Livelihoods. IWRM Short Course. University of the Western Cape, Cape Town, 29.9.08–2.10.08.

The World Bank (2008). Water-Africa. and Water data & statistics.www.worldbank.org.

UN Millennium Project Task Force on Water and Sanitation (2005). Health, Dignity and Development: What Will It Take. Millennium Project. Stockholm International Water Institute.

UNEP (2002). Global Water Supply and Sanitation 2000 Assessment Report.

UNEP (2007). Chapter 4. Freshwater-Africa. UNEP GEO-4: Global Environment Outlook.

WHO and UNICEF (2004). Joint Monitoring Programme Midterm Assessment. WHO/UNICEF Joint Monitoring Programme for Water Supply and Sanitation.

WHO and UNICEF (2008). Progress in Drinking-water and Sanitation—special focus on sanitation. WHO/UNICEF Joint Monitoring Programme for Water Supply and Sanitation.

Winpenny, J. (2003). Financing Water Infrastructure. World Water Council and Global Water Partnership.

Winpenny, J. (2008). Water and Sanitation MDGs—Status, Achievements & Missing Links. First African Water Week, Tunis, 26–28 March 2008.

World Water Council (2006). Water Resources Development in Africa. Africa Regional Document. 4th World Water Forum, Mexico 2006.

WSP Africa (2008). Water and Sanitation Program. The World Bank. www.wsp.org.

2

Groundwater resources in Africa

Yongxin Xu & Eberhard Braune
Department of Earth Sciences at the University of the Western Cape, South Africa

To be able to consider the sustainable utilization of groundwater in relation to water supply and sanitation, a general understanding of the varying characteristics and roles of groundwater resources across the continent is provided in this chapter.

2.1 CHARACTERISTICS OF GROUNDWATER RESOURCES

As the second largest continent, Africa lies across the equator with its southern tip, at Cape Agulhas near the Cape Town, at 35°S and the northern coast between the straits of Gibraltar and Tunis at about 35°~37°N. Due to its variety in geology, relief and climate, hydrogeological conditions are complicated and their elucidation has been held back by a breakdown of communication among various groundwater stakeholders and agencies that have hosted and managed the databases over the years. Early hydrogeological information on Africa, due to historical reasons, is largely kept within various European geological surveys, in particular BGS, BGR, BRGM and TNO. Investigations carried out by African countries themselves are not systematically reported and thus not made readily available, except in few countries. Groundwater occurrence in Africa is characterised by the large variety of geological formations, structures and the climatic differences that condition the regional hydrogeological settings. In general, Africa groundwater features can still be characterized by the four major hydrogeological domains that are weathered basement aquifers, aquifer systems associated with Great Rift Valley, major sedimentary basins and unconsolidated deposits along drainage systems (Figure 2.1). The detail of the four types of hydro-stratigraphic units are discussed below.

2.1.1 *Major types of aquifers*

The major aquifers systems in Africa include North Africa sedimentary basins, Kalahari and Karoo basins, Atlas and TMG systems, Nubian sandstone and Iullumeden systems, West and East basement aquifer systems and aquifer systems associated with great rift valley. In the following section, we will introduce a few main types of aquifers.

2.1.1.1 *Basement aquifers*
Basement aquifers are often referred to as fractured water-bearing igneous and metamorphic rocks with low storativity and poor permeability in the near surface regions of the Earth's crust. They are found extensively in sub-Saharan Africa in the form of usually fractured Precambrian crystalline rocks, which account for about 40% of all aquifers and maintains livelihood for more than 200 million people in Africa. Due to their ready availability and portability, the basement aquifers collectively play a vital role in meeting the MDGs on water in the southern Africa, west Africa and east Africa.

Groundwater storage of basement aquifers is largely recharged by rain that infiltrates through the regolith (i.e. weathered overburden) into the underlying bedrocks. For basement aquifers located in humid and sub humid tropical climatic sub regions, recharge to the underlying

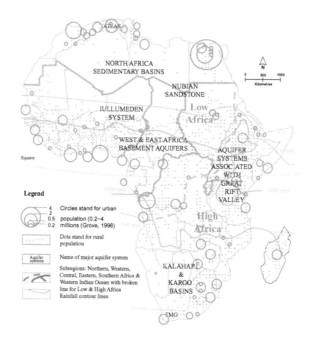

Figure 2.1.   Simplified hydrogeological domains of Africa. (See colour plate section).

fractured and unweathered bedrock is associated with the high storativity of the weathered overburden. The main source of recharge, in wet-dry and dry tropical climatic sub regions, is the drainage systems that are aligned along fracture systems. In arid and semi-arid regions, the storage is occasionally augmented through much thinner regolith by extraordinary floods.

It is recognized that a thick saturated regolith is essential (especially in humid regions) for adequate aquifer storage and available drawdown (Taylor and Howard, 2000; Chilton and Foster, 1995; and Acworth, 1987). Groundwater, within the lowermost unweathered basement rocks, is stored in interconnected systems of fractures, joints and fissures associated with regional tectonism (Reboucas, 1993). The hydraulic conductivity can vary, within the same rock mass, by orders of magnitude and over short distances. Water is generally stored and transmitted in fractures and fissures through a relatively impermeable matrix. Basement aquifers have very low transmissivity (T) values (i.e. geometric mean) ranging generally from 1 to 5 m^2/day, calculated in relation to a saturated thickness of the regolith varying from 12 m to 22 m (Chilton and Foster, 1995). The weathering (i.e. dissolution of minerals) and leaching processes tend to increase the porosity, permeability and specific yield, while the deposition of clay minerals (as products of the weathering processes) can cause a reduction in these hydrogeologic properties (Chilton and Foster, 1995; and Acworth, 1987).

Chemical weathering is the dominant process in the development of the weathered overburden on basement rocks (Acworth, 1987). Weathering extends far below the water table in certain areas where groundwater is the principal chemical reagent. The natural chemical quality of groundwater from basement aquifers in tropical regions of Africa is considered to be of acceptable quality (Chilton and Foster, 1995). The natural groundwater chemistry, being the product of various weathering processes, will exhibit vertical differences in chemical composition due to varying mineral assemblages at different stages of weathering and leaching in the regolith (Chilton and Foster, 1995). Furthermore, the groundwater quality varies over short distances due to complex groundwater flow patterns and weathering processes.

As the variation of the basement aquifer profile is largely dependent on local climate conditions, groundwater development/exploration in basement aquifers is notoriously complex, especially where a thinner weathered overburden is present. The weathered overburden is usually the main groundwater storage compartment, although boreholes may be developed in the underlying fractured bedrock. Basement aquifers in humid regions are characterized by saturated regolith thickness in excess of 10 m and shallow groundwater levels (Chilton and Foster, 1995). Similar aquifers in arid regions are characterized by relatively thin saturated regolith generally present above deeper groundwater levels. This necessitates the drilling of deep boreholes to intercept structural features and contact zones at depth within the unweathered bedrock. In contrast to the acceptable yield and quantity for basement aquifers located in humid and sub-humid tropical climatic sub-regions, the yield and quality for basement aquifers located in wet-dry and dry tropical climatic sub regions have generally been poor (Reboucas, 1993).

Gustafson and Krasny (1994) gave an account of the occurrence, properties and importance of basement aquifers in particular Third World countries. For a more detail treatment of the basement aquifer in Africa, the reader is referred to Acworth (1987), Wright and Burgess (1992), Olofsson (1993), Reboucas (1993), Gustafson and Krasny (1994), Chilton and Foster (1995), Lloyd (1999).

2.1.1.2 *Aquifer systems associated with Great Rift Valley*

Aquifers associated with Great Rift Valley are complex lava systems which are mainly distributed in the east Africa, especially in Ethiopia (see chapter 6 for more detail). It is interesting to notice that the aquifer systems seem to provide the headwater for three major river basins, i.e., Nile River, Congo-Zaire River and Zambezi River. This part of Africa's high land is recharged by precipitation in mountainous areas and groundwater is often discharged to surface water bodies forming regional lakes, which in turn become the sources of the big rivers.

2.1.1.3 *Sedimentary basins*

The sedimentary basins consist of the North Africa basins, including Quaternary basins, and southern Africa (or sub-Saharan) basins, mainly made up by Karoo and Kalahari basins. They are dominantly medium to low yielding shallow aquifers with a few transboundary systems significant at regional scale.

In addition, unconsolidated sediments occur in and along drainage systems, which make up significant aquifers for urban and community water supplies in the vicinity of the sedimentary aquifers.

Among these sedimentary basins, a most significant and risky one is the Atlantic coastal aquifer system. In the densely populated coastal areas along the Gulf of Guinea, high population densities and growth rates, are leading to rapid urbanization and economic development. The coastal plains are characterized by fast-growing future mega-cities, including the national capitals and the peripheries of Abidjan, Accra, Lome, Cotonou and Lagos, with intermediate to densely populated provincial towns, and semi-urban areas. Furthermore the coastal zone contains important fisheries, coastal lagoons and wetlands. These areas are usually also the heart of national industrial production.

The coastal aquifers in the Golf of Guinea (GOG), in common with the other transboundary coastal aquifers along the African coast, are essential to the sustained long term functioning of Large Marine Ecosystems (LME's). The aquifers in the GOG also support a series of critical terrestrial services, such as public water supply, coastal lagoons and mangroves.

2.1.2 *Groundwater resources in Africa*

Groundwater resources are identified as important source of water supply in many parts of Africa. While groundwater supplies are of obvious importance to many rural settlements, townships and cities in arid and semi-arid areas, they are also extensively utilized in humid areas. This is largely due to the fact that they require little or no treatment and can be cheaply developed.

At this stage no readily acceptable continent-wide assessment of groundwater resources exists. For instance, one figure cited by World Water Council (2006) states that groundwater resources account for about 15% of Africa's renewable water resources, amounting to about 810 billion m^3 per year. As pointed out by Braune and Xu (2009), this assessment of groundwater use has focused on its bulk water supply and may have produced a picture that did not come close to capturing groundwater's unique characteristics and potential role. In another extreme case, a groundwater storage of 5,5 million km^3 was estimated (UNESCO IHP, 2003), which would require an aquifer that must stretch over the entire continent and have the average thickness of close to 2 km below the ground if an average porosity of 10% is assumed for such an aquifer. In a modelling assessment, groundwater represents 51% of Africa's renewable water resources (Doell et al., 2008). Here, still clearly lies a challenge for hydrogeologists to improve these estimates.

Perhaps one good handle on improvement of our understanding of renewable resources would be to look at groundwater recharge estimation. Different methods—such as the hydrogeological modelling, tracer method, water balance approach, base flow, chloride mass balance, GIS and remote sensing approaches—have been used to estimate recharge rates in different parts of Africa. Although this information may be available somewhere, it remains impossible to characterise each of the geographic regions of Africa (Northern Africa, Western Africa, Central Africa, Eastern Africa and Southern Africa) with a single realistic recharge value that can be used for comparative purposes. This is due to the large size of each region with its diversity in geological settings, topography and climatic conditions.

Some expressions of regional recharge estimation, based on investigations at local scales, in different parts of Africa are summarised below:

Western Africa: Favreau et al. (2001) estimated a recharge of 10–30 mm/yr in a semi-arid south-western part of Niger; while Le Gal La Salle et al. (2001), using radiotracers method in the Iullemeden basin, estimated a recharge of 5 mm/yr in the unconfined continental terminal aquifer the semi-arid Niger. Using base flow method, Totin et al. (2008) calculated a value of 405 mm/yr in the coastal sedimentary basin of Benin.

Northern Africa: Wright, E.P. et al. (1982) gives a recharge of 10 to 20% of precipitation for the Kufra and Sirte basin (eastern Libya) while a range of 4 to 8 mm/yr was estimated in the Central Sudan (Abdalla, 2008).

Central Africa: Using isotope method, B. Ngounou Ngatcha et al. (2008) gives a recharge of 14 to 45 mm/yr for an unconfined aquifer in the vicinity of the Lac Chad Quaternary aquifer. Recharge information is not available for the Congo basin due to lack of hydrogeological investigations in this region. In volcanic region of Virunga, Christian (2007) find a recharge of 300 mm/yr.

Eastern Africa: Taylor, R. and Howard, K. (1995), modelling the groundwater flow in a regolithic unconfined aquifer in Uganda, calculated a recharge of 200 mm/yr while in Tanzania Onodera (1993) and Mjemah (2008) found similar recharge values in the upland of Tanzania and in Dar es Salam, respectively. 53% of precipitation (550 mm/yr) and 240.7 mm/yr we calculated in these sites characterised by sand and silt aquifer and Quaternary sand, respectively. In the volcanic terran of Ethiopia, Ayenew (2008) gives a range between 10 and 150 mm/yr which is about 10 to 20% of rainfall amount.

Southern Africa: In the Kalahari area, recharge have been estimated to 5 mm/yr in the Central Kalahari (de Vries et al., 2000), 1 mm/yr in the Eastern Kalahari (de Vries et al., 2000), and 8 mm/yr in the North-eastern Namibia (Wanke et al., 2008). In South Africa, the average recharge rate of the TMG is about 30 mm/yr. The highest recharge rate is 137 mm/yr associated with rainfall of 1842 mm/yr; the lowest is 0.7 mm/yr associated with 164 mm/yr. Percentage wise, this variation is from 0,28% to 12,6% of the mean annual precipitation (WRC, 2008).

In southern Africa, Namibia, Botswana and the northwestern part of South Africa have semi-arid and arid conditions. Groundwater storages in these areas are not recharged annually, but may be recharged by episodic heavy rainfall events that infiltrate through the soil into the underlying layers. As a consequence of the climate, the arid and semi-arid countries,

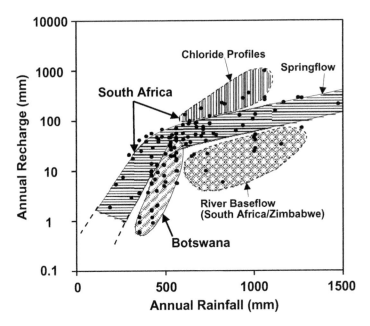

Figure 2.2.   Comparison of results of recharge studies in southern Africa (Beekman and Xu, 2003).

especially southern Angola, Namibia, Botswana and western South Africa, and western Zimbabwe, have lower recharge rates as shown in Figure 2.2. In these areas, groundwater recharge may be limited and is probably largely localized to line and point sources such are streambeds and dam basins respectively. Surface water resources are largely ephemeral, and most perennial rivers in these areas receive their recharge from humid areas. As a result, the groundwater resource has assumed great importance as the principal source of fresh water.

These relatively comprehensive local results can be used as basis for calibration of estimates made at regional scale. FAO (2005) estimated the Internal Renewable Water Resources (IRWR) of Africa, including groundwater resources per country. More recently a GIS based approach was used to evaluate recharge for the African continent (BRGM, 2005; BGR and UNESCO, 2008), with the following estimations:

a. Northern Africa: mostly <5–20 mm/yr (Sahara) and 5–100 (Maghreb and Mediterranean region).
b. Western Africa: 100–500 mm/yr (Guinea, Cote d'Ivoire, Ghana, Benin, Togo, and Nigeria); >500 mm/yr (Sierra Leone, Liberia and southern Nigeria), and mostly <5–20 mm/yr in the Saharan countries.
c. Central Africa: 300–500 mm/yr in the heart of the Congo Basin, the remaining part of the region is mostly recharged with 100–300 mm/yr.
d. Eastern Africa: 5–20 and 50–100 mm/yr, with a high recharge of 100–300 mm/yr surrounding the Victoria Lake.
e. Southern Africa: <5–50 mm/yr, except in Mozambique and some regions of Zambia and Angola where a range between 100 and 300 mm/yr is estimated.

Despite of the availability of these data, one should not attempt to characterise each of the geographic regions of Africa with a sole realistic recharge value. Foe example, by comparison with the more detailed southern Africa estimates, it seems that some regional recharge maps (BRGM, 2005; BGR and UNESCO, 2008; Doell et al., 2008) may overestimate recharge in some regions such as Namibia, where actual recharge does not occur annually, but only episodically. Further detailed estimation/calibration is still required in the different sub-regions of the continent.

Taking a more conceptual approach, salient hydrogeological features in Africa seem to suggest that recharge time scales can be grouped under three typical regions for purposes of resources planning (Figure 2.3). In terms of present conditions, the hydrologic function and distribution of groundwater is highly correlated with rainfall patterns. In more arid areas (Sudano-Sahelian regions and southern Africa) there seems to be little evidence to substantiate the claim that groundwater systems are connected to surface water bodies. In humid regions (Gulf of Guinea and Central Africa) aquifers tend to be connected to river systems and groundwater becomes a major factor determining baseflow (FAO, 2003a).

Large volumes of non-renewable, fossil groundwater resources (Figure 2.3), are a relatively unique feature of Africa, and are located in large sedimentary aquifer systems charged in past pluvial periods (Walling, 1996; Puri and Aureli, 2005). Several form large groundwater reservoirs but contain non-renewable resources that can only be exploited for limited periods when there is very good hydrogeological information (Struckmeier et al., 2006).

Although the groundwater recharge estimate does not fully represent what would be a sustainable resource use, it does provide an indication of what is termed replenished resources.

2.1.3 *Roles and functions*

Groundwater has long been recognized in its role as a geological agent. As water, and in human time scales, it has both a beneficial and a harmful side. It can be a source of production, growth & cooperation, but it can also serve as a source of destruction, poverty & dispute. It must be recognized that as a part of the hydrologic cycle, groundwater and its functioning form an important component in the overall hydrological and environmental system. Despite a serious lack of data, information and knowledge about such roles, some conceptual understanding has become apparent. Such a more conceptual picture of groundwater's role in Africa is presented in Table 2.1.

It cannot be overstressed that groundwater is of strategic importance for rural water supply. Most African countries rely to a large extent on groundwater for their drinking water supply, ranging from shallow hand-dug wells to deeper public supply boreholes. In sub-Saharan Africa 300 million people have no access to safe water supplies, the lowest water supply and

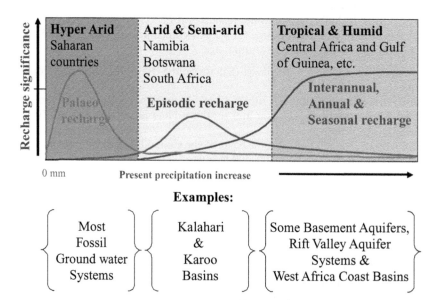

Figure 2.3.   A sketch of groundwater recharge scenarios under different climate conditions.

Table 2.1.   The role of groundwater in Africa.

Item	Roles of groundwater in sub-Saharan Africa		Example
1	High dependence for basic survival	Domestic supply (remote household, village & small town)	Namibia
		Sole source and security during droughts	Botswana
2	Sustainable rural livelihoods	Water supply for school, clinic and stock farming	South Africa
		Irrigation, subsistence-level cropping	Tanzania
		Inland fishery	Tanzania
		Community industries (brick making etc)	South Africa
3	Industrial development	Urban water supply	Benin
		Mining & industrial development	South Africa
4	Ecological services	Ecosystem sustainability (baseflow, wetland, etc)	Malawi
		Tourist development	South Africa

sanitation coverage of any region in the world,—approximately 80% of these live in rural areas (World Water Council, 2006). Therefore, significantly increasing the coverage of rural water supply in Africa is fundamental to achieving MDGs. The health and socio-economic implications of the above situation are enormous in terms of morbidity, mortality and sickness from water and sanitation related diseases such as cholera, children kept out of school and woman deprived of time for productive pursuits due to daily drudgery of fetching water and caring for sick family members. In addition, there is a strong link between the general health and poverty situation and the occurrence and management of HIV/AIDS (AMCOW, 2007).

Groundwater plays an important role in socio-economic development. However this value of groundwater is masked largely by the lack of sectoral integration. Because the links between users and the resource are often not apparent, and because many of the benefits associated with groundwater are public goods (such as environmental maintenance, health and poverty alleviation), the overall economic value of groundwater goes unrecognized.

Although no reliable, region-wide, statistics on groundwater use in Africa are known to exist, Braune et al. (2008) reported that throughout the continent there is very high dependence for domestic water-supply, rural livelihoods and livestock rearing, and increasingly for urban water supply at a range of scales. In addition, water security and development, and ecological services have received increasing attention in Africa.

2.1.3.1   *Water security and development*

Unlike food & energy security, water security reflects water's unique dichotomy. Grey and Sadoff (2008) defines the water security as a ratio of the availability of acceptable quantity & quality of water for health, livelihoods, ecosystems & production to the acceptable level of water-related risk to people, environment & economies. This definition can, for example, be meaningfully applied to the basic rural water supply security norm adopted in South Africa. Its 25 ℓ per person per day water service standard is much easier to be secured than 100 ℓ per person daily, under the same acceptable level of the water-related risk.

Agriculture has remained the most important sector in the African economic development. Only 7% of total arable land is presently irrigated (World Water Council, 2006). The mainstay remains the small farmer, with 70 percent of agricultural output coming from this group (UNESCO, 2001). With regard to groundwater use in agriculture there is very limited data available, but it appears that groundwater is used on only 1–2 million hectares of cropped area, directly contributing to the livelihoods of 1.5–3% of the rural population (Giordano, 2006). The figure includes many traditional and small scale farms of about 1 hectare, still the dominant farm type according to FAO. Despite the problems associated with lack of data or incomplete data, some of the data available do sketch a picture that agricultural groundwater use is important at local scales in parts of sub-Saharan Africa (Masiyandama and Giordano, 2007). The information is also adequate to show the relative sharp contrast to a rapid expansion in agricultural groundwater use in the last few decades, which has transformed rural economies in large parts of the developing world, in particular South Asia and North China. It appears that certain hydrogeologic and socioeconomic factors are currently interacting to deter more widespread and intensive use of groundwater for commercial irrigated agriculture (Foster et al., 2006):

Hydrogeologic Factors:

• the hydrological characteristics of the weathered crystalline basement and the deeper sedimentary aquifers.

Socioeconomic Factors:

• the relatively high capital cost of water well drilling;
• the very low levels of rural electrification, and elevated cost and intermittent supply of diesel fuel, for pumping groundwater;
• the lack of social tradition in irrigated crop cultivation, compared to rain-fed arable cropping and extensive livestock rearing.

In conjunction these factors mean that capital investment in irrigated agriculture remains a relatively high-risk venture in sub-Saharan Africa.

Large areas of savannah, semi-desert and desert areas in sub-Saharan Africa are typified by livestock, rather than crop production. In these areas, groundwater plays a critical role in the maintenance of the livestock economy, which itself is the basis of human survival of the poorest segments (Masiyandama and Giordano, 2007).

No inventory of urban groundwater dependence exists, but in a substantial number of larger towns and cities groundwater is critical to the continuity of the existing water-supply—playing a key strategic role during drought or other emergencies and an important supplementary role at other times. In a few cases the use of groundwater has evolved as part of planned urban water-supply development, but more often it has occurred in response to water shortage and/or service deficiency, and often through private initiative (e.g. Lusaka, Nairobi, Dar-es-Salaam, Addis Ababa, Kampala, Cape Town, Windhoek, Gaborone, Nouakchott, Dakar, Abidjan and probably elsewhere) (Foster et al., 2006). In southern Africa it is estimated that 36% of the urban population relies on groundwater. There are numerous examples in the region where groundwater significantly contributes to supply to bigger settlements (Molapo et al., 2000). The urban explosion, referred to above, is creating unprecedented challenges, among which provision of water and sanitation has been the most pressing. It is in these large unserved areas that groundwater is already playing a major, but mostly unrecorded role (Nyamba and Maseka, 2000). This situation is not going to change in the short to medium term and groundwater's role should be formalized, be it as bridging supply, conjunctive or sole source (Braune et al., 2008). The fact that urban use is widespread shows that it is not so much a question of availability and accessibility and economic feasibility as it is of economic means and political decision and will to develop it (Masiyandama and Giordano, 2007).

2.1.3.2   *Groundwater and health*

World-wide, humans have been benefiting from clean groundwater supplies. Africa has been no exception (Xu and Usher, 2006). Depending upon types of soil and rocks through which groundwater flows, groundwater may acquire a variety of chemical constituents. Groundwater with high concentration of fluoride and iron can have a geogenic or natural cause, while high concentration of nitrate and heavy metals is often associated with anthropogenic impacts. Neglecting these variations of groundwater chemistry due to either ignorance or lack of information can cause harmful or even detrimental effect to the community who relies on the bad quality water as their domestic source, often observed in rural area in Africa. One statistic suggests that about 3% of deaths in South Africa are linked to contaminated water.

Various measures should be considered in management of groundwater quality to prevent communities from drinking water with unacceptable water quality. These measures may include the provision of accurately mapped water quality information, proper borehole construction and protection zoning, and appropriate water treatment, including chlorination of water for drinking water purposes.

As a general safety measure, water for drinking purposes should be disinfected with chlorine, especially during and after floods. When drinking water is disinfected with chlorine an undesired byproduct, chloroform ($CHCl_3$), is formed. The concern over carcinogenic chloroform in drinking water would be quickly dispelled as Potency factor for oral route is classified under the category of B2, a very low value (6.1×10^{-3} (mg/kg/day)$^{-1}$). Suppose a 60 kg person drinking 2 ℓ of water every day for 70 years with a chloroform concentration of 0.01 mg/ℓ. According to Masters (1998), the chronic daily intake (CDI) is 0.000333 mg/kg/day. The incremental lifetime cancer risk is the product of CDI × Potency factor, i.e., 1.46×10^{-6}. So over a 70-year period the upper-bound estimate of the probability that the person will get cancer from this drinking water is about one in a million.

The benefit of proper management of groundwater for community health can be illustrated by a case study in South Africa (Nel et al., 2009). In September 2005 a typhoid and diarrhoea outbreak in the small South African town of Delmas caused a month long health crisis due to contaminated borehole water. A total of 3 000 people were diagnosed with diarrhoea, 561 with typhoid infections and 5 deaths occurred according to official figures. The community claims that more than 49 deaths were caused due to the typhoid and diarrhoea (Groenewald and Dibetle, 2005; Masinga, 2005).

With typhoid having a limited life span outside its host, a simple travel time protection zone for the community boreholes could have prevented the typhoid and diarrhoea outbreak. The cost of implementing groundwater protection zones for Delmas, should as a minimum need to include public participation, awareness building, user training, scientific delineation, monitoring, sampling and data evaluation. Analysing groundwater flow data together with microbial size, lifespan, pH conditions of soils would give a relatively good indication of the distance to be travelled by typhoid bacteria allowing it sufficient time for inactivation (Personne et al., 1998). The total capital cost over an arbitrary 10 year management period (5% inflation considered) to implement and manage this protection zone is estimated at 3 million South African Rand (ZAR) (about US $500 000).

The emergency cost incurred to solve the typhoid and diarrhoea outbreak included medical cost, human resources and the trucking in of clean water. The direct investment cost of the outbreak totals 3.2 million ZAR without consideration of the loss of income for the sick and deaths. When this cost is considered for the 5 official deaths the economic cost of the contamination event exceeds 9 million ZAR (about US $1 500 000).

The economic benefit of preventing contamination for the Delmas case therefore far outweighs the cost of solving and cleaning up the problem after it has happened. Especially, sensitive communities like Delmas are a concern, where typhoid and diarrhoea has hit HIV-positive people harder because of their compromised immune systems. This explains the much higher death rate experienced by the community. Management agencies must therefore take note of the social vulnerability of communities (Cutter et al., 2003; WHO, 2006) when evaluating environmental and economic risks and effects.

2.1.3.3   *Groundwater and poverty eradication*

Water development is closely linked to poverty reduction. Large scale water development projects have played a major role historically in poverty alleviation by providing food security, protection from floods and droughts, and expanded opportunities for employment. In particular, the development of irrigated agriculture has been a major engine for economic growth and poverty reduction. However this kind of development will not be accessible to the majority of Africa's poor, because poverty is deepest in the thinly populated rural areas and rapidly emerging new urban areas. Poverty alleviation in these areas needs to address enhanced livelihoods security, reduced health risk, reduced vulnerability to a variety of risks and pro-poor economic growth (Poverty-Environment Partnership, 2006). In all these, local groundwater resources have a major role to play. Overall, by enabling individuals to accumulate reserves, access to groundwater enables rural populations to reduce their vulnerability, not just to drought, but to the full range of economic and social hazards that generate much rural poverty (Burke and Moench, 2000).

2.1.3.4   *Ecological service*

There is still little appreciation for the ecological services offered by groundwater throughout Africa. As pointed out recently by Braune and Xu (2009), groundwater is essential in sustaining ecosystems and landscapes in humid regions and in supporting unique aquatic ecosystems in more arid regions and along coastal belts. Groundwater discharge in these regions often takes the form of springs, diffuse slow seepage at the ground surface or evapotransporation along riverine ripare. It is these springs and seepage flows that keep rivers flowing during dry periods. In many places in Africa, a system of natural springs provides sufficient water to create ponds, lakes or rivers. For the determination of the ecological reserve (or environmental flow requirements) under the South Africa Water Act of 1998, groundwater-fed baseflow is recognized as an important component of the flow requirements to keep such an aquatic system as a river, wetland and estuary functioning normally and healthily (Xu et al., 2000, Xu et al., 2002, Seward et al., 2006, Colvin et al., 2007). Besides the serious socio-economic consequences of the loss of a life-giving resource, particularly for vulnerable rural communities, there are clear signs of irreversible impacts on the environment through the drying up of whole landscapes, e.g. destruction of wetland and terrestrial ecosystems as well as migration of poor quality water (Brendonck, Pers. communication, 2007).

2.2   THE SUSTAINABLE GROUNDWATER UTILIZATION AND MANAGEMENT IN AFRICA

While a lot has been said in recent times about a water crisis in Africa, the management of groundwater resources has to date not yet featured predominantly in national and regional African water agendas. This is largely attributed to failure to recognise major groundwater-related issues and problems on the African continent, lack of appropriate valuing of groundwater, lack of appropriate regional development approaches/instruments, widespread degradation of groundwater resources, the human issues relating to the sustainable utilization of the common property resource, the transboundary nature of aquifer systems and the lack of information and information management relating groundwater resources management.

2.2.1   *Major groundwater-related issues and problems on the African continent*

Attention to the groundwater issues and problems shown below is seen as essential in moving towards a more sustainable utilization and management of groundwater on the African continent. Groundwater resources account for about 15% of Africa's renewable water resources. Its most important role is for community water supply, with an estimated 75% of the African population using groundwater as its main source of drinking water (ECA et al., 2000). At this stage 300 million people in sub-Saharan Africa still have no access to safe water supplies—approximately 80% of these live in rural areas. The health and socio-economic implications of the above situation are enormous. This makes groundwater a potentially highly strategic resource in Africa.

Key issues relating to its sustainable utilization include:

- The general lack at all levels of appropriately valuing groundwater, a natural resource that is life-giving, and by its nature at the doorstep of every community and could therefore make a major contribution to the achievement the Millennium Development Goals, Africa's Water Vision, and the SADC IWRM Vision and Mission.

- The lack of appropriate approaches for the planning, financing, developing and sustainably utilizing a resource that can normally only be exploited in the form of many, widely distributed, relatively small, individual sources, given Africa's largely hard-rock aquifer environment.
- The wide-spread degradation of groundwater and the resulting diminishing resource base, mainly because of pollution of underlying groundwater in both urban and rural areas, but also over-abstraction, particularly in higher-yielding aquifer systems. There are also strong indications of severe land degradation (soil and vegetation) in areas of poorly managed stock watering from groundwater.
- The well-known human challenges of equity, justice, power and governance. These are at the heart of the debate on any finite natural resource and particularly pressing for groundwater because of its unseen and poorly understood nature. The challenge is a major one in Africa, because of the much greater lack of information and knowledge about the resource and a general lack of appropriate governance structures to help achieve objectives of equity and sustainability.
- The transboundary nature of aquifer systems and the environmental and societal systems that depend on them. It can be foreseen that growing water scarcity will bring increasing water resources as well as related land resources degradation and the accompanying human conflicts, increasingly across international boundaries. Again, these problems are exacerbated through the unseen and poorly understood nature of groundwater.
- The lack of data, information and knowledge about groundwater and its functioning as a critical hydrological and environmental system component and the serious lack of institutional capacity in this regard. This problem is an underlying cause to most of the other groundwater issues mentioned above.

These key issues and problems particularly prevalent in Africa, but also elsewhere in the world, are discussed in slightly more detail below.

2.2.1.1   *Lack of appropriate valuing of groundwater*

The management of groundwater has to date failed to feature prominently in the national and regional water agendas in Africa, except for countries that are virtually dependent on underground water resources. The continued use of a 'private water' legal classification for groundwater is just one of the expressions of its lack of recognition.

At the heart of moving forward with a more sustainable utilization lies the need for a much clearer understanding and articulation of groundwater's role and contribution to national and regional development objectives, in particular:

- community water supply;
- public health;
- rural development;
- small farmer development;
- urban development;
- mining development;
- functioning of vital ecosystems.

Deeper understanding of these roles and relationships is bound to establish that groundwater has a major, still largely locked up, role to play in the fulfilment of the Africa Water Vision of

> "An Africa where there is an equitable and sustainable use and management of water resources for poverty alleviation, socio-economic development, regional cooperation and the environment."

2.2.1.2   *Lack of appropriate regional development approaches and instruments*

No appropriate planning approaches: Large surface storages are often inappropriate for a wide spread of the resulting benefits—and still the alternative solutions called for have not yet addressed groundwater in a systematic way. Planning processes and accompanying study and exploration investment, addressing groundwater systematically through all phases of planning, are not in place.

No appropriate financing approaches: There are obvious hurdles to be overcome in financing many small groundwater schemes with limited accountability and initial cost recovery, compared to large schemes where national governments are ultimately accountable. However, solutions are being explored all over the world and in related sectors, which need to be evaluated and pilot-tested in Africa.

No appropriate sustainable resource development approaches: South Africa's very successful approach of BOTT (Build, Operate, Train, Transfer), for example, utilized to achieve the rapid reduction in the large, country-wide water services backlogs, was not designed to cater for the special needs of water supply from local groundwater sources. The challenge lies in the internationally accepted principle that local and vulnerable groundwater resources need a much more participative approach to development, operation and protection than was possible in this, largely, top-down programme designed to achieve rapid results.

Lost opportunities as a result of such shortcomings not only relate to conjunctive use of surface and groundwater resources, but also for mainstreaming of drought risk management and for development planning, both requiring integrated land and water management approaches. The above present a particular challenge to AMCOW policy direction referred to in chapter 2 regarding the institutionalization of groundwater management by river basin organizations.

2.2.1.3   *Widespread degradation of groundwater resources*

In contrast to the potentially strategic role for groundwater outlined above, it has remained a poorly understood and managed resource. This has become a clear threat to sustainable water service delivery and meeting the Millennium Development Goals on water. Poor understanding and mismanagement is the norm rather than the exception. Examples of vital water supply sources and even whole aquifers being abandoned due to pollution are cited and polluted water served to communities through piped water supply systems leading to typhoid outbreaks and national repercussion (Xu and Usher, 2006). Because of its unique role in the landscape and its largely unseen nature, groundwater and the aquifers that host it are particularly vulnerable to these impacts. In addition, impacts can stay hidden for many years and by the time, they are discovered they are very difficult and costly to remediate.

This situation was brought to the international attention by a major UNEP/DEWA and UNESCO/IHP project undertaken in 11 African cities (Abidjan (Côte d'Ivoire), Dakar (Senegal), Niamey (Niger), Ouagadougou (Burkina Faso), Bamako (Mali), Cotonou (Benin), Keta (Ghana), Mombasa (Kenya), Addis Ababa (Ethiopia), Lusak (Zambia) and Cape Town (South Africa). The overall conclusion from all countries was that pollution of the vital underlying groundwater sources has reached critical levels.

A few recurring aspects that are common throughout the African continent are highlighted here:

- Rapid urban development has resulted in many informal settlements in the urban areas of Africa.
- In the bulk of these areas, groundwater from shallow wells is widely used as the source of water.
- There is a lack of formal domestic waste disposal, sanitation and sewerage/effluent systems and the pit latrines and other sources are often in close proximity to wells and seldom designed to minimize groundwater impacts.

- The result is that nitrate (values in excess of 1000 mg/ℓ recorded) and bacteriological contamination (values in excess of 106 total bacterial counts in certain wells and 106 faecal coliforms in surface waters around Addis Ababa) are the most pervasive parameters giving rise to poor groundwater quality.
- The health effects of this contamination appear to be significant based on data provided by the case studies from such as Mombassa, Lusaka and Addis Ababa.

The evidence of the groundwater quality deterioration in the above-mentioned communities is overwhelming. With few exceptions, groundwater protection strategies in these regions are rare, the communities are not always educated on the groundwater quality issues and significant public health risks to the poorest inhabitants exist. The risk of groundwater pollution should be determined by assessing (1) the vulnerability of the aquifer to pollution, (2) the loading of pollutants to which it may be subjected and (3) societal vulnerability associated with. It is suggested that the need for continued monitoring must be emphasized.

The pervasiveness of these results, and the correlation to results from rapidly urbanized areas elsewhere in the developing world show that groundwater protection strategies should be a very important consideration across the globe.

2.2.1.4 *The human issues relating to the sustainable utilization of the common property resource*

The 'common property resource' issues of economic efficiency, equity, safeguarding of basic needs and of avoiding harm to others are all particularly important for groundwater, because its essential role in meeting a most basic human need and because of its hidden and little understood nature.

The key instruments of local control and popular participation as well as orderly conflict resolution and an enabling environment are challenging to implement and need, first of all, political will.

2.2.1.5 *The issues relating to the transboundary nature of aquifer systems*

As geological formations do not recognise political boundaries, aquifer systems are transboundary by nature. Currently generally recognized transboundary aquifer systems in Africa include large sedimentary basins, e.g. some aquifer systems in Karoo and Kalahari basins in Sub-Saharan Africa and the aquifer systems containing the fossil water in the North Africa. Examples of the latter are the Nubian Sandstone and Iullumeden System where geological events trap water underground, cutting them off from both their sources of supply and their outlets. The transboundary aquifer systems have ecohydrological significance as interactions of groundwater with surface water are often dynamic. The different categories of transboundary issues emerging for large regional ecosystems (UNDP, 1999) are all relevant for groundwater and aquifer systems, i.e.:

- Regional/national issues with transboundary causes/sources;
- Transboundary issues with national causes/sources;
- National issues that are common to at least two or more countries and that require a common strategy and collective action to address;
- Issues that have transboundary implications, e.g. free basic water or drilling subsidies.

A particular issue is that the general lack of understanding about groundwater system functioning could introduce transboundary issues where they may not exist in reality.

Regarding the Atlantic coastal aquifer system, the inevitable urbanization of the coastal areas represents a strong strategic opportunity for regional socio-economic development. There is a call for establishment of regional cooperative framework for joint management and protection of the common natural resources in the context of sustainable integrated coastal area management.

The focus on Africa's transboundary aquifers was established in two international workshops (held in Tripoli, 1999 and 2002) sponsored by UNESCO/IHP under the Internationally Shared Aquifer Resources Management (ISARM) initiative. This resulted in significant scientific statements regarding the sound management of transboundary aquifers, as well as a preliminary inventory of about 40 main transboundary aquifers in Africa (including the Tano and the Keta basins).

Key drivers for an integrated system approach for the management of these transboundary aquifers are the Abidjan Convention with its emphasis on transboundary management of natural resources and the African water sector, through the African Ministers Council on Water (AMCOW) and the regional economic communities, e.g. ECOWAS and SADC, focus on Integrated Water Resource Management.

2.2.2   *The data and information management relating to groundwater resources management*

The lack of information and information management relating groundwater resources management is identified as a part of the major groundwater-related issues and problems. As this is the root cause of many other problems facing Africa, it is specially highlighted here.

The practical experience in many countries in Africa is that data collection is delegated to very junior staff members who are not well trained, not adequately supervised and the outcome from such data collection process needs a lot of groundtruthing or crosschecking before being accepted for interpretation.

In Malawi, for example, there are six levels of governing structures from bottom up, i.e., local Water Committee at a village level, Tribal authorities within a cluster of villages, District Office at township level, Regional Office at a city level and National Department based in the capital. Usually, data are collected by the assistants or technicians in the field, the lowest rank in the government setup, then such data are passed on to District Officers who then pass it on to Regional Officers who in turn on pass it on to Officers at Headquarter Offices (National Departmental Offices) in raw format. At the regional level hard copies of such data are made and kept but often are not calibrated or crosschecked before submission to the National Department. In general, the data are not readily available to the public.

In southern Africa, only South Africa, Botswana, Namibia and Mauritius appear to have well developed and well managed groundwater databases. Zimbabwe and Lesotho also have electronic databases, but these appear to be less well developed. Tanzania, Mozambique, Malawi and DRC appear to have only rudimentary groundwater information systems, and do not possess any electronic format database. Only Lesotho reports that they have external support for their database system. The national groundwater databases are usually stored and maintained by the relevant department in a government ministry. Zimbabwe has now established a quasi-statal entity, ZINWA (Zimbabwe Water Authority), which manages the groundwater sector and the database. Most countries indicate that some level of support for their database management would be valuable. Data is normally available only at a central level, but there may be certain procedures to be completed before data is made available. In South Africa, data is available readily to all areas, and it is not necessary to travel to the database host to obtain data.

With regard to the nature of the data available in the database, South Africa, Botswana, Namibia and Mauritius appear to have the most comprehensive data, with large scale (1:250,000 and 1:100,000) hydrogeological maps, water level and water chemistry data, and scientific studies on the major aquifer resources. Groundwater recharge rates are only well known in South Africa, Botswana and Mauritius. Second ranked countries with intermediate level systems would include Zimbabwe and Lesotho, while Tanzania, Malawi and DRC have only rudimentary groundwater data systems. Other data such as climatic and river flow data are generally available in all countries. All countries report that groundwater abstraction is generally not known, and typically, for private boreholes only the borehole user(s) know how much water is being pumped.

Apart from author's personal experience mentioned above, the rest of sub-Sahara Africa also has major problems regarding the data and information management in groundwater sector. Key issues can be summarised as follows:

- No systematic data base and resource assessment;
- Ad hoc groundwater exploration;
- Inadequate investment in groundwater information;
- Inadequate planning resulting in a lack of drive for improved information;
- Lack of capacity of the groundwater science sector to influence national/regional decision-making.

While poor and deteriorating hydrological networks and institutions, in general, were already seen as a major concern by AMCOW at its Pan-African Conference in 2003, it should be recognized that groundwater information services are the least established and in many African countries virtually non-existent.

2.2.3   *Implication of climate change*

Africa is likely to be the continent most vulnerable to climate change. Climate change has important implications for groundwater recharge. It also impacts the demand for groundwater resources related to spatial and temporal constraints on water supply provision imposed by recurring extreme climatic events.

Climate change impacts include an increase in average temperature (global warming), rise in sea-levels, rainfall pattern changes, and increased incidence of extreme weather events. Such impacts are already part of geological history, as vividly reflected in various sedimentary formations. Importantly, there is the probability that humans may not be able to cope with this global warming, given present environmental policies and management approaches. Major uncertainties regarding groundwater and climate change have a bearing on this:

- Relationship between precipitation and groundwater recharge: It is established that the relation between rainfall and recharge rate is not linear in Africa and there are several recharge mechanisms from annual recharge to episodic recharge;
- Impact of the climate change in the short term: Demand for water would change accordingly. Dependence on groundwater will be increasingly higher. Extreme events like floods would destroy on-site sanitation arrangements, causing pollution in the area affected;
- Impact in the long term: Recharge mechanism may be shifted from annual recharge to episodic one or vice versa. The recharge term of the water balance equation will be altered to accommodate either increased rainfall or decreased rainfall, and in either of these cases, the impact of climate change on groundwater recharge will not be linear, as shown above. The expected type of vegetation adaptation may cause a profile change at the root zone and consequently affect the moisture balance there as suggested by Hogan et al. (2004). The timing of recharge events for the same place may change and be out of phase and mismatched with other ecological processes.

Planning for climate change must involve consideration of climate related risks, including the uncertainty issues raised above. A key issue is a potentially high level of vulnerability of rural populations, particularly of women and children, to groundwater source threats in Africa. The problem has to be addressed through a better institutional arrangement. There is a need for responsibility and commitment on the part of local land and water users for the management of local resources, which is particularly important in a region characterized by high level of ethnic diversity and limited degree of political representation. The proper representation of rural communities, their concerns and the incorporation of successful indigenous arrangements for the management of groundwater resources in basin water management plans in the region is of great importance.

Local coping strategies are an important element of planning for adaptation. Wise land use, the protection and maintenance of groundwater systems and technical installations for the simple access to groundwater resources are keys to prevent groundwater contamination, ensure sustainability of economic investments, and groundwater availability during extreme conditions.

2.3   OUTLOOK FOR THE SUSTAINABLE UTILIZATION OF GROUNDWATER IN AFRICA

At this moment, a potential opportunity for much increased recognition of groundwater's role in Africa may lie in the political pressures, mentioned in the previous section, for the water sector to address growth and development comprehensively. This can be illustrated with a few proposed groundwater-related responses of the South African water sector to the 'Water for Growth and Development Strategy' (DWAF, 2008):

- Focus on multiple uses of water for poverty alleviation and improved livelihoods;
- Need groundwater for multiple uses to achieve spatial equity;
- Need subsidization of appropriate technology (for local resource development).

Rapid progress with such suggestions could be achieved if groundwater would be made an integral part of major existing country and regional development/water resources management initiatives. Examples are:

- the development of Integrated Water Resource Management and Water Efficiency Plans in all countries;
- the widely supported and implemented Rural Water Supply and Sanitation Initiative (RWSSI) in Africa, which, inter alia, promotes the utilization of wide range of appropriate technologies, including improved shallow wells and boreholes fitted with hand pumps, spring development and rain water harvesting as well as the maintenance of infrastructure by local communities (African Development Bank Group, 2005);
- devolution of water resources management to appropriate lower levels, allowing wide-spread stakeholder participation, which is starting to be implemented in many parts of Africa.

A major opportunity has arisen to take continent-wide action through a resolution taken by the African Ministers Council on Water (AMCOW) at its 6th Ordinary Session in Brazzaville in 2007, namely that AMCOW would become the custodian of a continent-wide strategic groundwater initiative.

The first two practical groundwater resolutions that AMCOW took, must be seen as very strategic in their emphasis of the need to create synergies with major existing initiatives, namely (AMCOW, 2007):

- Promote the institutionalisation of groundwater management by river basin organisations to ensure regional ownership of the initiative;
- Create synergy with the RWSSI to ensure groundwater's inclusion in resource assessment and the sustainable management of groundwater resources.

Very significantly, the southern Africa sub-region has already taken the lead with a process, together with AMCOW, towards a stakeholder-responsive broad-based programme and approach to capacity-building for the integration of groundwater resources management into the region's Integrated Water Resources Management (IWRM) initiative (SADC, 2009).

REFERENCES

Abdalla Osman, A.E. (2008). Groundwater discharge mechanism in semi-arid regions and the role of evapotranspiration, Hydrological Processes, vol. 22, no. 16, pp. 2993–3009 [17 page(s) (article)] (1 p.1/4).

Acworth, R.I. (1987). The development of crystalline basement aquifers in a tropical environment, *Quarter. J. Engineer. Geol.* 20 (1987), pp. 265–272.

Alabaster, G. (2008). Meeting the WATSAN MDGs in Africa—The urban sector. Presentation at First African Water Week, Tunis, 26–28 March 2008.

AMCOW (2007). Partnerships Towards Sustainable Utilization of Groundwater in Africa. Brochure prepared jointly by the UNESCO International Hydrological Programme, UNEP and the University of the Western Cape as input to the 6th Session of AMCOW, held in Brazzaville in May 2007.

Ayenew, T. (2008). The movement and occurrence of groundwater in the Ethiopian volcanic terrain. In Grounwater and Climate in Africa, International Conference, Kampala, Uganda, 24–18th June 2008.

Beekman, H. and Xu, Y. (2003). Review of Groundwater Recharge Estimation in Arid and Semi-Arid Southern Africa, In Y. Xu & H.E. Beekman (Eds) "Groundwater recharge estimation in Southern Africa", UNESCO IHP Series No. 64, published by UNESCO Paris. ISBN 92-9220-000-3.

Braune, E. and Xu, Y. (2009). The role of Ground Water in Sub-Saharan Africa, Issue Paper, Ground Water, doi:10.1111/j.1745-6584.2009.00557.x.

Braune, E., Hollingworth, B., Xu, Y., Nel, M., Mahed, G. and Solomon, H. (2008). Protocol for the Assessment of the Status of Sustainable Utilization and Management of Groundwater Resources—With Special Reference to Southern Africa. Pretoria. WRC Report No. TT 318/08, Water Research Commission: Pretoria.

Brendonck, Pers. communication, 2007.

BRGM (2005). Projet Réseau SIG-Afrique. Carte hydrogéologique de l'Afrique à l'échelle du 1/10 Million. BRGM/RP-54404—FR, Décembre 2005 (http://www.sigafrique.net/TravauxMethodologiques/EAU/Rapport_Technique_Hydro.pdf)

BGR and UNESCO (2008). Groundwater resources of Africa, UNESCO Paris.

Burke, J.J. and Moench, M.H. (2000). Groundwater and Society: Resources, Tensions and Opportunities. New York: United Nations.

De Vries, J.J. et al. (2000). Groundwater recharge in the Kalahari, with reference to paleo-hydrologic conditions. *Journal of Hydrology* 238, pp. 110–123.

Chilton, J. and Foster, S. (1995). P.J. Chilton and S.S.D. Foster, Hydrogeological characterisation and water supply potential of basement aquifers in tropical Africa. *Hydrogeol. J.* 3 (1995), pp. 36–49.

Colvin, C., Le Maitre, D., Saayman, I. and Hughes, S. (2007). Aquifer dependent ecosystems in Key hydrogeological typesettings in South Africa, WRC Report No. TT301/07, Pretoria, South Africa.

Cutter, S.L., Boruff, B.J. and Lynn, S.W. (2003). Social vulnerability to environmental hazards. *Soc Sci Q* 84(2): 242–261.

Doell, P., Fiedler, C. and Wilkinson, J. (2008). Diffuse groundwater recharge and groundwater withdrawals in Africa as estimated by global-scale water models, Ch3 WGII AR4, background Water-GAP results, Institute of Physical Geography, University of Frankfurt am Main, Germany.

DWAF (2008). Water for Growth and Development Strategy, Department of Water Affairs and forestry, Pretoria.

ECA, Organisation for African Unit and African Development Bank, (2000). Safeguarding life and development in Africa. A vision for water resources management in the 21st Century. African Caucus Presentation—Second World Water Forum, The Hague, The Netherlands, 18 March 2000.

FAO/AQUASTAT (1995). Water resources of African countries: a review. Revised by Jean Margat in 2001; revision by AQUASTAT and Jean Margat in 2005.

FAO (2003a). Groundwater Management—The Search for Practical Approaches. Water Report 25. Rome: FAO.

FAO (2005). Computation of renewable water resources by country, in km³/year, average. www.fao.org/nr/water/aquastat (accessed on Nov. 2007).

Favreau, G. et al. (2001). Groundwater recharge increase induced by land-use change: comparison of hydrodynamic and isotopic estimates in semiarid Niger. In Impact of human activity on groundwater dynamics. International symposium, Maastricht, 18–27 July 2001.

Giordano, M. (2006). Agricultural groundwater use and rural livelihoods in Sub-Saharan Africa: A first—cut assessment, *Hydrogeology Journal* 14, 310–318.

Grey, D. and Sadoff, C. (2008). Achieving Water Security in Africa: Investing in a Minimum Platform of Infrastructure & Institutions, presented at the plenary session during the First African Water Week 26~28 2008, Tunis.

Groenewald, Y. and Dibetle, M. (2005). Rage flares over typhoid 'spin'. Mail & Guardian, http://www.mg.co.za/.

Gustafson and Krasny (1994). Crystalline rock aquifers: their occurrence, use and importance. *Applied Hydrogeology* 2, 2:64–75.

Hogan, J.F., Phillips, F.M. and Scanlon, B.R. (2004). Groundwater recharge in a desert environment: The southwestern United States, Water Science and Application 9, *American Geophysical Union*, Washington, DC.

Le Gal La Salle, C. et al. (2001). Renewal rate estimation of groundwater based on radioactive tracers (3H, 14C) in an unconfined aquifer in a semi-arid area, Iullemeden Basin, Niger. *Journal of Hydrology*, Vol. 254, issues 1–4, pp. 145–156.

Lloyd (1999). Water resources of hard rock aquifers in arid and semi-arid zones. UNESCO Publication 58. Paris: UNESCO.

Masinga, S. (2005). Is government underestimating deaths in Delmas typhoid and diarrhoea outbreak? TAC Electronic Newsletter Sunday 18 September 2005, http://www.tac.org.za/newsletter/2005/ns 18_09_2005.htm.

Masiyandama, M. and Giordano, M. (2007). Sub-Saharan Africa: Opportunistic exploitation. In the Agricultural Groundwater Revolution: Opportunities and Threats to Development, ed. M. Giordano and K. Villholth, 79–99, Oxfordshire: CAB International.

Masters, G.M. (1998). Introduction to Environmental Engineering and Science (Second edition), ISBN 0-13-896549-8, Prentice-Hall, Inc. 1998, 1990.

Mjemah et al. (2008). Groundwater exploitation and recharge rate estimation of a quaternary sand aquifer in Dar-es-Salam area, Tanzania. In Groundwater and Climate in Africa, International Conference, Kampala, Uganda, 24–18th June, 2008.

Molapo, P., Pandey, S.K. and Puyoo, S. (2000). Groundwater Resource Management in the SADC Region: A field of regional cooperation, IAH 2000 Conference, Cape Town.

Nel, J.M., Yongxin Xu, Okke Batelaan and Luc Brendonck (2009). "Benefit and implementation of groundwater protection zoning in South Africa", Water Resource Manage (2009) DOI:10.1007/s11269-009-9415-4, Springer.

Nyamba, I.A. and Maseka, C. (2000). Groundwater pollution, landuse and environmental impacts on Lusaka Aquifer. In Groundwater: Past Achievements and Future Challenges, Proceedings of the XXX IAH congress, Cape Town, 26.11–1.12.2000. Sililo et al. (ed).

Ngounou Ngatcha et al. (2008). The state of understanding on groundwater recharge for the sustainable management of transboundary aquifer in the Lake Chad Basin.

Onodera, S. (1993). Estimation of a rapid recharge mechanism in the semi-arid upland, Tanzania, using soil water 18O and Cl. In Tracers in Hydrology (Proceesinf of the Yokohama Symposium, July 1993), IAHS Publ. No. 215, 1993.

Personne, J.C., Poty, F., Vaute, L. and Drogue, C. (1998). Survival, transport and dissemination of Escherichia coli and enterococcci in a fissured environment. Study of a flood in karstic aquifer. *J Appl Microbiol* 84: 431–438.

Poverty-Environment Partnership (2006). Linking Poverty Reduction and water Management. UNDP and the Stockholm Environment Institute.

Puri, S. and Aureli, A. (2005). Managing shared aquifer resources in Africa. Ground Water 43, No. 5: 661–668.

Reboucas, A. (1993). Groundwater development in the Precambrian shield of South America and west side of Africa, in *Hydrogeology of Hard Rocks*. Mem. 24th Congress IAH, Oslo, Oslo, Pt.2, pp. 1101–15.

SADC (2009). Outcomes of SADC/AMCOW Workshop on Groundwater for Development in the SADC IWRM Initiative. November 18–19 2008, Gaborone, Botswana. www.sadc-groundwater.org.

Stephen Foster, Albert Tuinhof and Hector Garduño, Task Manager: David Grey (World Bank—AFR), (2006). Groundwater Development in Sub-Saharan Africa: A Strategic Overview of Key Issues and Major Needs, http://siteresources.worldbank.org/INTWRD/Resources/WaterCP15.pdf)

Struckmeier, W.F., Gilbrich, W.H., Gun, Jvd, Maurer, T., Puri, S., Richts, A., Winter, P., Zaepke, M. (2006). WHYMAP and the world map of transboundary aquifer systems at the scale of 1:50,000,000 Special edition for the 4th World Water Forum, Mexico City, March 2006, BGR, Hannover and UNESCO, Paris.

Taylor, R. and Howard, K. (1995). Groundwater recharge to the regolith in Uganda. In Sustainability of water and sanitation systems, 21st WED Conference: Discussion paper, Kampala, Uganda.

Totin, H.S.V. et al. (2008). Groundwater recharge mechanisms and water management in the Coastal sedimentary basin of Benin. In Groundwater and Climate in Africa, International Conference, Kampala, Uganda, 24–18th June 2008.

UNEP (1996). Groundwater: A threatened resource, UNEP Environment Library No. 15, Nairobi, UNEP, 1996.

UNEP (2002). Assessing human vulnerability due to environmental change: concepts, issues, methods and case studies, UNEP/DEWA/RS.03-5. Unites Nations Environment Programme, Division of Early Warning and Assessment, P.O. Box 30552 Nairobi 00100, Kenya.

UNESCO (2001). Integrated drought management: Lessons for sub-Saharan Africa. Brochure with summary and recommendations of 1999 International Drought Conference, Pretoria, September.

UNESCO IHP (2003). World Water Resources at the Beginning of the 21st Century released by the Cambridge University Press and edited by I.A. Shiklomanov and J. Rodda.

WHO 2006. World Health Organization guidelines for drinking—water quality, third edition, incorporating first addendum. WHO Press, Geneva, Switzerland.

Walling, D.E. (1996). Hydrology and rivers. In the Physical Geography of Africa, W. Adams, A. Goudie, and E. Orme. Ed. Oxford: Oxford University Press.

Wanke, H. et al. (2008). Groudwater Recharge Assessment for the Kalahari Catchment of the North-eastern Namibia and North-western Botswana with a Regional-scale Water Balance Model. Water Resour Manage, 22:1143–1158.

Water Resources Commission (2008). Estimating Recharge of the Table Mountain Group Aquifer System. Groundwater Technical Brief, April 2008.

World Water Council (2006). Water Resources Development in Africa. Africa Regional Document. 4th World Water Forum, Mexico 2006.

Wright, E.P. et al. (1982). Hydrogeology of the Kufra and Sirte basin, eastern Libya. The Quaternary Journal of Engineering Geology, 15(2): 83–103.

Xu, Y. and Usher, B. eds. (2006). Grounmaitre D, dwater pollution in Africa. Taylor & Francis/Balkema. The Netherlands.

Xu, Y., Braune, E., Colvin, C., Le Maitre, D., Pietersen, K. and Hotom, T. (2000). Comprehensive determination of Resource Directed Measures for groundwater in South Africa. In Sililo et al. (ed). Groundwater: Past Achievements and Future Challenges, Proceedings of the XXX IAH congress, Cape Town, 26.11–1.12.2000.

Xu, Y., Colvin, C. van Tonder, G.J., Hughes, S., le Maitre, D., Zhang, J., Mafanya, T. and Braune, E. (2002). Towards the resources directed measures: groundwater component. WRC Report No. 1090-2/1/03.

3

Framework of best practices for groundwater supply and sanitation provision

Eberhard Braune & Yongxin Xu
Department of Earth Sciences at the University of the Western Cape, South Africa

3.1 INTRODUCTION

Best practices can be seen as guiding the local actions of groundwater utilization and management. To be efficient and effective, these best practices need to be applied in a broader environment and suitable framework for integrated water, sanitation and hygiene delivery. Logically, the IWRM approach should achieve this integration, but both sanitation and sustainable groundwater resource utilization do not yet fit comfortably under this approach. In urban water management the various elements, including water supply, flood protection and drainage, wastewater treatment and disposal, and maintenance of water-based amenities are already coming together and approaching sustainable service delivery. Still to find its place in this integration is the relatively new function of protecting receiving waters as recreational, aesthetic, environmental and ecological amenity. The South African water quality management policy is used here as example of how this can be achieved and how groundwater quality management can be included, also in rural and peri-urban areas. At least the elements of an overall framework for integrated water, sanitation and hygiene delivery are discussed below under the following headings:

- Integrated Water Resources Management (IWRM);
- Groundwater in IWRM;
- Integrated Water, Sanitation and Hygiene Delivery;
- Integrated service delivery in the urban environment;
- Integrated service delivery in the rural environment;
- A comprehensive approach to facilitate groundwater resource protection;
- Support for local actions.

3.2 INTEGRATED WATER RESOURCES MANAGEMENT

Water and IWRM have a major role to play in poverty alleviation in Africa. Many African countries have adopted Integrated Water Resources Management (IWRM) as the comprehensive and integrated approach to help unlock the full benefits of sustainable water management for poverty reduction and economic growth. The particular water management challenges that IWRM must address here, can best be illustrated by the Africa Water Vision (Economic Commission for Africa et al., 2000):

> "An Africa where there is an equitable and sustainable use and management of water resources for poverty alleviation, socio-economic development, regional cooperation, and the environment."

Integrated management approaches are required that are designed to change the way people view and use the resource. According to Burke and Moench (2000), this involves an appreciation of three effective levels of integration, ie integration within the hydrological cycle (the physical processes), integration across river basins and aquifers (spatial integration) and integration across the overall social and economic fabric at national and regional level.

In pursuing IWRM comprehensively, there is a need to recognize some overriding criteria that take account of social, economic and natural conditions (Global Water Partnership, 2000):

- Economic efficiency in water use: Because of the increasing scarcity of water and financial resources, the finite and vulnerable nature of water as a resource, and the increasing demands upon it, water must be used with maximum possible efficiency;
- Equity: The basic right for all people to have access to water of adequate quantity and quality for the sustenance of human wellbeing must be universally recognized;
- Environmental and ecological sustainability: The present use of the resource should be managed in a way that does not undermine the life-support system, thereby compromising use by future generations of the same resource.

Implementing an IWRM process is a question of getting the "three pillars" right: moving toward an enabling environment of appropriate policies, strategies and legislation for sustainable water resources development and management; putting in place the institutional framework through which the policies, strategies and legislation can be implemented; and setting up the management instruments required by these institutions to do their job (Global Water Partnership, 2000).

While embracing all the Millennium Development Goals in Africa, the initial water focus has been on Target 10: Halve by 2015, the proportion of people without sustainable access to safe drinking water and sanitation (UN Millennium Project Task Force on Water and Sanitation, 2005). This introduced a strong shift all over Africa from a water resource management and sustainability focus to service delivery. With this shift went a decentalization of service delivery, and often completely new institutions at national level, separate from the traditional water resource management institutions, to regulate and support this process.

Only through renewed political pressures for the water sector to address growth and development comprehensively, is there a new opportunity to realistically work towards IWRM in Africa. This has been illustrated in chapter 1 by some proposed key responses of the South African water sector to the 'Water for Growth and Development Strategy' (DWAF, 2008).

3.3   GROUNDWATER AND IWRM

Even in the contemporary accounts of 'integrated' water management, the unique nature of the groundwater system that underpins the whole resource base, is rarely discussed and addressed (Burke and Moench, 2000). Some of the particular challenges to groundwater resources management relate to its unique nature:

- hidden, unnoticed and less understood resource
- generally occurring, widely spread and locally accessible resource

Because the links between users and the resource are often not apparent, and because many of the benefits associated with groundwater are public goods (such as environmental maintenance, health and poverty alleviation), the overall economic value of groundwater goes unrecognized (Burke and Moench, 2000). This situation has led to lack of attention in general to groundwater by decision-makers and a lack of progress in formally treating it as part of IWRM.

Besides the valuation, there is the management issue. Because of its ubiquitous nature and relative ease of local access, we have widely distributed and generally dispersed abstraction points and have many stakeholders, who are involved in its development, use as well as misuse. This complicates the traditional national approaches to resource regulation and requires a very high degree of participative management. It also requires novel approaches to the systematic planning, financing and implementing of hundreds and even thousands of small, locally dispersed groundwater schemes (Braune et al., 2008 and Burke and Moench, 2000).

Burke and Moench (2000) make a plea for the development of effective approaches, which in most cases will require a long-term process through which viable national, regional and local systems can evolve. They suggest a strategic framework as an essential precursor and progressive instrument for effective groundwater management and provide a descriptive model change process in this regard, as illustrated in Figure 3.1 (Burke and Moench, 2000).

In it, the intended relationships between diverse sets of interventions or management approaches and the development goals are specified. It should also specify the relationship between the diverse set of interventions themselves (individual, community and government roles and educational, economic, legal and regulatory mechanisms)—how together they are to form an approach to management that is internally consistent. Foster (2006) comes to a similar conclusion and sees this approach to groundwater resource governance and information provision to be functioning at both the micro and macro level and calls the desired approach a 'top-down facilitation of local actions'.

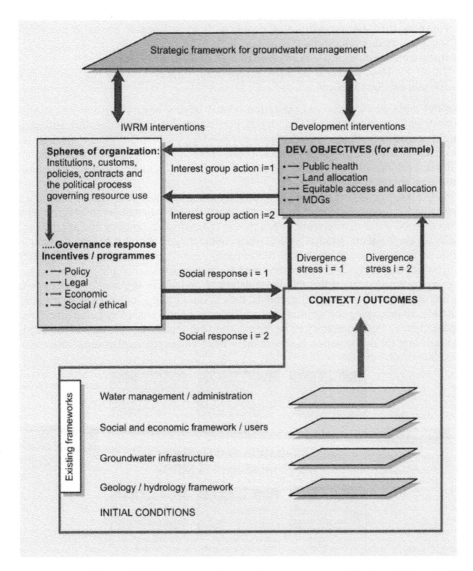

Figure 3.1.   Descriptive model change process in groundwater management (Burke and Moench, 2000).

Such a strategic, multi-stakeholder-driven approach still remains a major challenge in Africa for IWRM implementation as a whole. To make progress in this regard, the SADC Multi-Stakeholder Water Dialogue was introduced a few years ago (Global Water Partnership, 2008).

3.4   INTEGRATED WATER, SANITATION AND HYGIENE DELIVERY

By including water supply, sanitation and hygiene in the MDGs, the world community has acknowledged the importance of their promotion as development interventions and has set a series of goals and targets.

Goal 7: Ensure environmental sustainability

- Target 9: Integrate the principles of sustainable development into country policies and program and reverse the loss of environmental resources.
- Target 10:
  - Halve by 2015, the proportion of people without sustainable access to safe drinking water and basic sanitation.
  - Integrate sanitation into water resources management strategies.
  - Target 11: Have achieved by 2020, a significant improvement in the lives of at least 100 million slum dwellers.

Besides these direct water and sanitation related targets, improved water and sanitation also contributes to most other MDG targets, particularly for health, education and environmental sustainability (Figure 3.2).

An integrated approach should combine awareness-raising about hygiene behaviour and health impacts, improving access to appropriate sanitation facilities and water supplies, and institutional strengthening.

Cohesive programmes are required for protecting watersheds and ecosystems, protecting public health and ensuring sustainable ecosystems. Although IWRM attempts to integrate all these aspects of good water management, it has proven difficult to accomplish this in practice. Historically, an emphasis on technical aspects within this extremely complex system has been at the expense of the hydro-social and public health components (Schuster–Wallace et al., 2008). To move towards true integration will require transdisciplinary approaches and policies and institutional approaches that reflect the ubiquitous nature of, and competing uses for, water in ecosystems and social systems. Some of the required dimensions of integration are shown in Table 3.1.

There is an increasing recognition that an integrated service delivery is only possible if it happens close to the community to be served and with their active participation. The most advanced stage for this situation is service delivery by an elected local government.

Figure 3.2.   Sanitation and hygiene: Key ingredients in MDGs (Schuster–Wallace et al., 2008).

Table 3.1.   Integrated development.

Integrated Community Health Provision
Integrated Water Supply and Sanitation
Integrated Water Resources Management
Integrated Service Delivery

Water supply	**Resource protection**	**Sanitation provision**	**Health Provision**
Water services sector	Water resources sector	Health, Housing, Water services and Education sectors	Health sector
Water resources sector	Environment sector		
Local government	Local government	Local government	Provincial government
Local community	Local community	Local government	Local government
		Local community	Local community

EcoHealth: Examines the impact of environmental factors on health and well-being in order to identify opportunities to reduce associated human morbidity and mortality

EcoHydrology: Explores options for sustainable development of water resources that maintain essential ecosystem processes and services

Hydrosocial: An approach that includes not only the understanding of the hydrologic system, but also the social, political, cultural and economic systems that govern the flow of water through societies
(Schuster–Wallace et al., 2008)

Sustainable development: Improving the levels of service will require focus, not only on domestic use, but also on the productive use of water—a major step in sustainable development

Key principles for an integrated service delivery include (van Zyl, 2005):

- Core responsibility: Local Government is responsible and accountable for water services.
- Water services is an integrated business: water services cannot be rendered in isolation from development of other sectors. Co-ordination is necessary between all tiers of government and role players.
- Business approach: to ensure sustainability and viability of water services, the focus is not only on infrastructure, but also on effective and efficient management, proper institutional arrangements, sustainable delivery and financial management.
- Environmental integrity: the natural and health environment must be considered and protected in all development activities.
- Water and environment has economic value: the way in which water services are provided must reflect the growing scarcity of good quality water and must not undermine the long term sustainability and economic growth.
- Equitable allocation: the limited natural resources available should be equitable distributed among the regions, taking account of population and levels of development.

3.5   INTEGRATED SERVICE DELIVERY IN THE URBAN ENVIRONMENT

The provision of water supply, sanitation and drainage is a key requirement of the urbanization process. Where cities overlie productive aquifers, groundwater is invariably the first water resource to have been tapped. That underlying aquifers can provide a convenient and secure source of water for urban dwellers has long been appreciated. Less readily acknowledged is the use of the subsurface for other aspects of city development, such as wastewater disposal. These have different, and potentially conflicting objectives, that if not understood and managed on an integrated basis, can give rise to serious problems. The linkages of groundwater supply and wastewater disposal are illustrated in Figure 3.3 (Foster et al., 1998).

In most instances urbanization affects underlying groundwater in the following two ways:

- by radically changing patterns and rates of aquifer recharge;
- by adversely affecting the quality of groundwater.

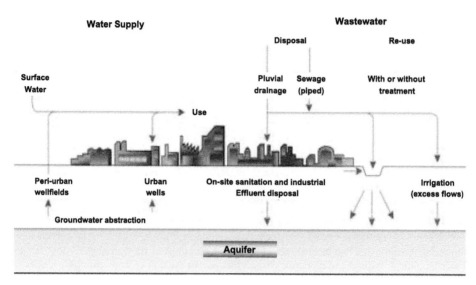

Figure 3.3.   Interaction of groundwater supply and wastewater supply disposal (Foster et al., 1998).

The effect on recharge arises both from modifications to the natural infiltration system, such as surface impermeabilization and changes in natural drainage, and from the introduction of a water service network, which is invariably associated with large volumes of water mains leakage and wastewater seepage.

The net effect of recharge on quality is generally adverse. Urbanization processes cause severe, but essentially diffuse pollution of groundwater by nitrogen and sulphur compounds and rising salinity levels. Relatively widespread groundwater contamination by petroleum products, chlorinated hydrocarbons and other synthetic compounds, and, on a more localized basis, by pathogenic bacteria and viruses, is also encountered.

Groundwater abstraction results in decline in aquifer water levels. Major changes in hydraulic head distribution within aquifers as a result of long-term over-abstraction can lead to the reversal of groundwater flow directions. This reversal can induce serious water quality deterioration, as a result of ingress of sea water, intrusion of other saline groundwater and induced leakage of polluted water form the surface.

Where groundwater is not heavily used, groundwater levels may eventually rise beyond their natural levels because of the additional urban recharge. In extreme cases the water table reaches the land surface and a health hazard may result because septic tanks malfunction and polluted water may accumulate in surface depressions.

The potential extent of the above-mentioned problems depends on the combination of the climatic conditions determining the groundwater recharge regime and the permeability of the subsurface material.

The acuteness and urgency of urban groundwater issues is greater in the urban areas of developing countries because of unplanned urban growth, often haphazard waste disposal and unregulated well water supply. The formulation of effective strategies and techniques for handling urban groundwater-related problems, geared to achieve healthy and sustainable results, will undoubtedly grow in importance in coming years (Tejada–Guibert and Maksimovic, 2004). A summary of the groundwater and sanitation-related risk is provided in Table 3.2.

Peri-urban areas lie at the interface between the rural and the urban environment. Typically, peri-urban areas are marginal to the to physical regulatory boundaries of the formal city, have uncertain or illegal land-tenure, are low income and high density, and lack formal recognition (Tejada–Guibert and Maksimovic, 2004). Most peri-urban areas are unserviced by

Table 3.2.   Integrated urban and rural groundwater supply and sanitation situation development.

	Urban	Peri-urban (informal)	Rural
Settlement density houses per ha		30–50+	<10
Development attention	***	**	*
Groundwater dependency	Often obtained from further afield	Often sole source	Sole source
Sanitation	Water-borne sewerage	Largely on-site sanitation	On-site sanitation
Groundwater contamination risk	*	***	**

piped water supply and sanitation coverage is even poorer. The challenges posed by peri-urban areas can be justifiably classified as social and institutional rather than technological, and requires an integrated solution.

In resource poor and low-population-density areas, on-site sanitation is preferred to off-site sanitation and groundwater is the main source for domestic use. Maintaining groundwater quality under these circumstances is becoming a critical livelihood intervention.

The sanitation option chosen has a major impact on the total water demand as well as on the pollution risk. While waste disposal options are available that require no water, it has been recognized that there are additional health benefits when about 20 litres per capita are provided for this purpose (Tejada–Guibert and Maksimovic, 2004). However, for a variety of reasons, the needs and current practices based on water-borne sanitation in modern urban settlements far exceeds this amount. An important logical effort is now going on into the development of less water-intensive forms of sanitation and into creating a resource out of "waste". Different kinds of toilets with low or no dilution and/or urine diversion are becoming important.

Conjunctive use of surface and groundwater is an important management option under these circumstances to use available resources more efficiently. It essentially takes advantage of the large storage capacity of the aquifer during periods when surface water flow is low or of untreatable quality, and utilizes surface water at other times. For hydraulically unconnected systems, the techniques involved are mainly operational, requiring careful design of water transmission system layouts and pipe sizing.

Urban water management facilitates the sustainable provision of specific services, including water supply, flood protection and drainage, wastewater treatment and disposal, and maintenance of water-based amenities. Urban water management has to incorporate the relatively new function of protecting receiving waters as recreational, aesthetic, environmental and ecological amenity.

3.6   INTEGRATED SERVICE DELIVERY IN THE RURAL ENVIRONMENT

The remaining backlog is largely in rural areas which will require much greater attention in future. A major problem is the misconception which regards small community water supply as merely a scaled down version of urban installations which require less engineering skill or ingenuity. The exact opposite may be the case as the technologies and methods selected must be integrated with the challenges of community involvement which is so essential in small scale schemes.

To plan successful small water supply schemes all possible water sources, infrastructure and management options must be considered and evaluated in terms of their sustainability. This is related to:

- affordability relative to the community's income;
- expected cost recovery and willingness of users to pay for system running costs;

- the appropriateness of the planned technology in terms of its design life, reliability, and the ability of the community to maintain and upgrade the system; and the long-term yield of the selected source (Sami and Murray, 1998).

Empowering people, reducing poverty, improving livelihoods and promoting economic growth requires local, appropriate and widely replicable solutions. The big challenge is that these are not just groundwater-technical solutions, but major societal and institutional ones. The fresh thinking generated through the shift in political priorities from water for basic needs to water for growth and development should help take groundwater in rural areas into a more sustainable utilization environment, requiring integrated management and assessment.

3.7   A FRAMEWORK TO FACILITATE GROUNDWATER MANAGEMENT

A framework for groundwater quality management is emerging in South Africa (DWAF, 2000). It has policy, strategy, institutional development and support provisions (Table 3.3).

Recently it was proposed that all groundwater support by the Department of Water Affairs and Forestry (DWAF) be incorporated into a broader framework of supporting local government in all facets of municipal water and sanitation services (Nel, 2005). The objective was formulated as: "Policy, regulation and support to ensure an equitable, sustainable and effective water service from groundwater." A concept of such a strategy framework is shown in Table 3.4. Brief comments are made on the various components within the framework.

A lot of thinking is required on how this responsibility and workload can be shared. Some examples, that can be developed further, are:

- Secondment of DWAF (national government water resources function) groundwater specialists to local government (an approach that Israel adopted about 20 years ago when local government functions were increased);
- Make this part of Catchment Management Agency responsibility in the longer term;

Table 3.3.   Strategy example framework for groundwater quality management (DWAF, 2000).

Policy principles
Sustainability and equity are recognised as central guiding principles in the protection, use, development, conservation, management and control of water resources.
    The special nature of groundwater must be recognised in implementing policy. Impacts on groundwater are often long term and irreversible. The precautionary principle must therefore be strictly applied when making decisions about groundwater.

Policy goals
The Department will achieve its mission through effecting three policy goals:
- to implement source-directed controls to prevent and minimise, at source, the impact of development on groundwater quality by imposing regulatory controls and by providing incentives;
- to implement resource-directed measures in order to manage such impacts as do inevitably occur in such a manner to protect the reserve and ensure suitability for beneficial purposes recognised by the National Water Act);
- to remedy groundwater quality where practicable to protect the reserve and ensure at least fitness for the purpose served by the remediation.

The institutional parts and the players
The Department will assume the leading role for groundwater quality management at national level and will rely on the following additional roleplayers: other national government departments; provincial and local government; the research community; the affected community; the regulated community.

(Continued)

Table 3.3.   (*Continued*)

Organisational strategies
The Department's organisational approach is based on centralised planning and decentralised implementation at regional and catchment level. Decentralised implementation of source-directed, resource-directed and remediation measures will be implemented by Catchment Management Agencies as part of a catchment management strategy for each catchment. Groundwater quality management will be devolved incrementally to this level.

Community participation
Community participation in water resource management, and more specifically in groundwater quality management, will be facilitated through formal structures such as Catchment Management Agencies and Water User Associations.

Support
The Department will seek to influence sectors that cannot be controlled by direct intervention or incentives. Protection of groundwater in rural and peri-urban areas cannot be achieved through the usual direct intervention or incentive-based instruments. For these sectors the Department will use the following instruments:
- Research and development to build capacity, to advance knowledge and understanding and to develop new and better ways of improving groundwater quality;
- Best Practice guidelines to educate and build the capacity of the community to regulate itself;
- Educational initiatives to raise the level of awareness and develop skills needed to empower communities to protect their groundwater supplies; and
- Extension services to advise and assist communities to implement groundwater protection programmes.

Table 3.4.   Strategy framework for supporting groundwater management at local level.

Management measures	Comment
Policies, standards and guidelines	These must clearly spell out how the desired service should be executed.
Regulations	Regulations can be considered for both the water service provider and for the groundwater technology service provider. The present initiative to register drillers and possibly other groundwater professionals could go a long way in providing a more sustainable groundwater infrastructure.
Training	Whatever knowledge and skills are necessary to convert guidelines and standards into practice, should as far as possible be translated into accredited training material and sustainable courses.
Awareness creation	Reaching the ideal state of capacitated local government will take a long time. This means that an ongoing, focused programme of awareness building will be necessary, done in consultation with the water services sector and their networks.
Information provision	There are vast amounts of knowledge, information and specific data, which could be of great benefit to local government if this was customised and readily accessible.
On-the-job support	This is, from experience, a major need and not easy to quantify. Instead of an ad-hoc approach a programmatic approach is crucial, probably linked to audits, which can provide a wider assessment of failures and successes.
Audit	As regulator we need to audit, but should see the audit as basis for corrective measures and for planning our support role.
Research and development	Once DWAF decides on a programmatic policy, regulation and support approach, R&D requirements will definitely emerge, which can be addressed through the Water Rresearch Commission and its networks.

- A drawdown facility through which local private groundwater professionals could support the DWAF officials in executing their support role;
- Establishment of a "Groundwater Trust" through which the private sector could take a much greater responsibility to ensure a healthy groundwater service provision. This is being done successfully in the United States.

The combination of above measures within the above discussed progressive instrument for effective groundwater resource management as part of IWRM can be seen, at least as the elements, of the desired "top-down facilitation of local actions" approach.

REFERENCES

Braune, E. and Xu, Y. (2008). Groundwater Management Issues In Southern Africa—An IWRM Perspective. Water SA Vol 34 No. 6.14–15 May 2008. GWP Information Note.

Braune, E., Hollingworth, B., Xu, Y., Nel, M., Mahed, G. and Solomon, H. (2008). Protocol for the Assessment of the Status of Sustainable Utilization and Management of Groundwater Resources—With Special Reference to Southern Africa. Pretoria. WRC Report No. TT 318/08, Water Research Commission: Pretoria.

Burke, J.J. and Moench, M.H. (2000). Groundwater and Society: Resources, Tensions and Opportunities. New York: United Nations.

DWAF (2000). Policy and Strategy for Groundwater Quality Management in South Africa. Water Quality Management Series. Department of Water Affairs and Forestry, Pretoria.

DWAF (2008). Department of Water Affairs and Forestry Seminar on Water for Growth and Development, 17.8.08. Stockholm Water Week 2008.

Economic Commission for Africa, Organisation for African Unity and African Development Bank (2000). Safeguarding Life and Development in Africa. A Vision for Water Resources Management in the 21st Century. African Caucus Presentations—Second World Water Forum, The Hague, The Netherlands, 18 March 2000.

Foster, S., Hirata, R., Gomes, D., D'Elia, M. and Paris, M. (2002). Groundwater Quality Protection. Washington: The World Bank.

Foster, S., Lawrence, A. and Morris, B. (1998). Groundwater in Urban Development—Assessing Management Needs and Formulating Policy Strategies. World Bank Technical Paper No. 390. Washington: The World Bank.

Foster, S., Tuinhof, A. and Garduño, H., Task Manager: David Grey (World Bank—AFR), (2006). Groundwater Development in Sub-Saharan Africa: A Strategic Overview of Key Issues and Major Needs.

Global Water Partnership (2000). Integrated Water Resources Management. GWP Catalyzing Change Series.

Global Water Partnership (2002). The Policy Guidance and Operational Tools on IWRM.

Global Water Partnership (2008). The SADC Multi-Stakeholder Water Dialogue. Maseru, Lesotho.

Nel, M. (2005). DWAF Groundwater Capacity to support Local Government. Submission on Water, Environment and Health (UNU-INWEH).

Safe Water as the Key to Global Health. United Nations University International Network.

Sami, K. and Murray, E.C. (1998). Guidelines for the Evaluation of Water Resources for Rural Development with Emphasis on Groundwater. WRC Report No. 677/1/98. Pretoria: Water Research Commission.

Schuster–Wallace, C.J., Grover, V.I., Adeel, Z., Confalonieri, U. and Elliott, S. (2008). Safe Water as the Key to global Health.

Tejada–Guibert, J.A. and Maksimovic, C. (2004). Urban Water Issues—An International Perspective. AGU Water Resources Monograph Series, Volume 16, 2004.

UN Millennium Project Task Force on Water and Sanitation (2005). Health, Dignity and Development: What Will It Take. Millennium Project. Stockholm International Water Institute.

Van Zyl, F. (2005). SA prioritizes water and sanitation delivery. in Engineering News Online, 8.12.05, Creamer Media (Pty) Ltd, Johannesburg, South Africa.

4

Best practice for groundwater quality protection

Yongxin Xu, Thokozani Kanyerere, Eberhard Braune & Jaco Nel
Department of Earth Sciences at the University of the Western Cape, South Africa

Phil Hobbs
Council for Scientific and Industrial Research, Pretoria, South Africa

Kenneth Bradbury
Wisconsin Geological and Natural History Survey, Madison, WI, USA

Nick Robins
British Geological Survey, Maclean Building, Wallingford, Oxfordshire, UK

4.1 GENERAL INTRODUCTION

A strategic framework discussed in Chapter 3 should be seen as an important premise on which international groundwater best practice is promoted and implemented. Currently, there are only limited references of what constitutes good groundwater management in Africa. What has been published addresses mainly resource quantity and quality, resource use and resource vulnerability (UNEP, 1996, Foster et al., 2007, Xu and Usher, 2006). For rural water supply, the reader may be referred to a guide book on developing groundwater (MacDonald et al., 2005). Other, more readily available indicators of good management could be expressions of the available information on groundwater, e.g. how up to date records are, how regularly published and how well expressed in predictive models. However, with the increasing devolution of water resources management to lower levels, in particular river basin organizations, clear guidance will also become increasingly important for the full spectrum of management actions.

A compendium of 'best practice' directions is probably the closest indicators of good management we will come to, given the complex management environment touched on above. An example of such an approach has been an evaluation of SADC groundwater management under IWRM principles (Braune et al., 2008). A very good summary of the scope and practice of groundwater management can be found in The World Bank Briefing Note Series on Groundwater Management (World Bank, 2002). A major step towards putting in place regional best practice has been the 'Guidelines for Groundwater Development in the SADC Region' (SADC WCSU, 2000).

In reality best practice has to adapt to local conditions for the intended benefit. The best practice ranges from borehole location, drilling, construction and equipment, environmental protection, community participation etc. In this chapter, the focus is placed on some key practices that all have a bearing on a sustainable groundwater supply. The sections that follow include borehole construction standards, demarcation of water source protection areas, management of on-site sanitation and other related measures.

4.2 SOUTH AFRICAN BOREHOLE GUIDELINES

Establishing a water supply borehole is a combination of science, trade and art with many and varied nuances that often surprise even the most experienced of geoscientists, geotechnicians and groundwater practitioners. The adage of never being too old to learn enjoys great application in this field, and has mostly to do with the relative inscrutability of the resource being targeted. It is sometimes knowledge of the devil in the detail that determines the difference between success and failure. The literature, both published and electronic, abounds with often exhaustive material on the topic. It would be easy to merely repeat the subject matter in this contribution which, due to length limitations, would necessarily reduce such repetition to largely generic material. Instead, this contribution discusses specific aspects and components that often represent the devil in the detail which, if not recognised or given attention, can lead to costly failure of the process. Included amongst these are contracts/agreements, the supervision of activities, consideration of materials and equipment, recognition of anomalous circumstances, safeguarding of the resource and borehole straightness considerations. If the information conveyed in this chapter contributes to the trouble-free and successful establishment of just a single safe water supply borehole, it will have served its purpose.

4.2.1   *Introduction*

The process of establishing a water supply borehole represents a significant investment in both financial terms and in faith. The financial terms relate to the cost of construction of the facility and, if successful, its subsequent proving and equipping. The faith aspect relates to the anticipation (hope?) that the borehole will be successful, an intangible factor that depends not only on the competence of the person(s) responsible for marking the site of the borehole, but also on the competence of the contractor(s) who will be responsible for seeing the process through to its conclusion. In both instances, competence can be measured by the knowledge, experience and integrity of the parties involved.

The contracted services are provided firstly by the drilling Contractor, whose task it is to sink and construct the borehole. If successful, a test pumping Contractor will be employed to determine the optimal yield of the borehole. Finally, an equipping Contractor will install a pump and fit the ancillary infrastructure to deliver the groundwater on surface. Ideally, these service providers will function under the supervision of a (preferably) experienced geoscientist or geotechnician who will gauge the terms of reference that inform each of these activities as integral parts of a consummate whole.

This contribution explores only a few of many aspects that represent the devil in the detail of establishing a successful and safe water supply borehole. It is the author's opinion that these receive less attention than they deserve, even though neglect thereof can lead to costly failure.

4.2.2   *Relevant literature*

It is appropriate to commence this contribution with a synopsis of relevant literature that informs the subject. The definitive text on borehole drilling and construction is considered by many in the groundwater industry to be the 1089-page publication Groundwater and Wells (Driscoll, 1986). This book has recently been released in its 3rd edition (Driscoll, 2008). Other publications that provide detailed and comprehensive information on the topic are the Manual of Water Well Construction Practices published by the United States Environmental Protection Agency (EPA, 1975), and the AWWA Standard for Water Wells published by the American Water Works Association (AWWA, 2006). Another useful source is the publication Minimum Construction Requirements for Water Bores in Australia (LWBC, 2003) funded by the Australian Land and Water Biodiversity Committee (LWBC). This publication is clear in its aim to ensure, amongst others, "… that the very large investment in bore construction be protected by proper construction methods." The explosion in electronic knowledge brokering represented by the world wide web and the internet has itself spawned freely available literature on the topic (e.g. FCI, 2004; Lifewater, 2004) to those with access to e-communication services.

In a regional context, the Water Sector Coordination Unit of the Southern African Development Community (SADC) commissioned the development of minimum common standards for groundwater development in the SADC region (SADC, 2000a). This project also incorporated country-based situation analysis reports (SADC, 2000b) which report and discuss documentation that informs standards, guidelines and manuals for groundwater development as exist for SADC member states. In South Africa, the Department of Water Affairs and Forestry (DWAF) produced guidelines for groundwater resource development in support of the community water supply and sanitation programme (Hobbs and Marais, 1997), whilst Sami and Murray (1998) also discuss borehole design, construction and development in a rural water supply context. The South African Bureau of Standards (SABS) developed a 9-part set of South African National Standards for the groundwater industry, of which Part 2 (SANS, 2003a) has direct relevance to this topic. It is evident, therefore, that the subject is comprehensively covered in readily available literature. As a consequence, this contribution refrains from repeating especially generic subject matter that can be sourced elsewhere.

4.2.3 *Contracts and agreements*

The process of establishing a water supply borehole is as much a science and a trade as an art. The science and trade find expression in knowing what drilling method and technique to employ in specific hydrogeologic conditions and employing it 'by the book'. The art finds expression in the practices (often and typically innovative) employed to achieve success when the science and trade fails.

Two parties are instrumental in the process. The Owner (Client/Purchaser) who commissions a Contractor to perform the work. The Owner may choose to appoint a third party, the Representative, to supervise and instruct the Contractor on their behalf. Table 4.1 outlines the possible inter-relationship between, and associated implications for, the parties concerned.

A key element that underpins the process is the professional relationship that exists between the Contractor and the Representative. Ideally, this relationship is based on mutual trust in the competencies of the respective parties. Further, that this relationship fosters their functioning as a team whose common goal is the establishment of a serviceable and efficient groundwater

Table 4.1. The possible inter-relationship between, and associated implications for, the parties concerned.

Owner/Contractor/Representative permutation	Implication
Owner employs a Contractor directly, and does not employ a Representative	Contractor fulfils the dual role of Contractor and Representative. Owner accepts that the Contractor will deliver a serviceable product. Owner liable for cost of the contracted service.
Owner employs a Contractor and a Representative separately	Representative assumes responsibility for Contractor delivering a serviceable product. Owner liable for cost of the contracted and representation services.
Owner employs a Representative, who in turn employs the necessary Contractor(s)	Representative assumes responsibility for Contractor delivering a serviceable product. Representative liable for cost of the contracted service(s). Owner liable for cost of the representation and associated services.

supply facility. Although perhaps a contradiction in this context, it is an absolute imperative that the drilling and construction activity, and also the test pumping activity for that matter, be underpinned by a legal contract document that clearly spells out the terms and conditions of the service(s) to be provided. This aspect of the process normally finds a place towards the end of texts on the subject, when in fact it is one of the first aspects to be given attention.

There are numerous standard 'off the shelf' contract documents available, for example that provided in Driscoll (1986) and the Standard Form Drilling Agreement obtainable from the Borehole Water Association of Southern Africa (BWA, 2009). Driscoll (1986) devotes 27 pages to a discussion of borehole drilling and construction specifications and contract problems. To the author's knowledge, the document that covers this aspect the most comprehensively is that of the United States EPA (1975).

It is often what does not appear in the contract or agreement that gives rise to acrimony between parties in the event of something going wrong. An unsuccessful borehole is often cause for unhappiness due to the considerable yet fruitless expense it represents. In most instances, such circumstances also do not signify neglect or the lack of applying due diligence by any party concerned. It is simply beyond reasonable expectation for any party involved with the establishment of a water supply borehole to guarantee its expected or required water-producing success. This is part and parcel of the relatively inscrutable nature of the resource mentioned previously.

4.2.4   *Supervision of activities*

Although full time supervision of the process represents the ideal, the LWBC (2003) recommends that "The client, or representative, should be on site for a substantial amount of the construction period ...". With reference to the tasks of the project coordinator in a drilling programme, the FCI (2004) states that "Actually you should be in the field most of the time during drilling". Unfortunately the financial cost and availability of personnel invariably reduces this service to shorter and often infrequent visits to the site of operations. Seen against a background where drilling contractors "... are not renowned for the quality of their paperwork". (Entrepeneur.com, 2009), these circumstances place an added burden on the Contractor(s) to report as comprehensively and honestly as possible on their operational activities. This will vary from accurately recording time-based parameters such as bit penetration rate during drilling and water level drawdown/recovery during test pumping, to event-based parameters such as water strike depth and yield during drilling, and equipment breakdowns during test pumping.

Perhaps more important than any other aspect, reporting will also include recording circumstances that may compromise the serviceability of the facility at the time or in the future, and communicating such to the Owner or Representative at the earliest opportunity. This will typically include circumstances that constitute a material defect or impediment that compromises the successful commissioning of the facility for its intended purpose. Due to the nature of the facility, such defects and impediments are typically 'invisible', even to the trained eye, from surface. A prime example of this is where equipment, e.g. drill bit, spanner, pump, etc., is lost or left behind in the borehole.

A non-negotiable clause in a contract or agreement must therefore stipulate that the Contractor vacating the site of operations leaves such site, and in particular specific installations such as a borehole, in the same condition as prevailed when first occupying the site/installation. In the case of the drilling Contractor, this might simply entail proving that the borehole is capable of accepting a pump of a given diameter to a given depth, since it is not generally incumbent on this service provider to prove the potential yield of the borehole. Apart from the similar obligation imposed on the next 'occupying' Contractor, it is also in the interest of the latter to establish the exact condition of the installation when first occupying the site. In the case of the test pumping Contractor, this will entail proving that the borehole is capable of accepting a pump of a given diameter to a specified depth both before and after test pumping of the borehole. The initial proving will establish the serviceability of the borehole for test pumping purposes, whereas the final proving will establish the serviceability of the facility for equipping. The given diameter must be the same or greater than that of the pump which will be used for testing the borehole. The SADC (2000a) document recommends that "… the maximum pump diameter should not be more than 80% of the finished diameter of the borehole at the depth where pump installation is planned". In the case of both the drilling Contractor and the test pumping Contractor, this proving is readily achieved with a straightness/alignment test (Section 4.2.7).

4.2.5   *Materials and equipment*

Countries such as Australia and America which enjoy a longer history of mechanised water borehole drilling than any on the African continent, have developed standards for materials and practices that are specific to this industry. The American Society for Testing and Materials (ASTM) standards D2855-96 (ASTM, 2002) and F480-06b (ASTM, 2002), and the Australian AS/NZS 1477:1999 standard for PVC pipes and fittings for pressure applications, serve as examples. In a regional context, the SADC (2000a) document refers to the SABS 966:1998 standard for uPVC pipe as having the minimum characteristics for use as borehole casing. The SANS 10299-2:2003 (SANS, 2003a) standard, however, does not reference a specification in this regard; it only specifies threads, and then in accordance with ASTM F480 (ASTM, undated) or BS 879-2:1988 (BS, 1988). The SAPPMA (2006) technical manual provides information (some of which is replicated in Table 4.2) on uPVC and mPVC pipe as produced in South Africa, but does not refer to the collapse strength from external pressure of this pipe beyond stating that PVC "cannot withstand crushing". It is worth noting that some casing/screen manufacturers may use specially manufactured uPVC pipe with 'odd' specifications to those shown, e.g. Class 12 equivalent (8.5 mm sidewall thickness) pipe with an outside diameter of 168 or 186 mm.

Perhaps few aspects illustrate the devil in the detail adage better than factors which influence the application and use of PVC (polyvinyl chloride) pipe as casing in the construction of a borehole. The groundwater industry has traditionally favoured uPVC (unplasticized PVC) pressure pipe. Technological developments in pipe manufacturing processes have, however, produced mPVC (modified PVC) and oPVC (oriented PVC) pressure pipe with thinner sidewall thicknesses than uPVC pipe of the same class. This characteristic recognises the superior mechanical property of mPVC and oPVC pipe to withstand internal pressure. In the borehole

Table 4.2.   Some information from the SAPPMA (2006) technical manual.

	Class	9		12		16		20	
		\multicolumn{8}{Inside diameter and wall thickness (WT) (mm)}							
	Type	uPVC	mPVC	uPVC	mPVC	uPVC	mPVC	uPVC	mPVC
Outside diameter (OD) (mm)	OD 125	116	118	113	116	109	114	104	111
	WT	4.5	3.5	6.0	4.5	8.0	5.5	10.5	7.0
	OD 140	129	133	126	130	122	127	118	124
	WT	5.5	3.5	7.0	5.0	9.0	6.5	11.0	8.0
	OD 160	148	151	144	149	139	145	134	142
	WT	6.0	4.5	8.0	5.5	10.5	7.5	13.0	9.0
	OD 200	185	190	180	186	174	182	168	178
	WT	7.5	5.0	10.0	7.0	13.0	9.0	16.0	11.0
	OD 250	231	237	225	233	218	227	210	222
	WT	9.5	6.5	12.5	8.5	16.0	11.5	20.0	14.0

application context, however, it is rather the ability of pipe to withstand external pressure such as that created by substantial differences in the water level inside the cased borehole compared to those outside the cased borehole during pumping, that is important. Under these circumstances, uPVC pipe has enjoyed favour due to its superior strength, stiffness and resistance to external pressure than the other types of PVC pipe. This does not mean that mPVC pipe is unsuitable for use as borehole casing. It is probable that a higher class (greater wall thickness) mPVC pipe exists which offers greater strength, stiffness and ability to withstand external pressure than a lower class uPVC pipe. The interested reader can find more information in this regard in Vinidex (2004) and PIPA (2007a; 2007b). It is these circumstances, however, which indicate that an awareness and appreciation for the various types of uPVC pipe and their mechanical properties is important.

Another aspect regarding the use of PVC pipe as borehole casing that warrants discussion, is the joining of sections of this pipe. The SANS 10299-2:2003 (SANS, 2003a) standard allows only screw-threaded joints, forbidding solvent welded socket joints. The SADC (2000) document also states that "… threaded couplings are required …". This is in contrast to Driscoll (1986) and the EPA (1975) and LWBC (2003) documents, which sanction both solvent welded and threaded-and-coupled joints. The LWBC (2003) document is clear both in the type of cement to be used and that it should be "… applied evenly to both spigot [male] and socket [female] ends, applied to the socket end first". The Lifewater (2004) document recommends roughening the joint surfaces with sandpaper prior to the application of solvent cement. In regard to supporting the joint while the solvent cement cures, the LWBC (2003) document instructs that "… only stainless steel screws are to be used", and that "Care must be taken to ensure the screws do not protrude internally". No mention is made of using pop rivets to secure the joint. Since general purpose 'blind' pop rivets are typically manufactured from aluminium, the process of passivation whereby corrosion of the rivet is mitigated by the formation of a protective aluminium oxide coating implies a greater longevity and, consequently, less risk of it corroding away entirely to leave a hole in the joint. Any concern that may exist for the use of pop rivets in securing joints must therefore relate to the internal protrusion of the 'buck-tail' (deformed) end. This end may be sheared off by any equipment inserted or extracted from the borehole, leading to the possibility that the hole occupied by the mandrel could let material such as fine sand or contaminant enter the bore. Under these circumstances, the use of pop rivets is not encouraged so as to eliminate this devil in the detail.

The requirement that the pumping rate maintained during the constant discharge testing of a borehole should not vary by more than ~5% over the duration of the pumping period (EPA, 1975) often precipitates the preferred use of a positive displacement type pump for this activity (Hobbs and Marais, 1997; SADC, 2000). It can be argued that specification of the requirement itself is sufficient, irrespective of how and with what equipment it is achieved, so long as indisputable proof of meeting the requirement is provided.

4.2.6   *Anomalous circumstances*

Every so often during borehole drilling and construction, anomalous circumstances arise that warrant caution and which, if not heeded, could have severe consequences for the successful completion of the facility. An example hereof is where an inordinate amount of material is returned from the borehole during drilling, compared to the theoretical volume of the borehole as a function of drilling diameter and depth of advancement. This is more likely to occur during rotary percussion drilling (even if a surfactant is used) than during direct circulation mud rotary drilling. It is also often accompanied by a slower penetration rate than might be expected for ostensibly 'soft rock' strata, since the material produced is often fine grained, i.e. sandy or silty material as opposed to the rock cuttings/chips obtained from hard, competent strata. It is the author's experience that glauconitic sandstone in particular is prone to such 'behaviour'. These circumstances indicate the likely development of cavernous conditions that might be accompanied by caving or slumping of the borehole sidewall.

This will almost certainly impact on the final construction of the borehole by reducing the depth of the facility or preventing the insertion of casing to the required depth.

Table 4.3 serves to quantify this phenomenon by relating metres drilled and diameter to the theoretical volume of material that should be returned from a bore intersecting fine- to medium–grained soft rock sandstone using the air rotary percussion drilling method. Noticeably more material than indicated would be anomalous and sufficient reason to investigate the circumstances and assess the situation.

4.2.7 *Straightness and verticality*

Borehole straightness (alignment) and verticality (plumbness) are not synonymous concepts. A perfectly vertical borehole is necessarily straight, but a perfectly straight borehole is not necessarily vertical. These concepts are illustrated in Figure 4.1. The consensus of opinion in the groundwater industry is that straightness, within limits of course, is more important than verticality. In the author's experience, it is only the diviner who has targeted a 'water vein' of a specific geometry at a presupposed depth who finds in non-verticality a ready excuse for a failed borehole. Nevertheless, straightness is more likely to be achieved if the drilling Contractor uses a drill collar and stabilizer rods at the working end of the drill string, i.e. immediately behind/above the drill bit, than if these are absent.

The determination of borehole straightness is much more readily achieved than verticality. It merely requires lowering a device (often referred to as a dummy) of specific dimensions under its own weight down a borehole and bringing it back to surface. The free and unfettered movement of the device in its passage down and back up the borehole is sufficient proof of straightness.

The dimensions of a borehole dummy comprise a specified length and a variable diameter determined by the smallest diameter of the borehole being surveyed for straightness. The literature sources consulted specify different lengths, e.g. SANS (2003a) specifies a length of at least 5 m, Hobbs and Marais (1997) and SADC (2000) specify 6 m, whereas EPA (1975), Driscoll (1986) and LWBC (2003) specify 12 m. The latter dummy straddles two standard lengths (6.1 m) of casing when centred over a casing joint. It therefore provides a much stricter survey than the shorter dummies, which straddle only ~50% of a standard length of casing when centred over a casing joint. The SANS 10299-2:2003 (SANS, 2003a) standard specifies the most lenient straightness requirement, and the outcome must be gauged against this knowledge.

The second dimension, that of diameter, defines the tolerance between the outside diameter of the dummy and the inside diameter of the borehole being surveyed. In this regard, the EPA (1975) offers two tolerances, namely ≤13 mm for a borehole diameter ≤254 mm, and ≤25 mm for a borehole diameter ≥305 mm. The LWBC (2003) distinguishes between a tolerance of 20% when using a rigid dummy and 15% when using a test 'dolly', where a dummy is simply a length of pipe (Figure 4.1) and a test 'dolly' a rigid tube forming an axis onto

Table 4.3. Relation among metres drilled and diameter to the theoretical volume of material that should be returned from a bore intersecting fine- to medium–grained soft rock sandstone using the air rotary percussion drilling method.

Strata	Nominal bore diameter (mm)	Volume per metre (l)	Packing coefficient	Equivalent volume	
				(L)	(Buckets)*
Fine- to medium-	165	21	0.8 (80%)	26	~1.3
grained soft rock	204	33		41	~2
sandstone	254	51		64	~3

* 1 bucket = 20 L.

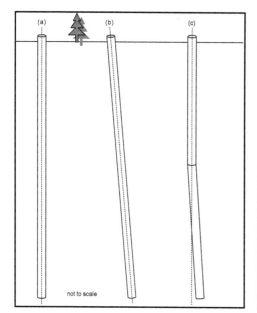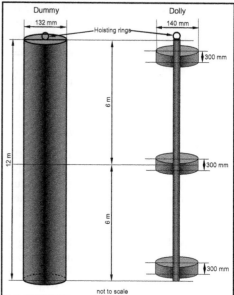

Figure 4.1.   A diagrammatic representation of a borehole that is straight and plumb (a), straight but not plumb (b), and neither straight nor plumb (c); b, diagrammatic representation of a dummy and a test dolly as are suitable for surveying the straightness of a 165 mm nominal diameter borehole according to LWBC (2003) specifications. The 300 mm long pipe sections on the dolly derive from SADC (2000).

which three pipe sections, one at each end and one in the centre, are attached (Figure 4.1). Both Hobbs and Marais (1997) and the SADC (2000) specify a tolerance of 10 mm less than the smallest finished diameter of the borehole being surveyed. On a 165 mm nominal diameter bore, this translates to a diameter of 145 mm, which is similar to that prescribed by the LWBC (2003) as shown in Figure 4.1 for a 'dolly' test.

4.2.8   *Protecting the resource*

It is an absolute imperative in potable water supply applications that a borehole installation not only provides clean and safe drinking water, but also safeguards the resource from possible contamination (Xu and Usher, 2006). This is especially relevant in a rural community water supply context (Braune and Xu, 2006). Whilst the facility cannot secure this for the surrounding environment, access to the resource via the borehole itself is a possibility that must be eliminated. This is achieved by grouting the annular space (annulus) between the borehole sidewall and the surface casing with impervious material, typically neat cement, a sand/cement mixture or concrete, to form a sanitary seal. The construction of a proper sanitary seal must be a mandatory component of every water supply borehole. As with other aspects discussed in this chapter, the literature shows significant variation in the recommended specifications for and design of sanitary seals. The following recommendations sourced from the literature (EPA, 1975; Driscoll, 1986; Ferro and Bouman, 1987; SADC, 2000a; LWBC, 2003; Lifewater, 2004), are put forward as representing 'best practice' in this regard.

   The primary consideration is the depth to which the sanitary seal must extend below surface. Most of the source literature agrees on a minimum depth of 5 m, but all qualify this on the basis of the nature and extent of the near surface material and the proximity of compromising facilities such as on-site sanitation infrastructure. The devil in the detail of this aspect is the requirement that the annular space be sufficiently large and open to receive the grout mixture without restriction for the entire length of the sanitary seal, being completely filled with grout on completion. Further, that the sanitary seal rests on some form of underlying annular fill material. In the case of open borehole completions, this will generally be bed-rock into which the surface casing is embedded. Since drill cuttings accumulating around the borehole on surface are easily flushed into the annulus during drilling, this must be flushed clean before introducing the grout. An annular space of 50 to 100 mm is considered suffi-cient (Driscoll, 1986). This necessarily requires careful consideration of the size of the bore-hole and final surface casing. For example, a nominal 204 mm diameter borehole fitted with 160 mm outside diameter PVC casing (Section 4.2.5) will provide a barely acceptable annular space of roughly 45 mm.

Grout mixtures regarded as suitable include cement/water, cement/bentonite/water and cement/sand/water. The use of old cement, recognised by its lumpy and clotted nature, is unacceptable. Simply pouring the grout slurry evenly into the annulus is only acceptable where the grout interval is visibly open, dry and does not exceed some 9 m in depth below sur-face. It is preferable to introduce the grout mixture from the bottom upwards using a tremie pipe with a nominal diameter of 38 mm and a funnel. The tremie pipe is withdrawn as the annulus fills with grout. The mixing ratios presented in Table 4.4 define ingredient quantities for the respective grout mixtures (after EPA, 1975; Driscoll, 1986; LWBC, 2003).

A borehole which is no longer in use (abandoned or decommissioned) and has been left open, similarly represents an unacceptable risk and potentially lethal threat in its environ-ment. It provides an excellent pathway for pollutants to enter the subsurface and contaminate the groundwater resource. Further, erosion of the borehole at surface might enlarge the open-ing sufficiently for a person to fall into. As recently as 2 November 2008, a 3-year old boy died under tragic circumstances after falling into an abandoned borehole on a smallholding near Polokwane/Pietersburg, South Africa (Louw–Carstens, 2008a; 2008b). The abandon-ment of a borehole is addressed in most of the literature sources referenced in Section 4.2.2. The SANS 10299-9:2003 (SANS, 2003b) standard is specific to this activity. As in the case of sanitary seals, the following recommendations are put forward as representing 'best practice' in this regard. Most of the source literature appears to borrow substantially from the AWWA (2006) recommendations for sealing abandoned boreholes.

The objective is to re-establish as closely as reasonably possible, the hydrogeologic condi-tions that existed before construction of the borehole. This necessarily requires knowledge

Table 4.4.   Relation among grout mixtures, mixing ratios of ingredient quantities and grout volume yields per length of annulus filled.

Grout mixture (per 40 kg bag of cement)	Volume of water (L)	Quantity of bentonite (kg)	Quantity of clean sand* (kg)	Volume of grout made (L)	Length of annulus filled** (m)
Cement/water	20	–	–	~33	3.0
	25	–	–	~38	3.5
	30	–	–	~43	3.9
Cement/bentonite/ water***	40	0.8 (2%)	–	~53	4.7
	50	1.6 (4%)	–	~63	5.7
	60	2.4 (6%)	–	~74	6.7
Cement/sand/water	20	–	60	~60	5.7

* Preferably a fine- to medium-grained sand (0.1–1.0 mm) washed free of clay or organic material.
** Annular space = 50 mm plus 25% allowance for washout or loss of grout into formation.
*** Best practice recommends mixing bentonite and water before adding cement (Driscoll, 1986; LWBC, 2003).

of at least the depth, thickness and water-bearing properties of the strata intersected. In the absence of a drilling record (Section 4.2.4), some of this information may be re-constructed from a downhole geophysical survey and camera inspection of the borehole. Unfortunately the expense and relative scarcity of such a service are prohibitive factors in such an application. Prior to sealing, the borehole must in any event be checked for depth (plumbed) and any obstructions (Section 4.2.7) that may compromise the objective. It is advisable to also remove any casing, and especially surface casing, from the borehole. Boreholes that intersect an unconfined or semi-unconfined aquifer must be sealed by placing cement grout or concrete from the bottom upward in a manner that prevents segregation of the material. In the case of a semi-confined or confined aquifer, the borehole can be filled with clean sand, gravel or aggregate to within 5 m of surface, and sealing the remainder with cement grout or concrete to surface. Finally, the methods and materials used to achieve successful abandonment, as well as the various dimensional parameters such as quantities, depth intervals, etc. must be recorded for future reference.

4.2.9   *Safety*

Unfortunately, loss of human life is not associated only with abandoned boreholes. In an assessment of fatalities among humanitarian workers in Africa, Sheik et al. (2000) report that "Unintentional violence was in some cases related to carelessness, such as running into a spinning airplane propeller or being killed during borehole drilling". Although the nature of the latter instances is not reported, the blanket association thereof with carelessness (as opposed to accident) is questionable. Nevertheless, the nature of any work which involves the handling and lifting of heavy equipment and material poses a safety risk, and borehole drilling is no exception. Although it is the responsibility of the drilling Contractor to ensure that basic safety measures such as cordoning off the work area and the wearing of hardhats, gloves and steel-toe footwear are adhered to, the duty rests on all parties concerned to ensure that this is enforced. The greater concern, especially when working in rural areas where the sinking of a borehole enjoys substantial spectator value, is for the safety of the local populace who may crowd the site and encroach on the work area.

4.2.10   *Conclusions*

This contribution touches on only a few aspects associated with the establishment of a water supply borehole, yet even in these the considerable complexity and variation that exists in the process is highlighted. Further, the measure of inconsistency that exists in regard to standards, specifications and even guidelines is apparent. These circumstances impose considerable obligations on the various service providers involved in the process to ensure that a serviceable groundwater supply facility is delivered. More so, they draw attention to the need for a uniform and consistent set of guidelines that may serve as a definitive reference for geoscientists, geotechnicians and groundwater practitioners operating on the African continent. The Australian (LWBC, 2003) example shows that this does not need to be a mammoth and hugely detailed document. The devil in this detail, however, is reaching an acceptable measure of consensus within and across the groundwater industry.

4.3   GROUNDWATER PROTECTION GUIDELINES

Once groundwater is tapped as source of water supply, its protection merges as a top priority for the sustainable utilization. On-site sanitation can contaminate nearby water points, a commonsense statement but a piece of advice that is often overlooked. The worst situation is that of a pit latrine just up-hydraulic gradient from a shallow well or borehole. Commonsense also tells us that the wellhead should be protected from surface waters and fenced to avoid access by animals. Simple though these rules might be there remains an important task for hydrogeologists in particular and field workers in general to get these messages across to villagers, their Chief, Headman and Water Committee. Disease risks from drinking water infected with faecal coli and viral contaminants include hepatitis, poliomyelitis, diarrhoeal diseases, typhoid, dysentery and cholera.

While the resource protection entails the assessment of aquifer vulnerability and pollutant loading at the scale of an aquifer, the source protection requires the best practice measures that can be realistically implemented around water source points. The latter would be implemented based on the differentiated principle (Xu and Reyders, 1995). A high level of the protection measures should be accorded to aquifers of the high risk associated with the heavily loading contaminants and high vulnerability. In addition the strategic value of the aquifer should also be considered as one of important requirements for accordance of high level protection. This section will introduce three important measures of groundwater source protection including safe distance, borehole protection (wellhead protection) zoning and spring protection.

4.3.1   *Basic approaches*

To guarantee a good quality water supply entails effort in many aspects ranging from borehole siting, borehole construction and pump installation to demarcation of protection zones around a borehole or wellfield (borefield). This section places an emphasis on the latter. Though protection zoning around a well or borehole is widely practised in some developed countries, it proved difficult in its adoption locally in southern Africa and in other parts of the continent. Taking into account the developing nature of socio-economic infrastructures in Africa, this section first introduces a simple minimum distance (or safe distance) as an important measure against well/borehole contamination for Africa. Then the concept of a protection zone is presented for possible implementation wherever feasible.

4.3.2   *Concept of safe distance*

The safe distance or called minimum distance or sometimes termed optimum distance was loosely defined as separation distance between drinking water supply wells and sources of potential or existing pollution. It provides minimum protection measures against bacteria and viruses propagating from on-site sanitation systems. In an Africa context, these pollution sources could include cattle kraals, drinking troughs (feedlot); graveyards (cemetery sites) etc in rural areas. The concept of the safe distance has, for example, been adapted in South Africa in an attempt to deal with the negative impact of pit latrines on groundwater quality. A technical guideline was thus developed by the national Department of Water Affairs and Forestry in South Africa (Xu and Braune, 1995) with "rule of thumb" estimates of the safe distance. Minimum distances of between 10 and 50 m are prescribed between water point and latrine or other point source of pollution. The guideline identifies three factors that dictate the optimum separation between latrine and water point: the depth to watertable, the composition of the soil and the characteristics of the aquifer. The guideline was followed by practical protocols aimed at field operators (DWAF, 1997 and DWAF, 2003).

In certain areas where hydrogeological conditions are known to water professionals, a set of formulae can be used for estimation of the minimum distance (Xu and Braune, 1995). Actual values of the separation distance to be used for specific sites depend on such parameters as aquifer thickness, porosity, hydraulic gradient, average pumping rate and duration.

Another typical example is a minimum distance of 10 m between borehole and pit latrine for basement aquifers, as proposed in the **DFID (UK) ARGOSS Project** (Lawrence et al., 2001).

The concept of the minimum distance is related to a travel time that would ensure the decay of degradable contaminants like bacteria and viruses. It may also delay and dilute encroachment of non-degradable contaminants such as nitrate but cannot prevent their encroachment in the long-term. An alternative would require a much comprehensive zoning equivalent to the catchment area of the borehole concerned.

In short, the minimum separation, together with proper borehole construction, forms the core of a first tier protection strategy to ensure that potential sources of contaminants, such as inappropriate sanitation and poor borehole construction are dealt with immediately.

4.3.3   *Stepwise procedure from case studies*

The link between sanitation and water source has been demonstrated repeatedly with faecal coli contamination of rural community drinking water supplies. The DFID (UK) ARGOSS Project (Lawrence et al., 2001) highlighted the transport of faecal bacteria in groundwater using case study examples from Uganda and Bangladesh. A key product from this work is a set of guidelines with an accompanying set of rules for determining the optimum distance to separate pit latrine from water source in a range of hydrogeological conditions (see Table 4.5).

There are four steps to the assessment:

1. collect information on the location and design of existing water points and effluent disposal units;
2. assess attenuation potential in the unsaturated zone;
3. assess attenuation potential below the water table;
4. assess attenuation potential due to natural groundwater transport.

The easiest way to demonstrate the methodology is by example from Uganda:

- STEP 1 A village is in low relief weathered basement strata, clayey near surface but otherwise granular down to 20 to 30 m below which is fractured bedrock. The water table is between 5 and 10 m below ground level. Long term average annual rainfall is 1000 mm.
- STEP 2 As the water table is shallow the unsaturated zone cannot offer any useful protection above the water table despite the presence of a shallow clay horizon in the upper zone of weathering, because the clay may be fractured or impersistent. Besides the base of the pit latrines may be below the base of the clay unit.
- STEP 3 Shallow dug wells are likely to allow contact with the shallow groundwater just below the water table where contamination is likely to be greatest. In the case of drilled boreholes, well screen needs to be at least 10 m below the water table to afford protection from nearby pit latrines. If it is less than 10 m proceed to Step 4.

Table 4.5.   Horizontal separation guide (after Lawrence et al., 2001).

Rock type	Porosity	Kh: Kv ratio	Likely permeability (m/d)	Is horizontal separation feasible?	Separation needed to reduce pathogen arrival at water point (m)
Silt	0.1–0.2	10	0.01–0.1	Yes	Several
Fine silty sand	0.1–0.2	10	0.1–10	Yes	Several
Weathered basement	0.05–0.2	0.01–10	0.01–10	Yes	Several
Medium sand	0.2–0.3	1	10–100	Difficult	Tens of hundreds
Gravel	0.2–0.3	1	100–1000	No	Hundreds
Fractured rock	0.01	1	High	No	Hundreds

- STEP 4 The Table suggests a separation between latrine and water point of 'several metres'. Erring on the conservative side a suggested separation of 10 m will allow for any higher permeable zones that may be present in the weathered zone.

As a practical guide assume the unsaturated zone and the saturated zone provide inadequate protection for water points and go straight to Step 4. Use the table as a guide to obtain the recommended horizontal separation.

The pollution hazards imposed by the point sources would be greatly reduced by application of the minimum distance during water borehole siting stage. Thanks to its simplicity and no need for complicated calculation, the minimum distance concept, as an immediate, basic protection measure, should be applied throughout Africa where no better protection measures are implemented.

If the safe distance is regarded as the first line of defence against contaminants attack, a protection zone would be a comprehensive approach to deal with the encroachment of both degradable and persistent contaminants toward a production well or borehole.

4.3.4 *Zoning approach*

The core of the protection zoning approach is the wellhead protection or borehole protection zoning adopted in many developed countries, which is illustrated by Figure 4.2. Figure 4.2 shows the basic concept of wellhead protection based on delineating the capture zone, or zone of contributing for a specific well (borehole). In this simple concept, the wellhead protection area, or WHPA, is the surface expression of the region contributing water to the well. In the simple porous-media flow system illustrated in Figure 4.2 the WHPA is a symmetrical area extending from just downgradient of the well to an upgradient groundwater divide.

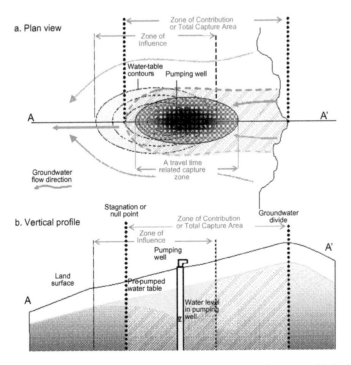

Figure 4.2.   A sketch illustrating basic concept of a wellhead protection area with both plan view in (a) and cross section in (b) (after US EPA, 1987). (See colour plate section).

The delineation of a wellhead protection zone is the process of determining what land or geographic area should be included in a protection zone program. This area of land is then managed to minimize the potential of groundwater contamination by human activities that occur on the land surface or in the subsurface.

Contamination of groundwater sources has been observed world-wide, and it is becoming self-evident that concentrated human activity will lead to even more groundwater contamination. In general it has been shown that contamination of drinking water occurs where three main components exist:

1. A potential source of contamination,
2. An underlying aquifer, and
3. A pathway for transfer between the two.

This pathway can be either indirectly through the soil, or directly through man-made structures which intersect the water table such as boreholes, trenches and quarries. The size and shape of the borehole protection zone depends upon the hydrogeologic characteristics of the aquifer system, and the design and operational characteristics of the boreholes (or wells) used to pump water from the aquifer system.

Delineation of groundwater source protection zones aims to reduce the effect from these different components, based on the fact that all contaminants do not pose the same risk to the aquifer users, and that the contamination flow path in the aquifer, might not affect users far away or in another part of the aquifer. Commonly, zones or areas are delineated to achieve the following levels of protection (Figure 4.3) (Javandel and Tsang, 1986; Chave et al., 2006; Nel et al., 2009):

- A Wellhead Operational Zone immediately adjacent to the site of the borehole or wellfield to prevent rapid ingress of contaminants or damage to the borehole (also referred to as the 'Accident Prevention Zone').
- An Inner Protection Zone based on the time expected to be needed for a reduction in pathogen presence to an acceptable level (often referred to as the 'Microbial Protection Area').
- An Outer Protection Zone based on the time expected to be needed for dilution and effective attenuation of slowly degrading substances to an acceptable level. A further consideration

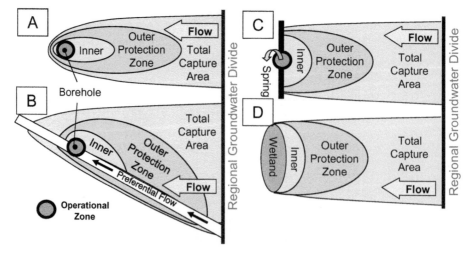

Figure 4.3.   A sketch illustrating protection zones under four hypothetical hydrogeological settings. (See colour plate section).

in the delineation of this zone is sometimes also the time needed to identify and implement remedial intervention for persistent contaminants.

• A further, much larger zone sometimes covers the total catchment area of a particular abstraction where all water will eventually reach the abstraction point. This is designed to avoid long term degradation of quality.

The number of zones defined to cover these functions varies between countries, usually ranging from 2 to 4. By placing some form of regulatory control on activities taking place on land which overlies vulnerable aquifers, their impact on the quality (and in some cases quantity) of the abstracted water can be minimised. The concept can be applied to currently utilized groundwater and to unused aquifers, which might be needed at some time in the future. Whatever number of zones are to be implemented, one can not overemphasise the importance of a sanitary seal in the vicinity of a borehole as pointed out in Section 4.2.8. A typical operational zone where headworks allow drainage of washing water and other spillage away from the wellhead is advisable (Figure 4.4). The zone may also include some water pumping facilities of fixed nature, often built on or attached to headworks including the pump house and some other traditional equipments as illustrated by the Libyan example in Figure 19.1 of Chapter 19 in Part Two of this book.

It is important to provide guidance on activities which are either acceptable, unacceptable or need to be controlled in various protection zones. In some countries such lists are very extensive and specific. In others, general guidance is issued. With each protection zone comes land use constraints. These constraints are of increasing strictness moving from the outer protection zone to the wellhead operational zone. Table 4.6 gives a list of typical land use constraints associated with each zone.

The extent of groundwater contamination is often indicated by the distribution and nature of land uses in an area. From this information, the environmental and public health significance of any such contamination can be assessed (NWQMS, 1995; Usher et al., 2004). For example, areas of concentrated manufacturing industries are often associated with contamination of heavy metalsand organic compounds. Areas of intensive horticulture have been related to excessive pesticide and fertilizer contamination. Landfills generate leachate which can result in groundwater contamination from a range of contaminants.

Figure 4.4.   A typical wellhead protection layout in Malawi.

Table 4.6.   Land use constraints for protection zones (Jolly and Reynders, 1993; Xu and Braune, 1995; Foster et al., 2002; Nel et al., 2009).

Zone	Land use constraint
Wellhead operational zone	All constraints of inner protection zone and outer protection zone Agriculture Traffic—both pedestrian and automotive
Inner protection zone	All constraints of outer protection zone Informal waste disposal Cattle kraals/Feedlots Sewage sludge Small settlements Pit latrines Mining Fuel storage Cemeteries Workshops Farm stables and sheds Roads and railways Parking lots
Outer protection zone	Hospitals Wastewater and sewage treatment facilities Solid waste sites Mass livestock Airports and military facilities Oil refineries Chemical plants and nuclear reactors Large informal settlements using pit latrines Storage of hazardous substances underground

4.3.5   *Protection approach in fractured rock aquifers*

Fractured rock aquifers are common on the African continent, and provide water supplies for hundreds of millions of people. For the purposes of this chapter, a fracture is defined as any hydraulically conductive discontinuity or break in a rock mass, without regard to origin. Although most geologic materials contain fractures at some scale, fractured rock aquifers are water-bearing formations in which most groundwater transport is through a fracture network rather than through a porous matrix. Almost any rock type can contain fractures, and the lithologies that generally form fractured aquifers include crystalline rocks (granite, basalt), metamorphic rocks (schist, gneiss, quartzite), and fine-grained carbonate rocks (limestone, dolomite). In Africa and elsewhere the igneous and metamorphic rocks commonly occurring near the land surface are often called basement rocks. Such rocks, composed of interlocking mineral grains, have very low matrix or primary porosity. Fractures in these rocks form through weathering processes and structural stresses, and the resulting fracture porosity is called secondary porosity. Limestone and dolomite are sedimentary rocks that also can have significant secondary porosity. Groundwater movement through these rocks can widen the fractures by chemical solution to produce karst features (conduits, caves, and sinkholes). Fractured rock aquifers therefore span a continuum from the sparse, small-aperture fractures often found in crystalline rocks such as granite to large bedding-plane fractures and conduit features found in carbonate rocks such as limestone.

Protection of fractured-rock aquifers requires recognition of the hydraulic and geologic characteristics that make these settings so vulnerable to contamination. Such characteristics can include flow through interconnected preferential pathways (fractures, karst conduits, macropores), high flow velocities, low effective porosity, rapid and large changes in hydraulic head or water levels, rapid recharge, and potentially little attenuation of contaminants. Fractured aquifers are generally more vulnerable to contamination than are porous granular aquifers composed of sandstone or sand and gravel. Groundwater movement through fractured aquifers can be very rapid—tens to hundreds of meters per day compared to only a few millimeters or centimeters per day in many granular aquifers. In addition, fractured rocks usually offer less opportunity for natural filtration or degradation of contaminants than do porous rocks. Consequently, many fractured aquifers can transport contaminants rapidly for long distances with little attenuation of contaminants.

The concept of groundwater protection zoning covers a spectrum of activities from general protection of large regions to delineation and protection of detailed contributing areas for specific wells or springs. In all cases, protection is a two-step process, including technical delineation of critical areas followed by the application of zoning or land-use controls to protect those areas. Most groundwater protection plans must strike a balance between the availability of technical resources and data and the need to accomplish meaningful protection in the face of uncertainty and complexity. The goals of groundwater protection must also be clear, and can differentiate between regional groundwater protection, protection of specific wells and springs, and protection of "environmental" groundwater needs such as stream baseflow and maintenance of wetlands.

The key to aquifer protection is a basic understanding of local hydrogeology. Even in complex geologic environments, the basic principles of groundwater flow still hold. Water moves from higher to lower hydraulic head through the most permeable parts of the geologic framework. Recharge depends on the characteristics of the uppermost geologic and soil layers and on the distribution of precipitation and runoff. Surface water features (lakes, rivers, oceans) and geologic structures (faults, unconformities, intrusions) form important boundary conditions. All protection methods, from simple calculations to sophisticated models, must begin with these basic conceptual models.

4.3.5.1   *Index methods for regional groundwater protection*

Most regional groundwater protection plans begin with development of vulnerability or susceptibility maps based on an index ranking method (Figure 4.5). A variety of such ranking methods have been developed, and Gogu and Dassargues (2000) provide a good summary of these techniques. Vulnerability mapping methods usually include overlay mapping of various geologic characteristics controlling groundwater vulnerability, such as depth to bedrock, depth to the water table, aquifer type, soil characteristics, and other factors. Figure 4.5 shows an example of an index mapping techniques applied to a fractured dolomite aquifer in a rural area of Wisconsin in the United States. In this area, thin soils overlie the fractured dolomite. These authors considered soil thickness to be the main control on vulnerability of the aquifer in this area, and the resulting map rates areas of thin soil (less than 1.5 meters thick) as most susceptible to contamination. Recently Vias and others (2006) proposed a vulnerability ranking scheme called the COP method (Concentration of flow, Overlying layers, Precipitation) and applied it to pilot sites in carbonate aquifers in southern Spain. Once vulnerability maps are produced local managers might then enact land-use controls or restrictions on the distribution of land-use activities having the potential to contaminate groundwater.

4.3.5.2   *Delineation of the wellhead protection*

Wellhead protection is essentially the two-step practice of, first, delineating the land surface area contributing water to a specific well (borehole) or wellfield (borefield) and, second, enacting zoning to minimize potentially polluting activities within this area. Application of wellhead protection in fractured rock aquifers remains a challenge as practical approaches for groundwater and wellhead protection in such environments range from simple (establishing a fixed-radius protection zone around a well, with low confidence) to complex (development

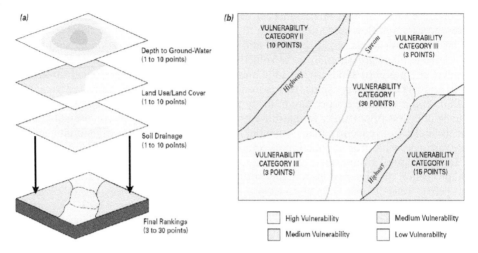

Figure 4.5.   Basic concept of an index method for mapping relative vulnerability to contamination. From Focazio et al. (2002). (See colour plate section).

of a groundwater flow model for a large region, with high confidence) and usually depend on the resources and data available. In 1991 the US Environmental Protection Agency published a manual for delineating wellhead protection areas in fractured rocks (USEPA, 1991). The basic concepts of wellhead protection have changed little since that time, although methodologies have improved due to recent advances in field instrumentation and computer techniques.

Many wellhead protection studies are based on analytical or numerical model simulations of groundwater flow to wells. Numerical groundwater flow models, such as the USGS MODFLOW code (McDonald and Harbaugh, 1988) are based on porous-media assumptions. Even so, such models can provide reasonable results in some densely fractured rock settings if used judiciously and if the fractured rock characteristics are considered as part of the conceptual model and numerical model construction. For example, Rayne and others (2001) applied a MODFLOW model to delineate contributing areas for municipal wells finished in fractured dolomite at Sturgeon Bay, Wisconsin. They incorporated near-horizontal bedding-plane fracture zones as discrete layers in a three-dimensional model, and calibrated the model to local water levels and groundwater discharges. The model-delineated contributing areas were nearly 10 km in length, and travel times from recharge to the wells were less than two years (Figure 4.6).

In sparsely-fractured aquifers, the contributing area for a well can be very complex, and porous media models may not be appropriate. Figure 4.7 (based on Bradbury and Muldoon, 1994) compares pathlines and time-related contributing areas for a well in a uniform porous medium (top) and a well in a sparsely fractured medium (bottom). Both wells are pumping steadily at equal rates in a regional flow field in which groundwater moves from left to right in each diagram. The oval-shaped WHPA in the upper diagram is typical of well capture in uniform porous media. The irregular WHPA in the lower diagram shows how the fracture pattern controls flow to the well. This hypothetical diagram, based on a numerical model, illustrates the potential error in using porous-media methods for delineating capture zones in sparsely fractured aquifers. Worthington and others (2002) provide a real-world example of this problem from a fractured dolomite site at Walkerton, Ontario, where several people died following contamination of a community well. In this example a porous-media computer simulation predicted a 720 hour travel time to a pumping well, while tracer tests showed that actual flow times through bedding plane fractures ranged from only five to 26 hours.

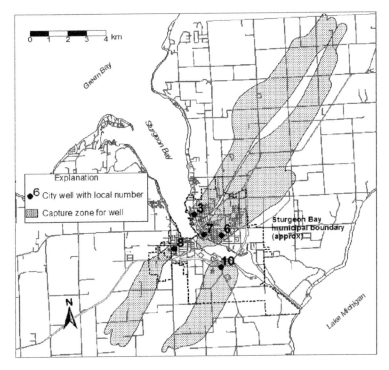

Figure 4.6.   Capture zones delineated using a groundwater flow model in a fractured dolomite aquifer, Sturgeon Bay, Wisconsin, USA. Travel times from the termination of the capture zones to the wells are approximately two years. After Rayne et al., 2001.

Bradbury and Muldoon (1994) and the Minnesota Department of Health (2005) discuss criteria for determining when fractured aquifers can be treated as porous media.

In regions or structural complexity or suspected conduit flow, water level contours can be poor predictors of groundwater flow direction. In such areas, investigators often apply delineation techniques based on structural analyses or utilize groundwater tracers. For example, Ginsberg and Palmer (2002) developed guidelines for delineation of contributing areas in steeply dipping limestone aquifers in the eastern United States, where structural dip exerts a major control on groundwater flow direction. Spangler (2002) demonstrated that dye tracing studies can be useful for delineating contributing areas for springs and wells in a karst terrain in Utah, and showed that porous media approaches can underestimate the size of contributing areas in these settings.

Given the complexities and data requirements for detailed delineations of capture zones for wells in fractured-rock settings, most practitioners use a step-by-step approach, moving from simple mapping techniques to (where appropriate) more complex numerical models (USEPA, 1991). The Minnesota (USA) Department of Health (2005) has developed a manual of suggested procedures for delineating contributing areas in fractured-rock and karst settings. Many of these methods begin with a calculated fixed radius approach but then extend the area to natural hydrogeologic boundaries along known fracture or fault traces.

Assessments of aquifer vulnerability should also include biological, geochemical, isotopic, and temperature data (USEPA, 1991). Such information is particularly valuable in fractured-rock settings because it offers a separate line of evidence from often-uncertain

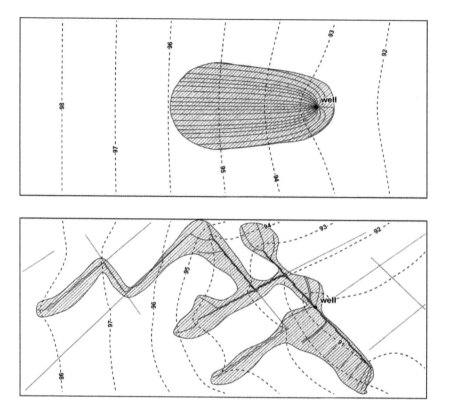

Figure 4.7.   Comparison of a contributing area in a uniform porous medium (top) with a contributing area in a sparsely fractured network (bottom). In both diagrams the hatched areas represent the contributing area for a well pumping at steady state in a uniform flow field. Regional groundwater from is from left to right. Time-related particle paths were generated using a numerical flow model. Dashed lines represent equipotentials. Grey lines in the lower diagram represent connected fractures.

hydraulic calculations. For example, the presence of detectable tritium (>1 T.U.) in a well is important evidence that the local groundwater is relatively "young" and vulnerable to contaminants originating at the land surface. Other geochemical constituents (e.g. nitrate, chloride) are inexpensive to sample and analyze and can also indicate recent inputs from the land surface. Isotopes of oxygen and hydrogen can also help discriminate recharge sources. Groundwater temperature changes often correlate with recharge events. And knowledge of biological contaminants (E. Coli, other bacteria, viruses) is essential for protection of public health and for design of water treatment facilities.

4.3.5.3   *Summary*

The most important considerations for protection zoning in fractured rock environments are, first, a clear understanding of the purpose and scale of the desired protection and second, a clear and accurate conceptual model of the local groundwater system. Regional contamination vulnerability or susceptibility maps developed using index methods are relatively inexpensive to construct and can be the basis for protection zoning that limits contamination sources in the most vulnerable areas. In fractured rock settings these schemes usually give high weight to bedrock depth and soil characteristics as controls on groundwater contamination, with areas of thin or very permeable soil being ranked as extremely vulnerable. Protection of specific wells requires knowledge of the hydrogeology of the region around the well and delineation of the area contributing groundwater to the well. Where rocks are highly fractured the flow to wells can often be approximated by porous-media models, the resulting capture zones are similar to those delineated for porous media. However, wells drawing water from sparsely fractured rocks can have irregular and complex contributing areas, and porous-media approaches and models will not be adequate and in fact can be very misleading. For these situations a conservative approach is often to extend the protected area in the directions of known or suspected conductive features. If resources are available, field studies, including tracer experiments can be used to test connections and determine flow paths to the well. Other geochemical indicators, such as temperature, basic chemical parameters, and isotopes, can be used to help verify the delineated areas.

4.3.6   *Spring and shallow well protection*

Springs and shallow wells are probably still the most common way of water supply to rural communities and even informal urban areas in Africa. The appropriate protection of these already available sources should be one of the highest priorities of any water supply and sanitation programme.

Spring protection is important aspect in modern hydrogeology. As pointed out by Pochon et al. (2008), various techniques including tracer experiment and vulnerability mapping have been applied to groundwater protection zone delineation in several fractured aquifer case studies but criteria for choosing one technique over another for protection zoning purpose are not readily available. Four major types of aquifers in Africa have been introduced previously. These aquifers and their combinations in places allow for a variety of spring occurrences. The most typical and generally observed springs include those located in a topographic depression, along a formation contact, major fault, joints, fracture zone and in the vicinity of Karst cavities as illustrated in Figure 4.8.

Use of springs in Africa has a long history (Xu and Usher, 2006). Community participation plays a crucial role in the sustainable use of this local resource. Mwami (1995) relates a story of

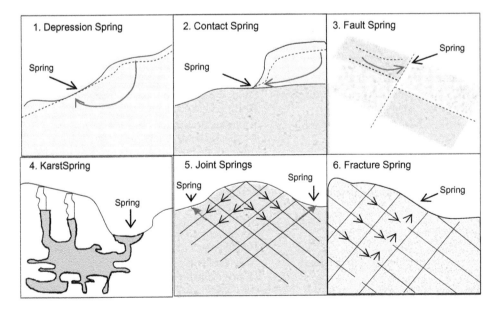

Figure 4.8.   Types of springs often observed in field (modified from Harvey, 2004).

successful spring protection through church-led community participation in Rukungiri in South Western Uganda. As the church there is a very strong indigenous organization, the church leaders are highly respected and their words are taken serious. As a result of the project, a total of 950 springs were protected in the area, serving 56,000 people and providing a 14% coverage.

To illustrate concept of spring protection, Figure 4.9 is used, in which spring protection requires a spring box with a discharge pipe outlet that can be piped to the point of use. The area between the spring and the spring box is backfilled with appropriate material—gravel to act as a filter medium, capped by a thin clay layer and overlain by sand and topsoil grassed over above. The sides of the infill should be contained within a wall of impermeable material such as clay. The filter medium should reach the upper part of the spring flow horizon under wet weather conditions, and the infill area fenced off. The use of a spring box here also facilitates chlorination of the spring water should this be required. For more detail information on designs, the reader may be referred to Howard et al. (2000).

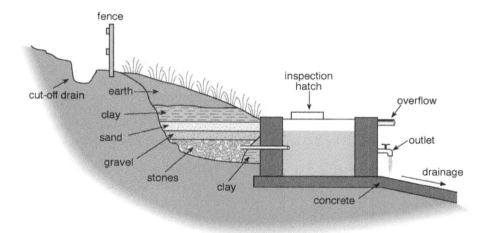

Figure 4.9.   Schematic spring protection cross-section (MacDonald et al., 2005).

Table 4.7.   Sanitary protection measures (after Lawrence et al., 2001).

Source	Sanitary measures	Details
Spring	Local protection works to prevent contamination	Spring eye protected with clay layer and undamaged Works kept in good order
	Surrounding area managed	Spring eye area fenced off Diversion ditches to take run off away Adequate drainage of waste water away from the site No ponding in vicinity
Dug well	Wellhead protected to prevent source contamination	Concrete apron is at least 1.5 m diameter No cracks in apron No ponding of water on the apron Sound join between apron and borehole casing Apron floor slopes away from borehole Handpump or windlass used for raising water
	Surrounding area managed	Area fenced off Diversion ditches to take run off away Adequate drainage of waste water away from the site No ponding in vicinity

Dug wells also require protection including a fence to prevent access by animals. The well lining should be impermeable within the top 1 to 2 m of the well and extend at least 0.3 m above ground level. Concrete rings are ideal for this purpose, with cement poured behind them to form an impermeable seal around the top of the well. If the well is already lined with brick or other material then an annulus can be dug down to a depth of 1 m and infilled with cement. An impermeable wall can then be built around the well to a height of 0.3 m which is founded on the cement seal.

In principle sanitary protection at headworks of groundwater supplies should inhibit contamination of the source by preventing pathways to develop that link the ground surface to the water table. A set of simple measures to protect sources is summarised in the Table 4.7.

If the spring, borehole or well is situated on sloping land then a diversion ditch dug above it will protect it from overland flow. Any abandoned wells or boreholes should either be infilled or properly capped.

4.4   DROUGHT PROOFING

4.4.1   *Basic concepts*

The concept of drought has been extensively researched, particularly from meteorological, sociological and agricultural perspectives. However, drought definitions are not consistent. The lack of universally accepted definitions and the inherent difficulties associated with separating causes from impacts creates confusion when describing drought (Calow et al., 1997).

A meteorological drought occurs when rainfall is abnormally low, i.e. less than a critical precipitation. A hydrological drought results when water supply falls below the minimum required for 'normal' functions, reflecting a deficit in the water balance. Similarly, in broad terms, drought also occurs when there is a deficit in water supply, due to insufficient rainfall and surface and sub-surface runoff and storage, so that demand cannot be met.

Groundwater drought is the phase during which groundwater levels are depleted by insufficient recharge so that the resource is no longer able to sustain normal demand and withdrawal. Theoretically, only when the meteorological drought is advanced does the groundwater drought occur, and only after the meteorological drought is over and recharge to aquifers occurs causing groundwater levels to rise will the groundwater drought end. Concurrent with the onset of groundwater drought, sources begin to fail. Indicators include failure of poorly maintained and overstressed pumps and additional pressure on the declining number of operating water points that a community can rely on. In practice it is rarely the total failure of the resource that leads a community to critical water shortage but their reduced access to the resource. Meteorological drought causes surface water sources to dry up, many of which may be ephemeral, although flow within sand river beds may still occur. A traditional drought coping strategy is the excavation of dug-outs in the sand rivers to sustain both community and its animals. A longer term water storage strategy involves the construction of sand river dams.

Past responses to drought have been reactive, e.g. the mobilisation of emergency borehole drilling, trucking water supplies, community feeding programmes and provision of animal feed (Clay et al., 1995). Responses need to take into account the sustainable management of all natural resources so that appropriate use of the available groundwater can be achieved to mitigate the impact of drought when surface waters are in deficit (Calow et al., 2009). This requires the development of appropriate groundwater management strategies for drought prone areas to mitigate the impact of drought on community life as well as associated natural systems and resources.

There are numerous coping strategies that can be invoked, each designed to lessen the impact of drought on community livelihood and wellbeing. Drought coping strategies include:

Sand-river dams: An impermeable wall is placed across a suitably constricted section of a sand river behind which sediment and water can accumulate. The water in the sand below

2 metres depth cannot evaporate and is stored in the sand following periods of flood. The filling of the reservoir with sand behind a newly constructed dam takes place in about 5 to 7 years. Mixed coarse to fine-grained sands and gravels are deposited by the initial flood, sediments tending to fine upward as the flood subsides (Hussey, 2007). With subsequent floods the finer top layer is eroded away by the new flood and more sand is deposited, so increasing the thickness of the mixed coarse to fine grained sands and gravels.

When the water level in the riverbed decreases following a period of flood, the groundwater in the banks may slowly discharge to the riverbed. This process can occur in ephemeral rivers. As the thickness of sand in the storage dam increases so the water storage potential in the sand also increases. By raising the level of the sand in the river, the 'base level' of the groundwater flow may also be raised, potentially increasing the amount of water that is stored in the banks and the volume that can be withdrawn. The amount of water stored behind a sand storage dam is thus higher than just the amount of water stored in the sand behind the dam. Typical yields from a sand dam are between 2 and 10 m³/day throughout a normal dry season. This is additional water which a community can use to its best advantage with virtually no impact on other local demands for water.

Shallow wells in bank deposits adjacent to sand river dams can draw on the enhanced storage created by the dam. This can enable facilities such as community gardens. Whereas these shallow wells draw on the enhanced storage of the dam, a well or shallow borehole placed below the dam will draw on leakage from it, either under the dam foot or in fractures and cracks in the underlying bedrock. These wells or boreholes actually increase the ultimate yield of the dam. The most important conditions for the application of sand storage dams are:

- Coarse sand and gravels material in the riverbed more than 2 metres thick.
- Poorly permeable layer underneath the proposed dam site.
- Short periods of high intensity storm-event rainfall, that can rework remove silt from the deposits within the sand dam aquifer.
- Organised community to participate in constructing and maintaining the dam and the sustainable use of water abstracted.

Monitoring pump status and repair: This is an essential exercise that is lacking in many countries in Africa and hinders reliable supply. Villagers can be empowered to report the status of their water points in order that repairs be undertaken. However, no programme of systematic maintenance can be effective without monitoring and the two need to be performed together to be effective. Very few if any communities normally monitor water levels or abstraction rates on a systematic basis.

Reserve deep boreholes for emergency use: Deep drilling at hydrogeologically favoured sites between a group of villages has been advocated by some as a valuable form of drought proofing. The boreholes would be opened and equipped only at times of water stress in the surrounding villages. Although it is focussed partly on livelihood, it in fact provides only domestic water as villagers cannot transport other than domestic requirements. Consequently this form of intervention is not cost effective. Reserve and isolated deep boreholes are also open to abuse as enterprising farmers will be tempted to tap the potential of the borehole during normal times in order to provide water for stock watering or irrigation of fodder crops.

Well deepening: Programmes of well deepening and of drilling through the bottom of hand dug wells to increase access to deeper groundwater are only successful if good quality groundwater is available at depth. Although relatively deep aquifers are sourced for domestic wellfield supply in some countries, the aquifers elsewhere—such as the weathered Basement— are, for the most part, relatively shallow and their transmissivity declines rapidly with depth.

Rehabilitation: Repair of broken infrastructure is a valuable form of drought proofing. Effort spent in redressing the operational status of a water point is money well spent and by far a cheaper option than attempting emergency intervention should drought again arise. Rehabilitation of the water points in a village community allows that village to maintain its livelihoods at times of water stress. It is important to retain accessible technology, i.e. repairs to hand pumps rather than installation of motorised pumps for which there is unlikely to be any fuel. There also needs to be some form of borehole failure assessment as breakdown may be due to aquifer resource depletion.

Community empowered monitoring: A simple and cost effective coping strategy is empowerment of a community. One effective means is to equip a community with the means of monitoring its own infrastructure. A length of string and weight is enough to measure the depth to water in its village wells—best done at dawn each day before the daily drawdown on the aquifer commences. This may not be possible in the majority of boreholes where there is no access to the water table from the ground surface. Measurement of the number of buckets drawn from a well provides a useful indicator of demand and how demand might change in response to changed conditions. Other indicators that can easily be self measured include number of cattle owned within a village, or quantity of vegetables taken from a community garden. These data combined with health indicators such as the number of patients dealt with by the local clinic can provide valuable insight into the overall status of the village not just in terms of its food security but its wellbeing and its water security status.

Community empowered water point maintenance: Empowerment of a village to self maintain its infrastructure can take a variety of forms including the appointment of a water committee which might chose to elect a keeper of the community well or wells. A small tariff might be applied against income within the village which allows the purchase of spare parts for pumps and pays the local mechanic to undertake repair as required. It can also be used to maintain the integrity of the sanitary slab at well/borehole top to minimise pollution. Such structures are widespread throughout many of the semi-arid savanah lands of Africa.

Small scale irrigation: Small scale irrigation of community gardens has been widely advocated as a means of improving livelihood and in making rural communities more resistant to external shocks. A variety of water sources may be available, for example, a spring used for stock watering and dirty water left to soak away downstream might be captured for use both for stock and garden watering. Clearly the garden has to be fenced in order to keep out animals, and some reticulation may be required to bring water to the garden. In addition a community garden needs to be accessible to the community and needs to be in close proximity to the village and under its guard from possible predators.

Rainwater harvesting: Tin roofs with gutters are suitable for rainwater collection. Downspouts draining to containers can collect a significant quantity of water to maintain a garden in the dry season at, for example, a district clinic or school.

4.4.2   *Types of drought*

There are several types of drought defined in the literature. Agricultural drought exists when soil moisture is depleted to the wilt point such that rain-fed crop and pasture yields are reduced and some susceptible crops die before cropping. A meteorological drought, the cause of agricultural drought, is the persistent failure of the rains. Typically one failed rain season causes failure of the crops, the second consecutive failed rain season causes death of animals, and subsequent to this the death of people. However, in the meantime livelihood has suffered and intervention is required. Various early warning systems are in place to predict meteorological drought in order that ameliorating strategies can be put in place. However, these concentrate more on food security than they do on water security and are in any case generally inadequate.

Drought is a relative concept, defined in terms of a deviation from the norm. In semi-arid lands it is critical to wellbeing whereas in European terms it may mean restriction in water use such as hose pipe bans with little real impact either on community and wealth. However, drought in Africa is potentially life threatening and should not be considered as unusual, it is a recurring event that must be catered for within both national and regional strategies.

Table 4.8.   Livelihood approach.

Livelihood	Food first solution	Sustainable livelihood approach
Objective	Access to food	Secure and sustainable livelihood maximising access to all assets to support production, income and production
Priorities	Food first	Livelihood needs including water for income generation
Entitlements	Narrow entitlement base	Broad base including access to common pooled resources such as water
Vulnerability	Lack of food	Insecurity: exposure to shocks and stress including lack of ready access to water
Vulnerable groups	Based on social and medical criteria	Based on wider set of livelihood security indicators but including water availability and use
Coping strategies	Designed to maximise access to food	Specifically designed to preserve livelihoods and include trade-offs between expenditure on accessing water and preserving consumption, production and income
Measurement and monitoring	Narrow: food availability and access	Broad: livelihood security and sustainability, including water security indicators. Emphasis must be on local assessment
Supporting interventions	Food aid, food-for-work, food stamps etc	To protect livelihoods and assets, i.e. co-ordinated food and water interventions

But in defining drought, sufficient data must be gathered to help describe the normal conditions of the system and the critical conditions at which normality can no longer be sustained without external intervention. Internal intervention can include subsidised purchase of livestock to remove them from the arid pastures, promotion of drought resistant but nevertheless rain-fed crops, provision of free fertilizer, food aid and local medical assistance. External intervention is that which Government can no longer cope with and which requires emergency intervention by global NGOs, much of which is short term assistance, e.g. emergency and ill-defined drilling programmes, which do little to drought proof an area in the longer term.

Other terms used in water scarcity are:

- Water shortage—occurs when available water does not meet minimum needs,
- Water scarcity—relates water demand to availability and use,
- Water stress—caused by scarcity can result in conflict due to crop failure,
- Water security—reliable and secure access to water.

4.4.3   *Livelihood approach*

Livelihood is one of great concerns associated with drought strikes. The relationship between livelihood, food first solution and sustainable livelihood approach is outlined in Table 4.8.

4.5   ON-SITE SANITATION AND GROUNDWATER

There is an urgent need for a joint discussion of on-site sanitation and local groundwater supply. Both have the ultimate objective of enhancing community health. The World Health Organisation (WHO) provides one definition of sanitation as "the science and practice of effecting healthful and hygienic conditions", whilst in its 'Guidelines for Drinking-water Quality' it equally stresses health: drinking-water quality is an issue of concern for human health in developed and developing countries world-wide. The risks arise from infectious agents, toxic chemicals and radiological hazards. Experience highlights the value of preventative management approaches spanning from water resource to consumer" (quoted from McCarthy, 2008).

Throughout Africa, groundwater is by far the most important source for drinking and other domestic needs. And at the same time, across the continent, it is inadequate sanitation that presents the greatest risk to the quality of groundwater supplies. This section will therefore briefly look at various on-site sanitation practices, not comprehensively and in terms of their overall pros and cons, but only in terms of their impact on local groundwater supplies (Figure 4.10 and Table 4.9). This is not just a technical issue, but has to become part of the choice each community must make towards the most feasible and convenient option to provide necessary health protection.

The further discussion of on-site sanitation practice is taken largely from WHO (1992) and IRC (2004). Sanitation refers to excreta and waste water management as well as runoff water and solid and industrial waste. There are different technologies available to deal with sanitation and two main types can be distinguished:

- off-site sanitation which is appropriate for large scale exploitation, based on technical and economic feasibility studies (sewer networks, runoff water drains, etc);
- and on-site sanitation.

On-site sanitation can be defined as the whole of actions related to the treatment and disposal of domestic waste water that cannot be carried away by an off-site sanitation system because of low density of population. This could mean:

- individual on-site sanitation, when a house (plot) makes use of the soil as a treatment medium (example of soak-away, latrines, etc);
- grouped on-site sanitation (or semi off-site sanitation), when many individual houses are linked to a network leading to a treatment system, or small communities: grouped

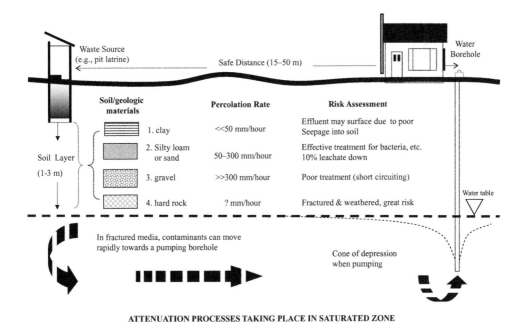

ATTENUATION PROCESSES TAKING PLACE IN SATURATED ZONE

Figure 4.10.   A sketch illustrating a on-site sanitation facility and its possible impact on groundwater resource.

Table 4.9. Types of on-site sanitation and their risk to underlying groundwater.

Type	Discription	Risk to underlying groundwater
Open defacation	Where there are no latrines people resort to defecation in the open. This may be indiscriminate or in special places for defecation generally accepted by the community, such as defecation fields, rubbish and manure heaps, or under trees	Will result in surface water pollution which can in turn impact shallow groundwater
Simple pit latrine	A simple wooden or concrete slab over a pit which may be 2 m or more in depth. The slab should be firmly supported on all sides and raised above the surrounding ground so that surface water cannot enter the pit. If the sides of the pit are liable to collapse they should be lined. A squat hole in the slab or a seat is provided so that the excrete fall directly into the pit	Does not need water to function; The urine plus rainfall plus flood water ingress can all contribute to percolation of contaminants to groundwater
Ventilated pit latrine	Fly and odour nuisance may be substantially reduced if the pit is ventilated by a pipe extending above the latrine roof, with fly-proof netting across the top. The inside of the superstructure is kept dark. Such latrines are known as ventilated improved pit (VIP) latrines	Same as simple pit latrine
Pour-flush latrines	A latrine may be fitted with a trap providing a water seal, which is cleared of faeces by pouring in sufficient quantities of water to wash the solids into the pit and replenish the water seal. A water seal prevents flies, mosquitos and odours reaching the latrine from the pit. The pit can be dug some distance from the latrine (which could be in a house) and connected with a pipe or covered drains. This system could be extended into a grouped sanitation system	Needs a source of water: 2–3 litres for each flush; Increased risk of groundwater contamination compared to a pit latrine
Single or double pit (for both pour-flush and pit latrines)	A common practice is to dig a second pit when the one in use is full to within half a metre of the slab. If the superstructure and slab are light and prefabricated they can be moved to a new pit. The first pit is then filled up with soil. After two years, faeces in the first pit will have completely decomposed and even the most persistent pathogens will have been destroyed. When another pit is required the contents of the first pit can be dug out (it is easier to dig than undisturbed soil) and the pit can be used again. The contents of the pit may be used as a soil conditioner	The general tendency is that more and more pits are dug, spreading and increasing the risk of groundwater contamination
ECOSAN latrines (ecological sanitation)	Ecosan latrines are latrines that ensure the recovery of waste by separating urine and faeces in view of their reuse in the fertilization of soil for agricultural purposes. Because of its its many benefits, including a low impact on groundwater, the Ecosan approach is discussed in greater detail as a 'best practice'	Recovery of urine and waste products; No pollution of groundwater

sanitation does not always use the soil as treatment medium (filtration beds, activated sludge are examples of purification systems at the end of the chain).

It is clear from the definition that the boundary between this and off-site sanitation is not always well-identified.

The most used on-site sanitation system in the world is the latrine. It is a site or a structure, located normally outside the house or building, destined to receive and store excreta and sometimes, to process it. This type of sanitation protects sensitive species of aquatic fauna by avoiding discharge concentrations in small waterways. Besides, it is cheaper than off-site sanitation, since the construction of a treatment plant is not required. They can improve living conditions of populations and solve sanitation needs in developing countries. Conventional water-born sewerage systems have been incapable of meeting populations' needs in these countries.

Crucial in any consideration of sanitation options is the question of proper construction and operation and maintenance. A non-functioning conventional sewage works can have a much higher impact on both surface- and groundwater than many pit latrines. Similarly a well chosen, constructed and maintained pit latrine can have minimal effect on local groundwater resources, whereas the same option, poorly executed, can be disasterous. Communities must understand this need of proper implementation in the same way as the upfront considerations like cost and cultural acceptability.

The need for an integrated approach to water supply and sanitation is stressed throughout this publication. An International Symposium on 'Coupling Sustainable Sanitation and Groundwater Protection' held in Hannover, Germany last year has put the focus on this issue and hopefully raised attention.

4.6 ECOSAN APPROACH FOR EFFECTIVE GROUNDWATER RESOURCES MANAGEMENT

Ecological sanitation (ecosan) is a new paradigm in the Africa sanitation sector. It recognizes human excreta and household wastewater not as waste but as resources that can be recovered, treated (where necessary), and reused. The concept ecosan focuses on reusing the waste in order to improve health issues, conserve water and other resources as well as protect aquatic ecosystems including groundwater resources. The ecosan approach places an emphasis on cleaning wastes or contaminated flow streams and shifts the concept from waste disposal to resource conservation and safe reuse (UNESCO, 2006). Ecosan is based on an ecosystem approach and treats human urine and faeces as a valuable resource to be recycled. It further shows that ecological sanitation is by no means untried. The needed action is to apply the ecological sanitation concept in existing toilets and wastes in communities and reap the benefits (Esrey et al., 1998).

4.6.1 *Historical context of Ecosan*

The current sanitation paradigm is failing the world, with the poor suffering most, threatening the integrity of fresh water supplies, including groundwater resources. The problems with conventional sanitation are crucial and an alternative approach is imperative. Ecosan provides a guide for improved sanitation by widening access to safe water and sanitation. Ecosan also offers a path out of extreme poverty by providing cheap manure to improve food security in communities. It is a cost effective approach to poverty reduction (UN, 2006). Sub-Saharan Africa remains the area of greatest concern. It is a region of the world where, over the period of 1990–2004, the number of people without access to safe drinking water increased 23% and the number of people without access to basic sanitation services increased by over 30%. Alternative approaches such as ecosan are needed to ease the situation (WHO & UNICEF, 2006).

4.6.2   *Main objective of Ecosan projects*

The main objective of an ecological sanitation system is to protect and promote human health by providing a clean environment and breaking the cycle of disease. In order to be sustainable, an Ecosan system has to be economically viable, socially acceptable, technically and institutionally appropriate, but it should also protect the environment and the natural resources such as groundwater. When improving an existing and/or designing a new Ecosan system, sustainability criteria related to the mentioned aspects need to be taken into consideration (SuSanA, 2007). Generally, objectives of ecological sanitation projects aim at:

- Reducing the health risks related to sanitation, contaminated water and waste
- Preventing the pollution of surface and groundwater
- Preventing the degradation of soil fertility
- Optimizing the management of nutrients and water resources.

4.6.3   *The philosophy about Ecosan*

Ecological sanitation (Ecosan) offers a new philosophy of dealing with what is presently regarded as waste and wastewater. Ecosan is based on the systematic implementation of reuse and recycling of nutrients and water as a hygienically safe, closed-loop and holistic alternative to conventional sanitation solutions. Ecosan systems enable the recovery of nutrients from human faeces and urine for the benefit of agriculture, thus helping to preserve soil fertility, assure food security for future generations, minimize water pollution and recover bioenergy. Ecosan ensures that water is used economically and is recycled in a safe way to the greatest possible extent for purposes such as irrigation or groundwater recharge (Esrey et al., 1998). Ecosan's focus is to:

- Destroy pathogens through flow stream separation, containment and specific treatment
- Conserve resource through a reduced use of potable water as a transport medium for human waste and by recovering wastewater for irrigation
- Reduce/stop wastewater discharges to the environment thereby protecting groundwater
- Close the resource loops through productive use of nutrients contained in excreta.

The new paradigm in sanitation is based on ecosystem approaches. Sanitation systems are part of several cycles, of which the most important cycles are the pathogen-water-nutrient- and energy cycle. In order to ensure public health, sanitation approaches primarily aim at interrupting the life cycle of pathogens. In addition, the new approach is recognizing human excreta and water from households not as a waste but as a resource that could be made available for reuse, especially considering that human excreta and manure from husbandry play an essential role in building healthy soils and are providing valuable nutrients for plants.

Ecosan systems restore the natural balance between quantities of nutrients excreted by one person yearly and that required to produce their food (7.5 kg nitrate, phosphorous and potassium to produce 250 kg of grain). In this case ecosan has potential to save limited resources especially in resource poor communities.

4.6.4   *Ecosan and groundwater quality*

With regard to fresh water and mineral resources, Ecosan protects groundwater resources from being polluted by wastes. Ecosan offers desirable solutions, in line with the Bellagio Principles as formulated by the Water Supply and Sanitation Collaborative Council) (UNESCO, 2006). Ecosan approaches strive to enable recovery of nutrients, organic material and water discharged in conventional sanitation systems. Recycling organic material contributes to safeguarding soil fertility and improving its structure and water retention capacity, while providing a natural alternative to chemical fertilizers. The process protects the groundwater from being contaminated by chemicals from such wastes. Therefore, Ecosan contributes to enhancing groundwater quality by closing the flow cycle of chemicals from wastes from reaching the groundwater.

4.6.5 *The challenge for Ecosan*

Ecosan approach has potential to become a reliable alternative approach to protecting groundwater resources. However, transmission of relevant knowledge and the existance of well trained people remain a challenge for the science and principles of ecologically sustainable sanitation to be applied in communities to yield its full benefits, including the protection of groundwater resources.

4.7 COMMUNITY PARTICIPATION BEST PRACTICE

Groundwater plays an important role in our society and it impacts all aspects of human life. Because of its ubiquitous nature and relative ease of local access, we have widely distributed and generally dispersed abstraction points and have many stakeholders, who are involved in its development, use as well as misuse. This complicates the traditional national approaches to resource regulation and requires a very high degree of participative management. Without this, resource utilization can not be sustainable.

Water resource issues, especially environmental ones, have to involve land use. Largely due to historical reasons, land is owned differently and administered at various levels. Land tenures can vary significantly from one tribe to another and from a country to another. Many stakeholders including the local community may have to be involved in issues like how to manage and protect water resources for a small piece of land in a given community. Furthermore, to have smooth implementation of above identified best practice measures, one cannot ignore the community participation either. In fact, successful management of groundwater resources will often include tales of community participation. Some good cases of participation are provided in this book, for example in spring protection (Section 4.3.6) and field data collection in Benin (Cranel et al., this book). Community participation is the process of learning and coordination by which communities control and deal with data, technology, challenges and development. This is especially critical for any groundwater protection measure that has maintenance and long term sustainability as its objective.

Water user associations, starting at the village level, are a very important way to formalise this participation. The holistic management, stressed previously, of water, land and waste can be put into practice in this way. Key targets of community institutions should be church organisations, tribal authorities, schools and water user organisations, if applicable. These groups have tremendous influence on water source ownership, access and maintenance etc in rural Africa. Non-government-organisations have a major role to play as intermediates between communities and water resource managers and their service providers at different levels. The water science sector should also become much more responsive to community needs and their knowledge and contribution to sustainable development. Some examples of initiatives to formalize community participation and empowerment are:

- EMPOWERS Methods, Tools and Guidelines (Moriarty et al., 2007);
- Generic Public Participation Guidelines (DWAF, 2001);
- Sustainability Best Practice Guidelines for Rural Water Services (DWAF, 2004);
- Water User Group concept as a sustainable management system for hand pump wells (SKAT, 2004).

4.8   TOWARDS A COMPREHENSIVE APPROACH

Information about implementation of best practice is usually not as well recorded and published as the development of specific scientific elements contributing to it. Therefore one such case, of which the authors are aware, of implementation of a range of best practice is briefly highlighted here. It was an attempt towards a comprehensive approach, and while it got some things right it also suffered a number of fatal flaws. This is the NORAD-Assisted Programme for the Sustainable Development of Groundwater Sources under the Community Water and Sanitation Programme in South Africa (DWAF NORAD, 2004).

The programme, managed by the Department of Water Affairs and Forestry (DWAF) between 2000 and 2004, undertook a series of inter-related projects aimed at enhancing capacity of water services authorities and DWAF to promote and implement sustainable rural water supply schemes based on groundwater resources and appropriate technologies. The activities were piloted in three of the recently established local government districts, located in three of the nine provinces in South Africa.

The programme addressed a national need, because water services had recently been devolved from national to local government level and more than 60% water supply schemes country-wide were from groundwater sources.

The programme has had an excellent knowledge output in the form of a 'Toolkit for Water Services', with an overall framework, local level strategies and a variety of best practice documentation and software (Figure 4.11). It also included a number of reference sites where communities had implemented appropriate technologies.

4.8.1   *Toolkit for water services*

In the pilot areas of the respective district municipalities the programme had a major impact in introducing appropriate groundwater source management practices and training local personnel. The reference sites included not only groundwater, but also sanitation best practice.

Successes were due to a number of key factors:

• The development cooperation partner and donor, NORAD, insisted and worked hard to build sustainability into the programme;

Figure 4.11.   A cartoon of community mobilization (DWAF/NORAD, 2004).

- The programme was enthusiastically driven by the national government institution responsible for water resources, the Department of Water Affairs and Forestry;
- An excellent combination of the best water science (the Council for Geoscience and the Council for Scientific and Industrial Research) and strongest water NGO (Mvula Trust) was acting as implementing agent for the programme;
- NORAD helped to bring international experience into the programme, which was lacking due to South Africa's political isolation before 1994;
- With an inception and an implementation period, the programme stretched over 5 years (2000–2004).

However, the multiplication impact of this intensive programme was limited. Some key reasons for this are summarised below:

- Water Services was a completely new and powerful arm in DWAF, established in response to the post-1994 government priority to address the massive water services backlog in the country. They were deeply committed to a major water services infrastructure development programme (Masibambane programme) and had considerable interest, but little time for a learning initiative, which the DWAF/NORAD Programme was;
- DWAF's lead agent for the Programme was in its existing Water Resource Management arm, through its Directorate Geohydrology. Groundwater was not yet well established as a management responsibility, because under the old water legislation groundwater had been classified as 'private water', and only with the new act, the National Water Act, 1998, was groundwater classified as a 'significant water resource', the same legal status as surface water;
- Programme implementation happened through the DWAF regional offices. At this level, support from the Water Services component, which was the DWAF link to local government, was very difficult to secure under the prevailing priorities;
- District municipalities had no appropriate institutional structures to start accommodating groundwater source management in any sustainable way, and without active involvement of national government Water Services, it was not possible to bring about long-term institutional changes;
- The Programme lost its national champion even before it was completed. With the major restructuring of DWAF in 2003 to support a devolution of water resources management, as foreseen by the National Water Act, 1998, the Directorate Geohydrology was broken up and and functions were integrated, in the spirit of IWRM, into a number of other resource management and assessment functions;
- Without a champion there was no follow up of this strategic initiative.

Also considering discussion in the previous chapters, it is clear that the fatal flaw of this initiative was the lack of an integrated approach between water services provision and resource management.

REFERENCES

ASTM (2002). ASTM D2855-96. Standard practice for making solvent-cemented joints with poly (vinyl chloride) (PVC) pipe and fittings. DOI: 10.1520/D2855-96R02. ASTM International. West Conshohocken. PA. Accessed on 24/02/2009 at http://www.astm.org/Standards/D2855.htm.

AWWA (2006). A100-06: AWWA Standard for Water Wells. American Water Works Association. Available from http://www.awwa.org/Bookstore/

Bradbury, K.R. and Muldoon, M.A. (1994). Effects of fracture density and anisotropy on delineation of wellhead protection areas in fractured-rock aquifers. Journal of Applied Hydrogeology, 3/94. pp. 17–23.

Braune, E. and Xu, Y. (2006). A South African perspective on the protection of groundwater sources. In Xu, Y. and Usher, B. eds. (2006). Groundwater Pollution in Africa. Taylor & Francis/Balkema. pp. 341–349.

Braune, E., Hollingworth, B., Xu, Y., Nel, M., Mahed, G. and Solomon, H. (2008). Protocol for the Assessment of the Status of Sustainable Utilization and Management of Groundwater Resources—With

Special Reference to Southern Africa. Pretoria. WRC Report No. TT 318/08, Water Research Commission: Pretoria.

BS (1988). BS 879-2:1988. Water well casing. Specification for thermoplastics tubes for casing and slotted casing. British Standards. ISBN 0-580-16488-8.

BWA (2009). Standard form drilling agreement. Borehole Water Association of Southern Africa. Available at http://www.bwa.co.za/

Calow, R.C., Robins, N.S., MacDonald, A.M., Macdonald, D.M.J., Gibbs, B.R., Orpen, W.R.G., Mtembezeka, P., Andrews, A.J. and Appiah, S.O. (1997). Groundwater management in drought prone areas of Africa. International Journal of Water Resources Development, 13, 2, 241–261.

Calow, R.C., MacDonald, A.M., Nicol, A.L. and Robins, N.S. (2009). Ground water security and drought in Africa: linking availability, access and demand (In press). Ground Water.

Chave, P., Howard, G. and Schijven, J. (2006). Groundwater protection zones. In: Schmoll, O., Howard, G., Chilton, J. (eds) Protecting groundwater for health. World Health Organisattion, IWA Publishing, London.

Clay, E., Borton, J., Dhiri, J., Gonzalez, A.D.G. and Pandolfi, C. (1995). Evaluation of ODA's response to the 1991–1992 Southern African drought. ODA Evaluation Report EV 568. UK Department for International Development, London.

Driscoll, F.G. (ed.) (1986). *Groundwater and Wells*. 2nd edition. Johnson Division. St. Paul. Minnesota. ISBN 0-9616456-0-1.

Driscoll, F.G. (ed.) (2008). *Groundwater and Wells*. 3rd edition. Johnson Division. St. Paul. Minnesota. ISBN 0-9787793-0-4.

DWAF (1997). A protocol to manage the potential of groundwater contamination from on site sanitation. National Sanitation Co-ordination Office, Department of Water Affairs and Forestry, Pretoria. First edition published in 1997.

DWAF (2001). Generic Public Participation Guidelines. Department of Water Affairs and Forestry.

DWAF (2003). A protocol for the protection of groundwater from contamination from sanitation practices. National Sanitation Co-ordination Office, Department of Water Affairs and Forestry, Pretoria. Second edition published in 2003.

DWAF (2004). Sustainability Best Practice Guidelines for Rural Water Services. No. 7.1, Toolkit for Water Services. NORAD-assisted Programme for Sustainable Development of Groundwater Sources under the Community Water and Sanitation Programme in south Africa.

DWAF/NORAD (2004). NORAD-Assisted Programme for the Sustainable Development of Groundwater Sources under the Community Water and Sanitation Programme in South Africa. Pretotia: Department of Water Affairs and Forestry.

Entrepeneur.com (2009). Maintenance for wells and boreholes. Accessed on 16/02/2009 at https://www.entrepreneur.com/tradejournals/article/146063295.html.

EPA (1975). Manual of water well construction practices. Document 570975001 accessed on 22/02/2009 at http://nepis.epa.gov/EPA/html/Pubs/pubtitleOSWER.htm. Last updated on 15/10/2007.

Esrey, S.A., Gough, J., Rapaport, D., Sawyer, R., Simpson-Hébert, M., Vargas, J. and Uno Winblad (ed) (1998). Ecological Sanitation; Swedish International Development Cooperation Agency, Department for Natural Resources and the Environment, Stockholm, Sweden.

FCI (2004). Borehole and hand pump implementation, operation and maintenance. A manual for field staff of NGO's. Foundation (Stichting) Connect International in collaboration with WES Consultants (Uganda). Accessed on 16/02/2009 at http://www.connectinternational.nl/english/smartmodules/smart-tec/wells-boreholes/conventionaldrilling.

Ferro, B.P.A. and Bouman, D. (1987). Explanatory notes to the hydrogeological map of Mozambique scale 1:1,000,000. Ministry of Construction and Works. National Directorate for Water Affairs. Mozambique.

Focazio, M.J., Reilley, T.E., Rupert, M.G. and Helsel, D.R. (2002). Assessing ground-water vulnerability to contamination: Providing scientifically defensible information for decision makers. U.S. Geological Survey Circular 1224. (available online as http://pubs.usgs.gov/circ/2002/circ1224/)

Foster, S., Hirata, R. and Games, D. et al. (2002). Groundwater quality protection—a guide for water utilities, municipal on authorities, and environment agencies. Groundwater Management Advisory Team, The World Bank, Washington.

Foster, S., Tuinhof, A. and Garduño, H. Task Manager: David Grey (World Bank—AFR), (2006). Groundwater Development in Sub-Saharan Africa: A Strategic Overview of Key Issues and Major Needs, World Bank.

Foster, S., Hirata, R., Gomes, D., D'Elia, M. and Paris, M. (2007). Groundwater quality protection: a guide for water utilities, municipal authorities, and environment agencies. The World Bank Publication, ISBN 0-8213-4951-1, Whashington, DC 20433, USA.

Ginsberg, M. and Palmer, A. (2002). Delineation of Source-water Protection Areas in Karst Aquifers of the Ridge and Valley and Appalachian Plateaus Physiographic Provinces: Rules of Thumb for Estimating the Capture Zones of Springs and Wells. Publication EPA 816-R-02-015, U.S. Environmental Protection Agency. p. 41.

Gogu, R.C. and Dassargues, A. (2000). Current trends and future challenges in groundwater vulnerability assessment using overlay and index methods. Environmental Geology 39 (6): 549–559.

Gotkowitz, M.B. and Gaffield, S.J. (2006). Water-table and aquifer-susceptibility maps of Calumet County, Wisconsin. Miscellaneous Map 56, Wisconsin Geolgical and Natural History Survey, University of Wisconsin-Extension.

Harvey, C. (2004). 1.72, Groundwater Hydrology, Lecture Packet #6: Groundwater-Surface Water Interactions, http://www.myoops.org/cocw/mit/Civil-and-Environmental-Engineering/1-72Fall-2004/LectureNotes/index.htm.

Hobbs, P.J. and Marais, S.J. (1997). Minimum standards and guidelines for groundwater resource development for the community water supply and sanitation programme. 1st edition. Department of Water Affairs and Forestry. Pretoria. South Africa.

Howard, G., Mutabazi, R. and Nalubega, M. (2000). Rehabilitation of protected springs in Kampala, Uganda, Water, sanitation and hygiene: chanllenges of the Millennium, 26th WEDC Confereence, Dhaka, Bangladesh, 2000.

Hussey, S.W. (2007). Water From Sand Rivers, Guidelines for Abstraction. Water, Engineering and Development Centre, Loughborough, UK.

IRC (2004). What is on-site sanitation? A case study of latrines. A fact sheet prepared by CREPA, Burkina Faso. IRC International Water and Sanitation Centre, The Hague.

Javandel, I. and Tsang, C. (1986). Capture-zone type curves: A tool for aquifer cleanup, Ground Water, 24(5): 616–625.

Jolly, J.L. and Reynders, A.G. (1993). The protection of aquifers: a proposed classification and protection zoning system for South African conditions. An International Groundwater Convention Entitled 'Africa needs groundwater' at the University of the Witwatersrand, Johannesburg, South Africa, 6–8 September 1993.

Lawrence, A.R., Macdonald, D.M.G., Howard, A.G., Barrett, M.H., Pedley, S., Ahmed, K.M and Nalubega, M. (2001). Guidelines for assessing the risk to groundwater from on site sanitation. Technical Report British geological Survey CR/01/142. Available from http://www.bgs.ac.uk/hydrogeology/argoss/docs/ARGOSSManual_144.pdf.

Lifewater (2004). Drilling and well construction manual. Version http://www.lifewater.ca/ndexdril.htm. Accessed on 16/02/2009 at http://www.lifewater.ca/resources/drillingtutor.htm

Louw-Carstens, M. (2008a). Stryd om kind in gat te red (E. Battle to save child in hole). Beeld Newspaper. Accessed on 25/02/2009 at http://www.news24.com/Beeld/Suid-Afrika/0,,3-975_2420616,00.html.

Louw-Carstens, M. (2008b). Boorgat: 'Mens verwag dit nie' (E. Borehole: 'One doesn't expect it'). Beeld Newspaper. Accessed on 25/02/2009 at http://www.news24.com/Beeld/Suid-Afrika/0,3-975_2421032,00.html.

LWBC (2003). Minimum construction requirements for water bores in Australia. 2nd edition. Land and Water Biodiversity Committee. ISBN 1-9209-2009-0.

MacDonald, A., Davies, J., Calow, R. and Chilton, J. (2005). Developing groundwater—A guide for rural water supply, ITDG Publishing, UK, 2005, ISBN 1 85339 596X.

McCarthy, P. (2008). The insidious nature of groundwater contamination—the great need for protection. Presentation at International Symposium on 'Coupling Sustainable Sanitation and Groundwater Protection', Hannover, 14–17 October 2008.

McDonald, M.G. and Harbaugh, A.W. (1988). A modular three-dimensional finite difference groundwater flow model: Techniques of Water-Resources Investigations of the U.S. Geological Survey, Chapter A1, Book 6.

Minnesota Department of Health (2005). Guidance for delineating wellhead protection areas in fractured and solution-weathered bedrock in Minnesota. Report prepared by the Minnesota Department of Health, Drinking Water Protection Section, Source Water Protection Unit. p. 80.

Moriarty, P., Batchelor, C., Abd-Alhadi, F.T., Laban, P. and Fahmy, H. (2007). The EMPOWERS Approach to Water Governance: Guidelines, Methods, Tools. EMPOWERS Partnership, Amman, Jordan.

Mwami, J. (1995). Spring protection—sustainable water supply, Sustainability of water and sanitation systems, 21st WEDC conference, Kampala, Uganda, 1995.

Nel, J.M., Xu, Y., Batelaan, O. and Brendonck, L. (2009). "Benefit and implementation of groundwater protection zoning in South Africa", Water Resource Manage (2009) DOI 10.1007/s11269-009-9415-4, Springer.

NWQMS (1995). National water quality management strategy: guidelines for groundwater protection in Australia. Department of Primary Industries and Energy, Agriculture and Resource Management Council of Australia and New Zealand, Canberra.

PIPA (2007a). Industry guidelines: PVC pipe equivalence. Issue 1.0. Plastics Industry Pipe Association of Australia Limited. Available from http://www.pipa.com.au/PVCs.html#POP104.

PIPA (2007b). Industry guidelines: PVC pipes in bore casings. Issue 1.0. Plastics Industry Pipe Association of Australia Limited. Available from http://www.pipa.com.au/PVCs.html#POP105.

Pochon, A., Tripet, J.P., Kozel, R., Meylan, B., Sinreich, M. and Zwahlen, F. (2008). Groundwater protection in fractured media: a vulnerability-based approach for delineating protecion zones in Switzerland, Hydrogeology Journal. V.16., No 7. 1267–1281.

Rayne, T.W., Bradbury, K.R. and Muldoon, M.A. (2001). Delineation of capture zones for municipal wells in fractured dolomite, Sturgeon Bay, Wisconsin, USA. Hydrogeology Journal. V.9., No. 5. 432–450.

SADC (2000a). Standards and guidelines for the groundwater development in the SADC region. Report No. 2. Groundwater Consultants for SADC Water Sector Coordination Unit. Maseru. Lesotho.

SADC (2000b). Situation analysis report. Report No. 1. Groundwater Consultants for SADC Water Sector Coordination Unit. Maseru. Lesotho.

SADC WSCU (2000). Guidelines for Groundwater Development in the SADC Region, SADC Water Division, Gobarone, Botswana.

Sami, K. and Murray, E.C. (1998). Guidelines for the evaluation of water resources for rural development with an emphasis on groundwater. Report no. 677/1/98. Water Research Commission. Pretoria.

SANS (2003a). SANS 10299-2:2003. Development, maintenance and management of groundwater resources—Part 2: The design, construction and drilling of boreholes. Standards South Africa. ISBN 0-626-14790-5.

SANS (2003b). SANS 10299-9:2003. Development, maintenance and management of groundwater resources—Part 9: The decommissioning of water boreholes. Standards South Africa. ISBN 0-626-14803-0.

SAPPMA (2006). Technical manual. 1st edition. Southern African Plastic Pipe Manufacturers Association. Available at http://www.sappma.co.za/pdfs/SAPPMA_TM_LR.pdf.

Sheik, M., Gutierrez, M.I., Bolton, P., Spiegel, P., Thieren, M. and Burnham, G. (2000). Deaths among humanitarian workers. British Medical Journal. Volume 321. pp. 166–168.

SKAT (2004). The Shinyanga Experience: Water User Group concept as a sustainable management system for hand pump wells. Tanzania Ministry of Water and Livestock Development with the Government of the Netherlands and Swiss Centre for Development Cooperation in Technology and Management.

Spangler, L.E. (2002). Use of dye tracing to determine conduit flow paths within source-protection areas of a karst spring and wells in the Bear River Range, Northern Utah. In: Proceedings, U.S. Geological Survey Karst Interest Group, U.S. Geological SurveyWater-Resources Investigations Report 02-4174. 75–80.

SuSan, A. (2007). Joint roadmap for the promotion of sustainable sanitation in the UN's "International Year of Sanitation 2008"—draft".

UN (2006). Hashimoto Action Plan: Compendium of Actions: United Nations Secretary-General's Advisory Board on Water and Sanitation.

UNEP (1996). Groundwater: A threatened resource, UNEP Environment Library No. 15, Nairobi, UNEP, 1996.

UNESCO (2006). Capacity building for ecological sanitation: Concepts for ecologically sustainable sanitation in formal and continuing education, UNESCO Working Series SC-2006/WS/5, Paris, France.

USEPA (1987). Guidelines for Delineation of Wellhead Protection Areas: U.S. EPA Office of Ground-Water Protection, Chapters paginated separately.

USEPA (1991). Delineation of Wellhead Protection Areas in Fractured Rocks. U.S. EPA Technical Guidance Document, by K.R. Bradbury, M.A. Muldoon and A. Zaporozec. U.S. EPA Office of Groundwater, EPA 570/9-91-009, p. 144.

Usher, B.H., Pretorius, J.A. and Dennis, I. et al. (2004). Identification and prioritisation of groundwater contaminants and sources in South Africa's urban catchment. WRC report 1326/1/04. Water Research Commission, Pretoria.

Vias, J.M., Andreo, B., Perles, M.J., Carrasco, F., Vadillo, I. and Jimenez, P. (2006). Proposed method for groundwater vulnerability mapping in carbonate (karstic) aquifers: the COP method. Hydrogeology Journal, 14: 912–925.

Vinidex (2004). Technical Note: PVC pipes under external pressure. No. VX-TN-4F.2. Available at http://www.vinidex.com.au/vinidex/live/RESOURCES/IMAGES/technotes/TN-4F.2.2004PVCpipes underexternalpressure.pdf.

WHO (1992). A Guide to the Development of on-site Sanitation. Geneva: World Health Organization.

WHO and UNICEF (2006). Meeting the MDG drinking water and sanitation Target. The urban and rural challenge of the decade, Geneva, Switzerland.

Worthington, S.R.H., Smart, C.C. and Ruland, W.W. (2002). Assessment of groundwater velocities to the municipal wells at Walkerton. In: Ground and Water: Theory to Practice: Proceedings of the 55th Canadian Geotechnical and 3rd Joint IAH-CNC and CGS Groundwater Specialty Conferences, Niagara Falls, Ontario, October 20–23, 2002. Edited by D. Stolle, A.R. Piggott and J.J. Crowder and published by the Southern Ontario Section of the Canadian Geotechnical Society. pp. 1081–1086.

Xu, Y. and Braune, E. (1995). A guideline for groundwater protection for the Community Water Supply and Sanitation Programme. Department of Water Affairs and Forestry, Pretoria.

Xu, Y. and Reyders, A.G. (1995). A three-tier approach to protect groundwater resources in South Africa, ISSN 0378-4738, Water SA, Vol. 21 No. 3 July 1995.

Xu, Y. and Usher, B. (2006). Issues of groundwater pollution in Africa. In Xu, Y. and Usher, B. eds. (2006). *Groundwater Pollution in Africa*. Taylor & Francis/Balkema. pp. 3–9.

5

Summary and recommendations

Eberhard Braune & Yongxin Xu
Department of Earth Sciences at the University of the Western Cape, South Africa

Achieving sustainable utilization of groundwater resources in Africa requires good understanding of many influencing factors including resource characteristics, user characteristics and requirements, and legal and institutional arrangements in a given area. These factors can be analysed in terms of their strengths, weaknesses, opportunities and threats (SWOT) as shown, by way of example only, in Table 5.1. Some key areas that still need attention are highlighted under the heading of Comment in Table 5.1.

The issues of groundwater use in the water supply and sanitation environment will have to be addressed within a broader framework for sustainable groundwater utilization as part of IWRM. Specifics will be highlighted in the summary and recommendations below.

5.1 SUMMARY

Water supply and sanitation service delivery is at this stage one of the highest development priorities in Africa. It is explicitly addressed in the Millennium Development Goals (MDGs) and is seen as one of the keys to the alleviation of poverty on the continent.

Governments in Africa face a twin challenge in terms of water supply and sanitation. Firstly to close the gap in rural areas where only two in five people have access to water supply and fewer than one in five have access to sanitation, and secondly the urban population explosion at a rate unique in history, which is characterized by growing slums, unemployment, poor access to water, sanitation and health services.

In both these environments, groundwater resources have a major role to play and their sustainable utilization has become of strategic importance.

It is important to note that groundwater's role in water and sanitation provision is related to its ubiquitous nature which matches the spread-out nature of the demand in rural and informal urban areas.

Groundwater occurs locally throughout the continent, albeit in limited quantities and sometimes difficult to access, because of the nearly 80% hard-rock nature of aquifer systems and limited recharge in the drier parts.

Groundwater is particularly important to mitigate the impact of recurring droughts on community life as well as associated natural systems and resources. It also has an important conjunctive use role together with other water resources. Artificial recharge of groundwater resources has a variety of benefits and finds increasing application.

In contrast to this strategic role, groundwater has remained a poorly understood and managed resource. Past assessments of groundwater use in Africa (and world-wide) have focused on its bulk water supply role and contribution to total water supply and, in so doing, have continued to produce a picture that did not come close to capture groundwater's unique characteristics and potential role.

A major indication of the poor state of management is that pollution of vital underlying groundwater sources in African cities as well as in many rural communities has reached critical levels.

Table 5.1. SWOT analysis of groundwater use in Africa.

Factor	Strength	Weakness	Opportunities	Threat	Comment
Resources characteristics	Widely distributed	Limited recharge and storage	Advanced exploration technology, databases for groundwater auditing	Lack of research and capacity-building	Conjunctive use, artificial recharge and localised deep drilling
	Sole source for many small schemes	No procedures for macro-implementation	Potential of deep aquifers and artificial recharge	Poor management is seen as poor resource	
	Clean and protected water	Shallow water table, vulnerable to pollution	Community awareness; Protection zoning	Resource degradation (pollution; localized over-use)	
	Balancing storage as security against drought	Lack of information and knowledge	International/regional support for better assessment (e.g. UNESCO ISARM)	Unwillingness in sharing of data, capacity and benefits of common resource	
Legal and institutional arrangements	New policy	Groundwater still treated as private water in practice	Paradigm shift, International transboundary aquifer articles	Insufficient implemention; Top down approaches	Promoion of groundwater through IWRM at all levels
	IWRM	Poorly understood; no sectoral ownership	National IWRM Plans Legislative reform Basin organizations Water user associations	Lack of coordination No groundwater champion	
	National function	Lack of institutional arrangement	Groundwater as regional priority (AMCOW and AGWC)	Lack of political will	
User perspectives	Domestic supply	Poor resource protection and operation and maintenance	A major contribution to poverty alleviation, the highest regional priority	Poor management giving the resource a bad name	Valuation and establishment in practice of groundwater's key roles, in particular for poverty alleviation
	Food security	Small scale—the poorer farmers	Major groundwater role in Asia	High costs for drilling (USD100/m)	
	Drought mitigation	Emergency focus—	Strategic response due to threat of climate change	Continued lack of mainstreaming drought management	
	Agricultural development	Many other non-water impediments	Africa's most important economic sector	No financing mechanism for small scale farming	
	Industrial development	Lack of knowledge—major exploration investment needed	Community involvement	Poor regulation	
	Environmental use	Lack of understanding of groundwater's role and functioning	Education about groundwater's role in nature	Under- or over- valued	

Management weaknesses cover the full spectrum of essential policies, institutions and practices. There is a general lack of appropriate approaches and investment for the planning, financing, developing and sustainably utilizing a resource that can normally only be exploited in the form of many, widely distributed, relatively small, individual sources.

Critical shortcomings appear to be in the organizational framework, in the building of institutional capacity for groundwater and in a lack of progress in formally treating it as part of IWRM.

Importantly, groundwater monitoring and information services are virtually non-existent in many African countries.

Climate change, which is expected to have major impact on the African continent, will have both short-term and long-term impacts on water supply from groundwater. While there is still considerable uncertainty about the nature of these impacts, it is clear that best mitigation approach at present is to systematically address the well-known management shortcomings which groundwater is already experiencing on the continent. Its role in adaptation to drought risks will need to be urgently developed.

A number of major opportunities exist in the region, led by AMCOW, SADC and other players, to make rapid progress with a stakeholder-responsive broad-based programme and approach of capacity-building for groundwater resources management as part of Integrated Water Resources Management.

5.2 RECOMMENDATIONS

Sustainable water services from groundwater need a variety of local actions (best practices). To meet their development objectives, these actions must be implemented as part of an integrated approach to water, sanitation and hygiene delivery. This need for integration is specifically called for in Target 10 of Goal 7 of the MDGs.

The formulation of effective strategies and techniques for handling urban groundwater-related problems, geared to achieve healthy and sustainable results, needs particular attention in Africa, because of continuing unplanned urban growth, often haphazard waste disposal and unregulated well water supply.

Highest priority needs to be afforded to the informal urban environment as this represents the highest groundwater contamination risk, followed by the rural environment.

An IWRM framework will lie at the heart of this integration, but will have to accommodate the groundwater and the sanitation components much more explicitly and systematically than at present.

The field of groundwater resource protection will specifically have to be addressed more strategically. The South African Department of Water Affairs and Forestry Framework for Groundwater Quality Management can serve as an example.

Groundwater's ubiquitous nature and relative ease of local access complicates the traditional national approaches to resource regulation and requires a very high degree of participative management. Water user associations and other forms of community-based organizations are a very important way to formalise this participation (chapter 18 in this book). The holistic management stressed previously, of water, land and waste can be put into practice in this way.

Particular attention will have to be paid to supporting local functionaries, e.g. local government, from national and intermediate levels in a coordinated way which covers all facets of water and sanitation delivery.

Because protection zoning inevitably involves land-use decisions, which are outside the jurisdiction of water resources agencies, it is crucial to bring the appropriate institutions at the national, provincial and local level, in particularly local government, on board.

Best practices for groundwater resource utilization and management should always be introduced in a systematic way, based on area-wide assessments of aspects such as resource characteristics, aquifer vulnerability and protection requirements (chapters 6, 7, 8 and 9 in this book); Macro-planning of groundwater resources is essential for a more planned implementation of sustainable approaches (the need for this is clear from chapters 15 and 16).

Piloting of best practices in the local environment and the adoption of locally appropriate best practices is crucial in the achievement of sustainability (chapters 8 and 9 in this book).

Groundwater quality problems and their mitigation have been neglected and need integrated attention to achieve sustainable solutions (chapters 10, 11 and 12 in this book).

Groundwater sources for domestic water supply need to be afforded the highest level of protection. Protection zoning appropriate to the risk and implementation capacity needs to be pro-actively pursued with appropriate policy and best practice (section 4.3 and chapter 10 in this book). Catchment-wide planning is required to to undertake this systematically (chapters 8 and 11).

Failure of groundwater supplies misses the development objective and continues to give groundwater as a secure supply a bad name. This has to be addressed in both the design and construction of groundwater abstraction infrastructure (section 4.2) and in their operation and maintenance (chapter 13 in this book). Cost-effectivity of technology should receive particular attention (chapter 14 in this book).

Particular attention needs to be given to the sustainability of groundwater supplies during drought periods (section 4.4).

All management actions should be accompanied by appropriate monitoring to be able to assess the effectiveness of management approaches and plan for corrective action. Participation of key stakeholders, e.g. communities, in this monitoring is strongly recommended to achieve understanding and buy-in (chapters 18 and 19 in this book).

At the heart of moving forward with a more sustainable utilization lies the need for a much clearer understanding and articulation of groundwater's role and contribution to national and regional development objectives, in particular community water supply, public health, and rural and urban development.

As a major step towards achieving a sustainable service from groundwater sources, it is important to mainstream groundwater resources management and protection as part of the thrusts of existing major actors and programmes for water supply and sanitation delivery in Africa, e.g. African Development Bank, UNICEF, WHO and UN-Habitat.

Given that poor sanitation is known to have the biggest pollution impact on community water supplies, an integrated implementation of water and sanitation (the need for this is also clear from case studies in chapters 15 and 16) would be by far the most pro-active measure of local resource protection, besides its obvious benefits for community health.

II
Case studies

6

Groundwater dynamics in the East African Rift System

Tamiru Alemayehu Abiye
School of Geosciences, University of the Witwatersrand, Johannesburg, South Africa

ABSTRACT: The East African Rift System (EARS) is characterized by a complex array of aquifers, variable water quality and a series of lakes that vary in size and geomorphologic setting and which contain fragile ecosystems. The sub-region is strongly groundwater dependent, although the water may be of poor quality and is susceptible to drought. The lower elevation of the rift, with respect to the adjacent plateaux, allows the regional and local groundwater flow to converge towards the rift floor. The extensive flow paths provide an opportunity for water-rock interaction to occur and the rift contains highly evolved and mineralized groundwater. This chapter concentrates on the groundwater dynamics and hydrogeochemical nature of the EARS, with representative examples from the Ethiopian Rift, given that the rift aquifers are of considerable resource potential for water supply and sanitation. The main quality degradation is high salinity due to evaporation, intensive water-rock interaction and increase in fluoride due to evaporation and contact with thermal groundwater. Recent changes in some lakes have grave environmental consequences to fragile ecosystems, which demand an integrated basin-wide water management approach with particular emphasis on the groundwater system whilst protecting the sustainability of the surface water resources.

6.1 INTRODUCTION

The East African Rift System (EARS) is one of the most spectacular structures on Earth, and is characterized by extensional tectonics which represents an active crustal spreading zone. The rift is several thousand kilometres long involving aligned successions of tectonic basins (Chorowicz, 2005). On a broad scale, EARS can be identified as the eastern branch: Ethiopian Rift (Main Ethiopian Rift and Afar Rift including Djibouti), Kenya Rift, and Tanzania Rift and western branches: Uganda Rift, Malawi Rift and Mozambique Rift. The eastern branch runs from the Afar triple junction to the north of Tanzania in a NE-SW direction while the western branch runs from Uganda in the north to Mozambique in the south with N-S and NW-SE direction (Figure 6.1) having a left offset from the eastern branch.

The comprehensive understanding of the groundwater dynamics in EARS is important for sustainable development of the region. Numerous lakes, fed by streams from the adjacent rift shoulders and highlands, occupied the rift floor in the geologic past. Volcanic and tectonic processes were responsible for creating topographic features and fertile soils. Apart from the presence of rich volcanic soil with great agricultural potential, the basin contains abundant resources of industrial minerals, geothermal energy, and surface and ground water. The special feature of the EARS is the chain of lakes varying in shape, volume of water and origin within the arid and semi-arid environment of the Rift Valley. In such dry environment, groundwater plays paramount role in sustaining the surface water resources. The lakes are concentrated in the Rift due to its tectonic basinal setting. Natural lakes are environmental resources having substantial importance to human wellbeing by regulating climate, making the environment suitable for life to sustain and enhance environmental development.

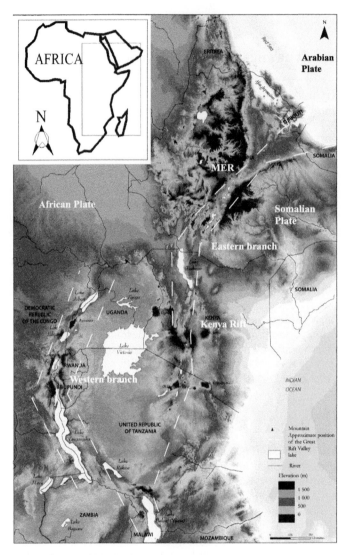

Figure 6.1.   Location of East African Rift System.

EARS provides vast land for agricultural activities which can be developed through water supply from groundwater.

EARS is characterized by high tectonic activity, where magma bodies are injected into the crust and where continental crustal fracturing provides paths by which surface water can penetrate freely to heating zones. These areas also provide permeable reservoirs in which large quantities of heat and hot fluid (steam or water) can be stored. Fracturing may also result in the escape of some of the heat to the surface, thus providing direct indications of the presence of a geothermal system. The result of this combination of factors is that a large number of hot springs, steam and hot grounds, which are direct indicators of the geothermal potential of the region, are abundant in the rift.

The eastern branch of EARS contains over 27 lakes (Ethiopia, 15; Kenya, 7 and Tanzania, 5) while the western branch hosts 12 larger lakes. The level of some of the lakes in the rift has changed through time while others are more or less constant in water level due to complex natural phenomena and man-made activities. Most lakes in the eastern branch show decreasing level while some show a water level increase due to natural causes. The Main Ethiopian Rift (MER) along with the Afar Rift represents the northern portion of the EARS and some examples from MER substantiate evidence for the groundwater dynamics. This chapter describes the status of groundwater dynamics in the EARS, which is intensively affected by volcanism and tectonics.

6.2   TECTONIC FRAMEWORK

The northern part of the EARS is characterized by the occurrence of the triple junction at Afar where three plates meet (African Plate, Arabian Plate and Somalian plate). In other words, three rifts meet at Afar: Red Sea Rift, Aden Rift and Ethiopian Rift. The western portion of the EARS is largely non-volcanic while the eastern branch is characterized by numerous Quaternary eruptive basaltic and acidic centres. The volcanoes are rooted on open fractures such as tension joints, tail cracks or tensional releasing bends (Korme et al., 1997). The southern part of the eastern rift, manly in Tanzania, contains abundant crystalline rocks and intrusive materials. In the eastern branch, the occurrence of basement rocks is rare, with the only localized out crop on the Kela horst of the central main Ethiopian rift. In the Eastern branch, the older volcanic rocks (Oligo-Miocene) outcrop on the Rift margins and on the highlands, while the recent volcanic products (Quaternary) cover the entire Rift. Large parts of the MER and Kenya Rift are dominantly covered by recent peralkaline acidic volcanic rocks (obsidian, ash, pumice and ignimbrite) and basaltic lava flows.

The Main Ethiopia Rift, with intense faulting in N–S, NE–SW, NNE–SSW direction, was developed through extension processes associated with mantle plume activities (Woldegabriel et al., 1990; Boccaletti et al., 1998; 1999; Acocella et al., 2002; Benvenuti et al., 2002). The repeated faulting resulted in numerous minor horst and graben structures with frequent occurrence of "rift-in-rift" structures. The frequent offset of the tectonic structures favoured the formation of lakes of variable sizes. The Eastern branch of the EARS is categorized as an active fault terrain with the expansion rate that reaches 5–10 mm/year (Skobelev et al., 2004). In the Afar region the expansion rate exceeds over 2 cm/year, which is an area where oceanic crust outcrops. Dalol depression, the hottest spot on Earth (mean temperature of 55°C) is located in the northern part of Afar where the elevation is 120 m below sea level. The width of the MER increases from the southern (30–60 km) to the central (65–80 km) and the northern (80–120 km) sectors, and is more than 200 km wide in the southern part of the Afar Rift (Mohr, 1962, 1967, 1987; Di Paola, 1972), ie, it funnels out to the NE direction. While the mean width for the Kenya Rift varies between 60 km and 80 km (Darling et al., 1995).

The central part of MER is characterized by active faulting with variable displacements along the strike of the boundary faults where the Rift floor lies at an elevation of 750 m. These margins are marked by high angle normal faults with large throws. The crustal extension rate that is determined from the plate tectonic model, geodetic and geological measurements indicate that the Ethiopian Rift is expanding at the rate of 1 or 2 cm per year. The presence of left lateral strike slip faults in the MER (Boccaletti et al., 1998) and the right lateral in Malawi (Delvaux et al., 1992) demonstrate the presence of tensional basins in the EARS that provide favourable place for groundwater storage. According to Chorowicz (2005), the Kenya Rift floor elevation rises from 1050 m in the north to 2100 m in the middle and steps down progressively southwards to 600 m at Lake Natron.

The western branch is characterized mainly by half grabens and structural high points that are represented by crossing points of different generation of faults and the rifting process in this part of the basin has been disturbed by uplift related to the Rwenzori horst block formation (Laerdal et al., 2002). The western branch is dominated by a series of en echelon structures linked to border faults. Down-faulting of the western rift valley and subsequent uplifting created significantly greater relief in the intra-arch basin. Chorowicz (2005) indicated that the

Malawi Rift is at an early stage of development where the tilted blocks dip in the same west-ward direction and accordingly, all extensional faults have the same eastward dip.

The tectonically and volcanically active MER offers different settings for the lakes. Figure 6.2 shows preferential alignment of the lakes based on the recharge condition, ground-water flow, river flow and volcano-tectonic structures where the rift acts as a main discharging zone for the regional groundwater flow. In the EARS the tectonic lineaments in the act both as conduits and barrier for the groundwater migration and rarely rivers cross the fault lines. In most of the cases, tectonic zones act as a boundary for the lakes.

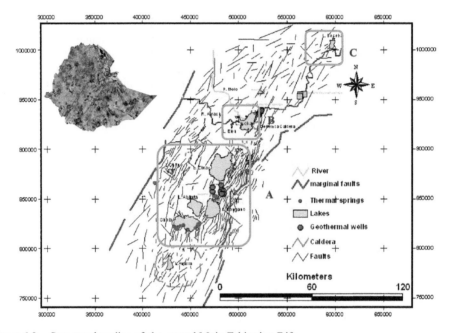

Figure 6.2.   Structural outline of the central Main Ethiopian Rift.

Figure 6.3.   Active extension faults within rhyolitic ignimbrites close to Lake Beseka, Ethiopia.

Figure 6.4.   Volcano-tectonic controlled lakes that lies in the major groundwater discharging zone, MER (Inset A in Figure 6.2).

The extensional structures provide suitable media for groundwater circulation in the region. Figure 6.3 depicts active extensional structures that act as a conduit for groundwater flow towards the lakes and sustain the lake ecosystems. This shows that lake level increases in EARS, in general, can be identified as non-climatic changes.

The tectonically well defined topography (Figure 6.4) shows the volcano-tectonically controlled Lakes in the MER. These lakes have intimate relation with groundwater where their survival and the chemistry is influenced by the groundwater input. Similar tectonic setting governs the occurrence of lakes in the EARS. In the EARS where rainfall is erratic, groundwater is vital in maintain the surface water system.

6.3   HYDROLOGY AND CLIMATE

The eastern branch of EARS contains over 25 large lakes while the western branch has over 11 lakes. The lakes are bigger in the western branch due to extensive tectonic grabens and more stable basins compared with the most active eastern branch of the EARS. The eastern branch contains more volcanic controlled lakes than the western branch. All lakes receive recharge from both rain and groundwater while some terminal lakes such as Abijata and Shalla in the MER (Ayenew, 1998, Alemayehu et al., 2006, Kebede et al., 2008) and Lake Chala in Kenya (Payne, 1970) do not have groundwater outflow. EARS is characterised by narrow belts of normal faulting, that give rise to graben-horst structures and act as groundwater discharge zone. This morphology provides a suitable place for the storage of large quantities of surface and groundwater fed by large numbers of rivers that drain toward the rift from the adjacent escarpments and plateaux. There are also numerous perennial rivers which are restricted to the rift. Such rivers contain several indicators from the groundwater inflow. The bigger rift rivers are the Awash, Omo (Ethiopia), Kerio, Turkuell (Kenya), Nyamgasani, Berarara, Ishasha and Semliki (Uganda). The rift is also characterised by wetlands that are sustained by groundwater.

The rainfall in EARS is controlled by the interannual oscillation of the Intertropical Convergence Zone (ITCZ) that regulates the moisture distribution derived from the Atlantic and Indian Oceans. Accordingly, the eastern branch of EARS is characterized by three distinct seasons: big rain season, small rain season and dry season. The western branch of EARS has a longer big rain season than the eastern part, which is dominated by a dry season. The MER and Kenya Rift are characterized by a semi-arid/arid climate with mean annual precipitation that ranges between 300 and 800 mm and mean annual temperature that ranges between 18 and 36°C. However, in Afar, the daily mean temperature can be as high as 45°C with a mean annual rainfall of less than 200 mm. The open water evaporation in large part of EARS reaches 2500 mm/year with a potential evapotranspiration (PET) of 1900 mm/year. The extreme case is for Afar where PET is over 3000 mm/year. In general, EARS is hot, dry with arid and semi-arid climate.

6.4   HYDROGEOLOGICAL FRAMEWORK

Due to the differences in mineralogy, texture and structure of the various volcanic rocks in the EARS, the water-bearing potential is also variable. Groundwater circulation and storage in the EARS aquifer depends on the type of porosity and permeability formed during and after the rock formation, and this is strongly conditioned by the tectonic and volcanic processes.

The groundwater flux between the higher rainfall areas of the adjacent plateaux and the drier rift floor aquifers depends on a number of interrelated factors including the presence of marginal grabens at the interface of the rift and the plateau; the presence of cross cutting transverse faults; and the distance of the rift valley aquifers from the bounding highlands. The groundwater connection between the EARS volcanic, volcano-lacustrine and alluvial aquifers and the bounding volcanic highland aquifers poses significant challenge due to the complex nature of groundwater flow, recharge and geochemical evolution in the volcanic and associated aquifers. The highland recharges appear in the center of the rift after deep circulation. Several fracture and contact springs characterize the rift escarpment from shallow circulation (Figure 6.5). This is also because the stratigraphy and tectonics of the interface zone is complex (rocks are laterally discontinuous, aquifer composition is heterogeneous and aquifer hydrodynamic characteristics are variable) (Kebede et al., 2008). Important hydrostructures in the Rift aquifers could be accounted for by:

- the occurrence of paleosols constituted by coarse alluvials, acidic lava flows;
- buried paleo-valleys;
- the occurrence of contraction joints within thick lava flows;
- the occurrence of thick residual soils as a product of weathering of volcanic rocks;
- permeable zones between the contact of the different generation of lava flows;
- secondary fractures due to faulting and weathering.

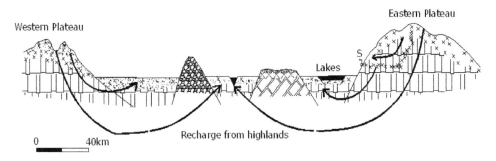

Figure 6.5. Conceptual regional groundwater circulation between the rift and the adjacent highlands.

Paleosols and several lava flow piles along the escarpment are known to generate perched aquifer system that are mainly marked by springs. EARS is an broad zone made of multi-layer aquifer system with variable groundwater potential. In most of the zones, groundwater emerges in the form of cold and thermal springs with variable magnitude. The MER, for example, is characterized by the occurrence of numerous cold springs which discharge at less than 2 ℓ/s in dry areas to 250 ℓ/s in high recharge zones adjacent to the escarpment (Alemayehu, 2003). The yield of such springs is directly controlled by the local rainfall and the presence of permeable hydrostructures linked to the regional groundwater flow system. The western branch is rich in highly productive springs due to the occurrence of high mean rainfall. In the eastern branch of the EARS, the potential aquifers are highly fractured and jointed basalts and ignimbrites. The weathered tuffs and paleosols are identified impermeable layers inhibiting vertical movement of groundwater, and these form perched water bodies in many parts of the EARS. Permeable alluvial and colluvial deposits associated with lacustrine soils form local shallow aquifers in the rift floor and along major river valleys. In the case of the Kenya Rift, productive aquifers exist within the fractured volcanics and along contacts between different lithological units (Gaciri et al., 1993; Olago et al., 2008). Unconsolidated sediments produce good quality water in the Rift valley of Malawi, (Water Aid, 2004).

The large part of the EARS, main water bearing structures are represented by fractures of tectonic origin through which surface water from lakes and rivers can easily infiltrate into the groundwater system. The hydrogeological characteristics of the volcanic rocks in EARS are subjected to varying degrees of secondary activities like weathering, fracturing and faulting which divide the area into a number of minor and major faults usually arranged in an en-echelon pattern. The secondary structures have a significant role in the circulation and distribution of groundwater in the rift. The main aquifers in the EARS are:

- Aquifers mainly characterized by primary porosity and permeability: lacustrine sediments, alluvials, and pyroclastic fall deposits. These deposits usually represent important aquifers in many places within EARS where there is dynamic recharge.
- Aquifers mainly characterized by secondary permeability due to fracturing and weathering: ignimbrites and welded tuffs, rhyolitic lava flow and massive basaltic lava flow.
- Aquifers characterized by both primary and secondary rock permeability: basaltic lava flows. Generally these rocks form moderate to good aquifers, in places of intense faulting and fracturing they are very good aquifers.

In the MER, for example, the transmissivity and hydraulic conductivity of rocks, dominated by both primary and secondary porosities, are about 2×10^{-3} m²/s with of 6×10^{-4} m/s respectively. However, wells located on fractured rocks have transmissivity of 8×10^{-2} m²/s and hydraulic conductivity of 4×10^{-4} m/s. The majority of faults act as conduits for groundwater flow, in some places, the open faults allow significant amount of preferential groundwater flow parallel and sub-parallel to the rift axis. In the EARS, the direction of groundwater flow is strongly, governed by the orientation of faults, which is often perpendicular to the regional groundwater contours in the highlands and escarpments. For easy circulation of groundwater from highland into the Rift Valley, the faults should be discontinuous or act as conduits. In most cases due to the extensional nature of the faults, groundwater can cross the faults which do not act as a barrier. It was found that these axial faults govern strongly the subsurface hydraulic connection of the rift lakes and the river-groundwater relations. In contrast to the high hydraulic conductivity of the rift fractured volcanics, some faults act as barrier for groundwater flow when they are filled with younger volcanic products. In Tanzania, groundwater flow is controlled by dipping of lithology and faults where localized flows are limited to weathered zones, while the bedrock is under the effect of regional flow (Mul et al., 2007).

Intensive faulting and crosscutting structures, mainly dykes and strike slip faults divide groundwater storage into different compartments. The occurrence of tensional basins and different generation of faults generates localized storage site. The extension fractures, which are quite abundant in the EARS play dynamic role in groundwater movement. In the Lake Beseka area of MER, submerged springs discharge as much as 500 ℓ/s of thermal groundwater into the lake. The springs were above the lake surface until the early nineties. Figure 6.6 shows the temporal increase of the Lake Beseka level derived from temporal satellite images. The water level and ambient temperature are plotted in Figure 6.7 for Lake Beseka, which shows a level

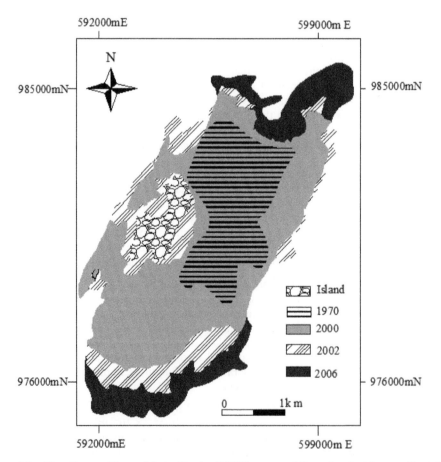

Figure 6.6.   Shoreline positions of Lake Beseka (MER) measured from Landsat images (Inset C in Figure 6.2).

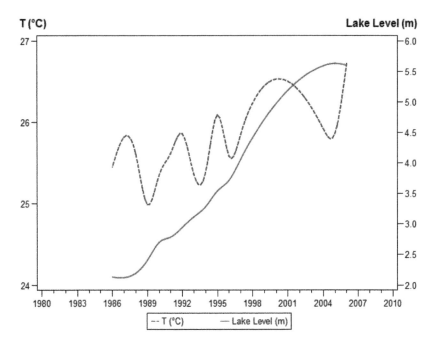

Figure 6.7.   Temporal variation of mean ambient temperature and mean lake level change of Lake Beseka.

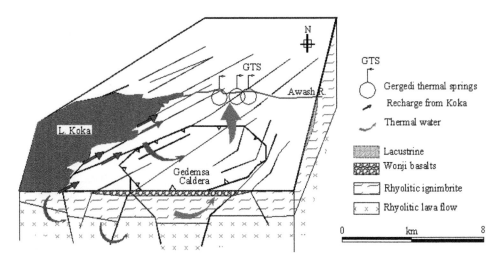

Figure 6.8.   3D conceptual model for the interaction of Lake Koka with the surrounding groundwater (Inset B in Figure 6.2).

increase with time as ambient temperature increases. The level increase is due to groundwater inflow facilitated by the active tectonics that act as a conduit rather than as a reservoir.

   Lakes also lose water into the groundwater system through extensional fractures. Lake Koka, straddling the western side of the Awash River in central MER, is located in an active tectonic area, and numerous high yield springs emerge at the downstream side. The lake is progressively losing water through fault lines. The Gedemssa volcano provides heat to the Gergedi thermal springs that are recharged by Lake Koka through active normal faults (Figure 6.8). Such interactions are dynamic in the EARS and are available throughout the rift. The overall discharge of the Gergedi thermal springs is over 2500 ℓ/s with a temperature of 42°C. Tectonic structures play a more important role as conduits for groundwater than as reservoirs. Kebede et al. (2008) interprets the isotopic composition of the thermal springs in the MER to indicate a meteoric source. This can indicate the recharging role of the fractures in the region. In general, the groundwater potential of the EARS is enormous except to some quality problems related to the natural source. Environmental degradation in the EARS is severely affecting the lakes ecosystem (Odada et al., 2003).

6.5   HYDROGEOCHEMICAL FEATURE

The hydrogeochemical characteristic of the EARS waters is complex due to the intensive faulting and volcanic processes that have taken place since Miocene. The chemistry of surface waters is influenced by discharging groundwater and strong evaporation and hence, most surface waters show alkaline pH and moderately high salinity. The effect is notable in the area where geothermal activity is highest due to the large geothermal gradient, for example, 17°C/100 m in the case of MER (Alemayehu and Vernier, 1997). In the EARS, the surface and groundwater reservoirs, which receive large amount of fresh water from highland rivers, are characterised by low salinity and mineral content, while those which have established direct contact with the deep hot water circulation are saline, highly mineralised and mostly alkaline.

There is a general progressive enrichment in the stable isotopes ($\delta^{18}O$ and δD) composition of the groundwaters from the mountains to the rift floor, except for the thermal waters. This trend suggests the progressive increase in the importance of evaporatively fractionated water infiltration to the groundwaters and the progressive decrease in groundwater inflow to the aquifers from the mountains (Kebede et al., 2008). Deep circulating groundwater that is recharged on the top of mountains which are part of the escarpment appears with lighter isotopes in the center of the rift. The plot constructed from 350 data points from MER (Figure 6.9) shows

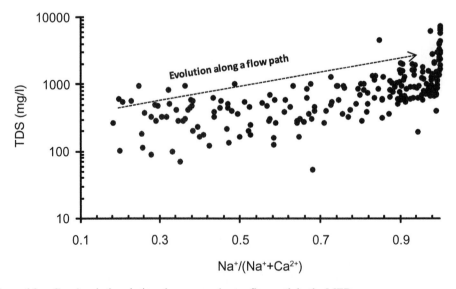

Figure 6.9.   Geochemical evolution along groundwater flow path in the MER.

evolution of water along the flow path where mineralized water with high sodium controls the geochemical system. This takes place because of the abundance of acidic rocks and loss of calcium through carbonate precipitation. This process also indicates the dominance of rock phase in the geochemical evolution. The southern part of the eastern rift in Tanzania contains alkaline type groundwater with low Ca, Mg and high Na (Water Aid, 2004).

Lakes that receive thermal springs tend to be saline and alkaline, such Lakes Abijata, Shalla and Beseka in the MER (Ayenew, 1998; Alemayehu et al., 2006); Naivasha (Darling et al., 1996) and Elementaita in Kenya Rift (Mwaura, 1999). The increase in salinity towards the axial part of the rift could be explained by the discharge of mineralized water after long water-rock interaction time in the volcanic aquifer. High lake mineralization is also related to the very high evaporation process of mineralized groundwater that joins the lake.

In the case of Lake Beseka (MER), electrical conductivity has fallen from 74 170 µS/cm to 7440 µS/cm between 1961 and 1991 corresponding to a change in size of the lake from 3 km^2 to 48 km^2 (Alemayehu et al., 2006) with the expansion rate of 1.5 km^2/year, which is strongly believed to be due to large groundwater influx through extensional tectonics that have modified the fracture permeability in the ignimbrites. Apart from the groundwater in the Afar region of Ethiopia, which is dominantly sodium chloride type due to the dominance of mafic volcanic rocks and deep groundwater circulation, the groundwater facies in the EARS can be categorized as sodium bicarbonate type of variable concentration. In general, two group of groundwater can be identified in EARS:

1. Weakly mineralised groundwater with near neutral pH, and calcium dominated.
2. Highly alkaline water which is enriched with fluoride, and sodium dominated.

EARS groundwater is characterized by high salinity due to a high degree of water-rock interaction, evaporation and the discharge of thermal water. Groundwater tends to evolve from Ca-HCO$_3$ type in the recharging zones to Na-HCO$_3$ water within the acid rocks of the Rift Valley. Salinity increases from the plateau volcanics that are characterized by high rainfall and lower evapotranspiration to regions of lower rainfall and higher evapotranspiration in the centre of the Rift Valley. High Na, HCO$_3$ and F characterize thermal groundwaters and alkaline lakes.

Fluoride occurs in both surface and groundwater. However, always the source is from groundwater that resides in the acidic volcanic. These waters offer a risk to health if consumed regularly. Obsidian and pumice contain large amounts of fluoride and weathering and hydrothermal activities release large amounts of it into groundwater. Fluorite occurs in igneous rocks worldwide, at an average concentration of 715 mg/kg (Hem, 1992). The substitution of hydroxyl ion by fluoride in clays is a likely removal mechanism (Von Damm and Edmond, 1984). Accordingly, formation of alumino-silicate minerals through reverse weathering, by evaporation and escape of CO$_2$ gases produces alkaline water (Na-HCO$_3$, Cl type) with pH > 9 in closed basins. The major thermal aquifers in the rift are found in the acidic rocks such as pyroclastic deposits and unwelded tuffs.

The fluoride concentration in the groundwater of the MER varies between 0.5 mg/ℓ and 98 mg/ℓ while in the lakes the concentration reaches 253 mg/ℓ (Ayenew, 1998; Alemayehu, 2000). High fluoride concentrations of up to 180 mg/ℓ are also a characteristic feature of the Kenyan Rift thermal waters (Clarke et al., 1990). Garciri et al., (1993) indicated that fluoride in the Kenya Rift occurs at concentrations in lakes between 13 and 2170 mg/ℓ, in groundwater between 0.9 and 180 mg/ℓ, and in thermal water between 1 and 110 mg/ℓ. The fluoride concentrations in Tanzania are lakes between 60 and 690 mg/ℓ, springs between 15 and 63 mg/ℓ and rivers between 12 and 26 mg/ℓ (Water Aid, 2004). In both cases, the sources are related to volcanic rocks. However, in the case of Malawi Rift (Msonda et al., 2007), the fluoride sources are related to weathering of biotite and hornblende in basement rocks. The spatial variation in fluoride concentration indicates that it is the discharge areas that are dominated by high fluoride rather than the recharge areas.

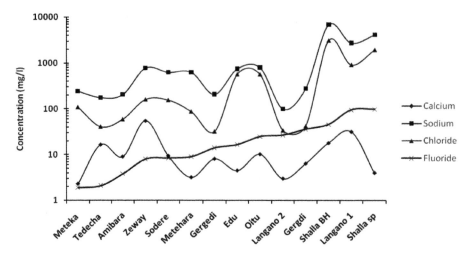

Figure 6.10.   Spatial distribution of fluoride in the central part of the MER.

The high fluoride concentration in the lakes of the EARS is attributed to the complete removal of calcium by carbonate precipitation, which is related to high salinity and alkaline environment. The fluoride is more chemically active at high temperatures and high temperature springs contain higher fluoride concentrations than the low temperature ones. Generally, high ambient temperature, alkaline rocks and low calcium concentrations favour the abundance of fluoride in waters of the Rift valley.

The variation in fluoride concentration in groundwater is the result of interaction between a number of factors:

- volcanic activity related to Rift formation that has generated rocks containing fluoride and thermal water, and
- chemical reaction involving calcium.

Figure 6.10 shows increases in fluoride concentration from NE to SE in the MER, and these are related to increases in the geothermal gradient and abundant fluoride bearing rocks, i.e. the fluoride concentration in surface and groundwater reflects the spatial variation of the environment of occurrence. The high groundwater temperature facilitates removal of Fluoride from rocks and concentrates in the groundwater.

The MER rocks contain high concentrations of fluoride. The Corbetti obsidians have 2500 ppm (UNDP, 1973), benmorites 818 to 2216 ppm, and rhyolites 1290 to 3138 ppm (Peccerillo et al., 1995). The pumices are known to have very high concentrations of fluoride and sodium. The weathering processes of these rocks could release large amount of fluoride into the groundwater. In the MER only, over 10 million people are living in high fluoride region. In the EARS, in general, over 50 million people are at risk due to the consumption of high fluoride water.

6.6   IMPLICATION ON WATER SUPPLY AND SANITATION

In the weathered volcanics and alluvial cover in the Rift, dug wells fitted with mono pumps are suitable for rural water supply. However, for urban and irrigation water supply, deep wells (300 to 400 m) are important. Springs are suitable for water supply along the rift escarpment. The wide range of ecosystems in the EARS are linked to the surface water and groundwater distribution. With recent increases in demand for water supply and irrigation, there is an increasing need to develop groundwater resources. EARS has a high human population and intense economic activities that require large quantities of freshwater to maintain. However, near absence of sanitation in the region is a critical concern to groundwater quality. Pollution and unsustainable exploitation of groundwater put groundwater at risk. EARS provides suitable media for groundwater provision except that some Quaternary acidic aquifers contain high fluoride that deteriorate the quality (the WHO drinking water guideline sets $F = 1.5$ mg/ℓ). The fluoride bearing acidic rocks can be excluded by locating bore holes in the basaltic aquifers. Some lakes are utilized for industrial mineral extraction, but most of them are used for large-scale irrigation, and the water quality and the lake ecosystem will continue to be degraded due to increasing human activities in the EARS. Lack of awareness and environmental regulations exacerbate the pollution effect, which could be easily controlled through the implementation of sustainable sanitation. Groundwater remains the key role player in improving the sanitation of the urban and rural communities. Improving sanitation and wastewater reuse or recycling provides sustainable solution for aquifer protection from where potable water is being pumped. Besides large scale irrigations in several countries, the water supply for rural communities in Africa can be provided through wise and knowledge based use of groundwater.

6.7   SUMMARY AND RECOMMENDATIONS

The western branch of EARS contains abundant relatively fresh groundwater compared to the eastern branch that contains a complex distribution of bicarbonate, sulphate and chloride type waters. The rift tectonics play a vital role as conduits through which rapid groundwater and surface water transport can occur. EARS is characterized by a high thermal anomaly and the elevated temperature of the volcanic aquifers that facilitates the removal of fluoride from the rocks due to the removal of calcium through carbonate precipitation. Extensive water-rock interaction allows the concentration of fluoride in the groundwater and evaporation of closed lakes facilitates the concentration of fluoride associated to alkaline pH condition. Surface waters receive fluoride from thermal waters in geothermally active part of EARS and the concentration gradually decreases towards the escarpment. The recent changes on lake levels may have grave environmental consequences on the fragile rift ecosystem, which demands an integrated basin wide water management practice with particular emphasis on the groundwater systems that guarantee the sustainability of the surface water resources. EARS can be considered as a groundwater rich zone (or rich in renewable groundwater) that can sustainably be used to alleviate the water stress in the region.

REFERENCES

Acocella, V., Gudmundsson, A. and Funiciello, R. (2002). Interaction and linkage of extension fractures and normal faults: examples from the rift zone of Iceland. Jour. of Structural Geology, 22: 1233–1246.

Alemayehu, A. (2000). Water pollution by natural inorganic chemicals in the central part of the Main Ethiopian Rift. Ethiopian Journal of Science, 23(2): 197–214.

Alemayehu, T. (2003). Controls on the occurrence of cold and thermal springs in central Ethiopia. African Geoscience Review Journal, 19(4): 245–251.

Alemayehu, T. and Vernier, A. (1997). Conceptual model for Boku hydrothermal area (Nazareth), Main Ethiopian Rift Valley. Ethiopian Jour. Science, 20(2): 283–291.

Alemayehu, T., Ayenew, T. and Kebede, S. (2006). Hydrochemical and lake level changes in the Ethiopian Rift. Jour. of Hydrology, 316: 290–300.

Ayenew, T. (1998). The Hydrological system of Lake District basin, Ethiopia. Ph.D. Thesis, ITC, Enschede, The Netherlands.

Benvenuti, M., Carnicelli, S., Belluomini, G., Dianelli, N., Grazia, S., Ferrari, G.A., Iasio, C., Sagri, M., Ventra, D., Atnafu, B. and Kebede, S. (2002). The Ziway-Shala lake basin (Main Ethiopian Rift, Ethiopia): a revision of the basin evolution with special reference to the Late Quaternary. Jour. of African Earth Sciences, 35: 247–269.

Boccaletti, M., Bonini, M., Mazzuoli, R. and Trua, T. (1999). Pliocene- Quaternary volcanism and faulting in the northern Main Ethiopian Rift. Acta Vulcanol., 11: 83–98.

Boccaletti, M., Bonini, M., Mazzuoli, R., Abebe, B., Piccardi, L. and Tortorici, L. (1998). Quaternary oblique extensional tectonics in the Ethiopian Rift (Horn of Africa). Tectonophysics, 287: 97–116.

Chorowicz, J. (2005). The East African rift system. Jour. of African Earth Sciences, 43: 379–410.

Darling, W.G., Berhanu, G. and Musa, K.A. (1996). Lake-groundwater relationships and fluid-rock interaction in the East African Rift Valley: isotopic evidence. Jour. African Earth Sciences, 22(4): 423–431.

Darling, W.G., Griesshaber, E., Andrews, J.N., Armannsson, H. and O'Nions, R.K. (1995). The origin of hydrothermal and other gases in the Kenya Rift valley. Geochimica et Cosmochimica Acta., 59(12): 2501–2512.

Delvaux, D., Levi, K., Kajara, R. and Sarota, J. (1992). Cenozoic paleostress and kinematic evolution of the Rukwa-North Malawi rift valley (EARS). Bulletin des Centres de Recherche Exploration-Production. Elf-Aquiaine, 16(2): 383–406.

Di Paola, G.M. (1972). The Ethiopian Rift Valley (between 7°00PN lat. and 8°40PN lat.). Bull. Volcanol., 36: 317–560.

Gaciri, S.J. and Davies, T.C. (1993). The occurrence and geochemistry of fluoride in some natural waters of Kenya. Jour. of Hydrology, 143: 395–412.

Kebede, S., Travi, Y., Alemayehu, T., Ayenew, T. and Tessema, Z. (2008). Groundwater origin and flow along selected transects in Ethiopian Rift volcanic aquifer. Hydrogeology Journal, 16: 55–73.

Korme, T., Chorwicz, J., Collet, B. and Bonavia, F. (1997). Volcanic vents rooted on extension fractures and their geodynamic implications in the Ethiopian Rift. Jour. Volcanol and Geothermal Research, 79: 205–222.

Laerdal, T. and Talbot, M.R. (2002). Basin neotectonics of lakes Edward and George, East African Rift. Palaeogeography, palaeoclimatology and palaeoecology, 187: 213–232.

Mohr, P.A. (1962). The Ethiopian Rift System. Bull. Geoph. Obse. Addid Ababa, 3(1): 1–50

Mohr, P.A. (1967). Major volcano-tectonic lineament in the Ethiopian Rift System. Nature, 213: 664–665.

Mohr, P.A. (1987). Patterns of faulting in the Ethiopian rift Valley. Tectonophysics, 143: 169–179.

Msonda, K.W.M., Masamba, W.R.L. and Fabiano, E. (2007). A study of fluoride groundwater occurrence in Nathenje, Lilongwe, Malawi. Physics and Chemistry of the Earth. 32: 1178–1184.

Mul, M.L., Mutiibwa, R.K., Foppen, J.W.A., Uhlenbrook, S. and Savenije, H.H.G. (2007). Identification of groundwater flow systems using geological mapping and chemical spring analysis in south Pare Mountains, Tanzania. Physics and chemistry of the Earth, 32: 1015–1022.

Mwaura, F. (1999). A spatio-chemical survey of geothermal springs in lake Elementaita, Kenya. International Journal of salt lake research, 8: 127–138.

Odada, E., Olago, D.O., Bugenyi, F., Kullndwa, K., Karimumuryango, J., West, K., Ntiba, M., Wandiga, S., Aloo-Obudho, P. and Achola, P. (2003). Environmental assessment of the East African Rift valley lakes. Aquat. Sci., 65: 254–271.

Olago, D.O., Opere, A. and Barongo, J. (2008). Holocene paleohydrology, groundwater and climate change in the Lake basin on the central Kenya Rift. Paper presented at the International conference on Groundwater & Climate in Kampala, 24–28 June, 2008.

Payne, B.R. (1970). Water balance of Lake Chala and its relation to groundwater from tritium and stable isotope data. Journal of Hydrology, 11: 47–58

Peccerillo, A., Yirgu, G. and Ayalew, D. (1995). Genesis of acid volcanics along the Main Ethiopian Rift: a case history, Gedemsa volcano. Ethiopian Jour. Science, 18(1): 23–50.

Skobelev, S.F., Hanon, M., Klerkx, J., Govorova, N.N., Lukina, N.V. and Kazmin, V.G. (2004). Active faults in Africa: a review. Tectonophysics, 380: 131–137.

Von Damm, K.L. and Edmond, J.M. (1984). Reverse weathering in the closed basin lakes of the Ethiopian rift. American Jour. Science, 284: 835–862.

UNDP, (1973). Technical Report, Investigation of geothermal resource for power development, Ethiopia, Geology, Geochemistry and Hydrology of thermal springs of the East African Rift system within Ethiopian. Ethiopian Institute of Geological Surveys, Addis Ababa, Ethiopia.

Water Aid, (2004). Groundwater Quality in Tanzania, in Malawi. Information sheet. BGS, UK.

Woldegabriel, G., Aronson, J.L. and Walter, R.C. (1990). Geology, geochronology and rift basin development in the central sector of the Main Ethiopian Rift. Geol. Soc. Am. Bull., 102: 439–458.

7

Aquifer vulnerability and its implication for community water supply of Porto-Novo region (South–East of Benin)

A. Alassane
Département des Sciences de la Terre, Faculté des Sciences et Techniques,
Université d'Abomey-Calavi; Cotonou, Bénin

A. Faye
Département de Géologie, Faculté des Sciences et Techniques,
Université Cheikh Anta Diop de Dakar, Sénégal

M. Boukari
Département des Sciences de la Terre, Faculté des Sciences et Techniques,
Université d'Abomey-Calavi; Cotonou, Bénin

S. Cissé Faye
Département de Géologie, Faculté des Sciences et Techniques,
Université Cheikh Anta Diop de Dakar, Sénégal

ABSTRACT: This study presents an estimation of the vulnerability to pollution of the Mio-Pliocian and Quaternary aquifer in Porto-Novo region (South–East of Benin), part of the coastal sedimentary basin of Benin. The DRASTIC and GOD methods based on the characterisation of seven parameters for DRASTIC and three parameters for GOD were used in a GIS environment with ArcView software. The physico-chemical parameters of the groundwater have been evaluated in order to compare the results of the two methods and the derived pollution status. The results show that DRASTIC is the better method given the local conditions. The vulnerability map indicates that the coastal plain (South of Porto-Novo lagoon) has high and very high vulnerability. But, this zone is not more contaminated because there is little human activity. In the North of Porto-Novo lagoon, the aquifer vulnerability is medium and low, but the nitrate contents are highest, with concentrations reaching 334 mg/ℓ. The medium, high and very high degrees of vulnerability represent 80% of the mapped areas and 99.95% do not have a natural hydrogeological protection. Measures should be taken to protect the quality of the groundwater which is critical to the wellbeing of the people living in this rural zone.

7.1 INTRODUCTION

Groundwater vulnerability to pollution mapping enables the identification of high risk contamination zones, regardless of the pollutant type. Soil occupation plans should incorporate groundwater vulnerability data and appropriate water supply protection measures should be adopted.

In Porto-Novo region the densities of population reach 5000 to 7000 inhabitants/km² (RGPH, 2002) on the Continental Terminal and Quaternary formations which contain phreatic groundwater. In these conditions, the variety of domestic and economic activities (industry, agriculture, fuel and lubricant storage), associated with the production of solid and liquid wastes, represents a serious pollution risk for this water resource. The coastal plain (South of Porto-Novo lagoon), is not much populated, but this zone is being occupied now by the populations coming from Cotonou.

The purpose of this study is to evaluate and map the vulnerability of the sub-surface aquifer in Porto-Novo and surrounding area using both the DRASTIC and GOD methods and compare the results with perceived hydrogeological protection indicated by nitrate concentrations.

The dual approach used for the DRASTIC and GOD methods was chosen for the comparison and reliability of results.

7.2   SITE DESCRIPTION

The study area, some 410 km² in area, is located in the South–East of Benin, between latitudes 06° 20′ N and 06° 35′ N and longitudes 02° 26′ E and 02° 48′ E. It covers the southern portion of the Sakété plateau and the coastal plain between this plateau and the Atlantic Ocean (Figure 7.1) and includes the communes of Porto-Novo, Adjarra, Akpro-Missérété, Adjohoun, Dangbo and Sèmè-Kpodji.

Basic information on hydrography, geomorphology and geology of Porto-Novo region is shown in Figure 7.2. Figure 7.3 shows a hydrogeological section of the Continental Terminal plateau (a) and the coastal plain (b) showing the aquifer levels (Alassane, 2004).

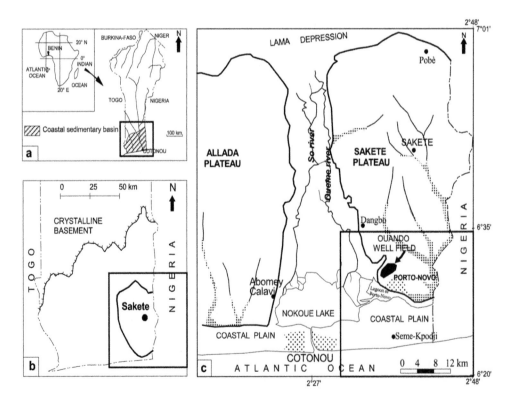

Figure 7.1.   Location of study zone: the coastal sedimentary basin in Benin (a), Sakété plateau in the coastal sedimentary basin (b) and Porto-Novo region (c).

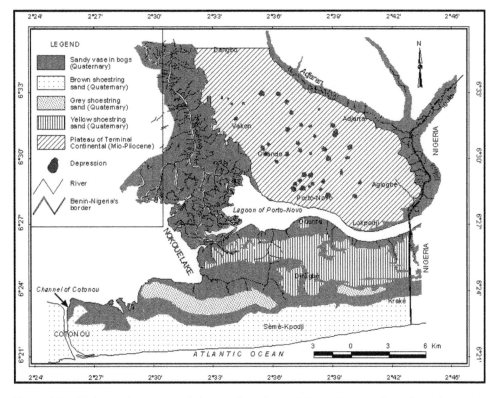

Figure 7.2. Hydrography, geomorphology and geology of Porto-Novo region. (See colour plate section).

7.3 DATA REQUIREMENTS

Data needed to support the evaluation included:

- Topographic maps at 1: 50000, R. of Dahomey, sheets 4a and 2c Porto-Novo, (National Geographic Institute, Paris, 1963);
- The pedagogical map of southern Benin (Volkoff, 1970);
- Data on static groundwater levels measured in July 2001, April and August 2002, February 2005 and August 2007 and existing data for February 1988 and December 1990;
- Hydroclimatical data collected by the Agency for Safety of Air Navigation (ASECNA) at its stations of Cotonou airport, Ouando (Porto Novo) and Sèmè-Kpodji from 1955 to 2000;
- Drilling and borehole logs;
- Existing data on hydraulic conductivity of the aquifer;
- Software for data treatment (ArcView, Excel, etc.).

7.3.1 *DRASTIC and GOD methods*

There is no absolute method for assessing groundwater vulnerability to pollution; several methods have been developed. The DRASTIC and GOD methods were selected because:

- The **DRASTIC** method is widely used for vulnerability mapping, it is standardized and takes into account many parameters;

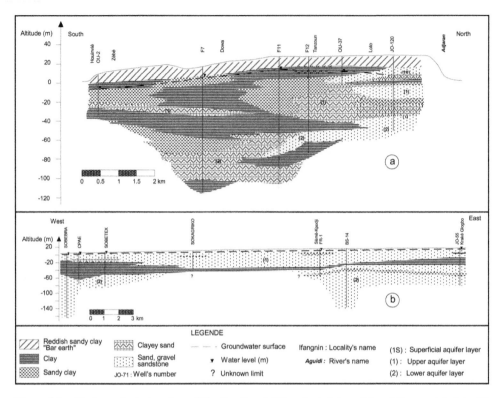

Figure 7.3.   Hydrogeological section of the Continental Terminal plateau (a) and the coastal plain (b) showing the aquifer levels (Alassane, 2004).

– The density of the study points is moderate and for this reason the numerical quotation method is recommended (Zaporozec and Vrba, 1994);
– The available data are related to **DRASTIC** and **GOD** parameters.

The DRASTIC method was developed by Aller et al. (1987) and is based on the evaluation of seven parameters which determine the water infiltration and its movement in the soil. It concerns the depth to water (D), the total recharge (R), aquifer media (A), soil media (S), topography (T), the impact of the unsaturated zone (I) and hydraulic conductivity of the aquifer (C). Each parameter has a fixed weight that reflects its relative importance to vulnerability (Table 7.1). The most significant parameters (depth of water and impact of the unsaturated zone) have a weight of 5 and the least significant (topography) a weight of 1. In addition, each parameter is assigned a rating between 1 and 10 depending on local conditions; the conditions of less vulnerability provide low rating and vice versa. Next, the local index of vulnerability (ID) is computed through multiplication of the value attributed to each parameter by its relative weight, and summing all seven products:

$$ID = (Dc \times Dp) + (Rc \times Rp) + (Ac \times Ap) + (Sc \times Sp) + (Tc \times Tp) + (Ic \times Ip) + (Cc \times Cp)$$

where Dc is the symbol of the parameter D (Depth) and Dp its weight.

The minimum DRASTIC index is 23 and the maximum is 226. The classes of vulnerability proposed by Aller et al. (1987) are presented in Table 7.2.

The GOD method is designed in 1987 by Foster (Murat et al., 2000). It is based on the evaluation of three settings: Groundwater occurrence (the aquifer type), Overall aquifer

Table 7.1.   DRASTIC settings and their respective weights.

Parameters	Weight
The water depth	5
Net recharge	4
Aquifer settings	3
Nature of soil	2
Topography	1
Nature of the unsaturated zone	5
Hydraulic conductivity	3

Table 7.2.   DRASTIC index intervals and classes of vulnerability.

Interval	DRASTIC class
<85	Very low vulnerability
85–114	Low vulnerability
114–145	Medium vulnerability
145–175	High vulnerability
>175	Very high vulnerability

Table 7.3.   GOD index intervals and corresponding classes.

Interval	GOD class
0–0.1	Very low vulnerability
0.0–0.3	Low vulnerability
0.3–0.5	Medium vulnerability
0.5–0.7	High vulnerability
0.7–1	Very high vulnerability

class (lithology and porosity of the aquifer) and Depth to water table. In GOD method, the parameters have the same weight. The vulnerability index (IG) is obtained using the following equation:

$IG = Cg \times Co \times Cd$,   where Cg = symbol of the type of aquifer, Co = symbol of lithology, Cd = symbol of the depth to water.

Five classes of vulnerability are defined from index 0 to 1 as indicated in Table 7.3 (Aubre et al., 1990).

7.3.2   *Evaluation of DRASTIC and GOD parameters*

Data were collected in 2001, 2002, and 2007 to update and complete the existing data sets and included georeferenced static groundwater levels and water physico-chemical characteristics.

Depth to water: This parameter was evaluated by interpolation of static groundwater level data measured in July 2001, April and August 2002, February 2005 and August 2007. The measurements were taken in 99 wells and data were entered in a Geographic Information System (GIS) environment using ArcView 3.2 for interpolation by the Inverse Distance Weighted (IDW) method (Longley et al., 2005). Vulnerability ratings were entered according to the DRASTIC and GOD methodologies.

Total recharge: The estimation of recharge was based on hydroclimatical data measured by ASECNA at Cotonou, Porto-Novo and Sèmè-Kpodji stations and the Thornthwaite water balance equation: $P = ETR + R + I$, with P = Rainfall; ETR = Actual Evapotranspiration; R = Runoff; I = Infiltration (Total recharge).

The average annual rainfall between 1952 and 1995 was 1309 mm in Cotonou, 1396 mm in Porto-Novo and 1490 mm in Sèmè-Kpodji. The actual evapotranspiration (ETR) was calculated by the Coutagne method and runoff (R) by the Tixéront-Berkaloff formula (Maliki, 1993). The estimation results indicate that total recharge values are between 10.2 and 17.8 cm (Table 7.4).

Aquifer media and impact of vadose zone: To evaluate these parameters, the lithological description of boreholes and piezometers was interrogated. The lithologic units based on DRASTIC and GOD classifications and ratings were assigned to each unit according to intervals determined by each method. Aquifer media consists of sand and gravel and vadose zone lithology facies, i.e. clay and sandy clay, silt, sand, gravel.

Soil type: Volkoff (1970) soil map was used to identify different soil types and regroup them according to the DRASTIC classification. The soil types, after classification, are: loam, clayed loam, sandy loam and sand. The ratings were then assigned to each soil type.

Topography: The topographic map (1/50000), Republic of Dahomey, sheets 4a and 2c, Porto-Novo, enabled slope to be calculated. It shows that the relief is almost flat in the coastal plain and is characterized by slopes of 0.12 to 0.8%. The plateau shows flat areas, characterized by low slopes (0.02 to 0.7%), but higher slopes are observed towards depressions and river valleys (1 to 6%).

Hydraulic conductivity: Pumping tests and data from the following sources were used:

- Bouzid (1991) in brown and gray sand cords ($K = 15 \times 10^{-4}\,m \cdot s^{-1}$);
- Nissaku Co LTD (1992 & 1994) in the yellow sand cords ($K = 4.85 \times 10^{-4}\,m \cdot s^{-1}$) and at the plateau (permeability map);
- Tastet (1977) concerning the granular deposits in the marshy areas of the coastal plain (3.3×10^{-4} to $4.7 \times 10^{-4}\,m \cdot s^{-1}$).

After evaluation, the parameters were entered into the ArcView 3.2 GIS environment and interpolation was made by the IDW method to generate spatial distribution maps. The ratings were then introduced and vulnerability maps generated.

7.3.3   *Validation of vulnerability maps and determination of aquifer protection*

The physicochemical characteristics of the aquifer were studied in order to be able to compare the appropriateness of the results from both the DRASTIC and GOD methods. The physical parameters taken into account are the temperature, conductivity and pH. These were measured in situ in 61 wells and boreholes in April and August 2002 and August 2007. The chemical parameters measured were nitrate concentrations and these were used as pollution indicators. The chromatographic method (liquid phase) was used to determine nitrate and nitrites. Samples were directly collected into polyethylene bottles and preserved at 4°C.

After validation of the DRASTIC map, the degree of aquifer protection was estimated according to the method of Ministère de l'Agriculture, des Pêcheries et de l'Alimentation au

Table 7.4.   Total recharge estimation in mm (1952–1995).

Parameters	Cotonou-Airport	Porto-Novo	Sèmè-Kpodji
Rainfall (mm)	1309	1396	1489
ETR (mm)	945.6	986.5	1019
Runoff (mm)	249.8	275.2	368.0
Total recharge (mm)	113.6	134.4	102.6

Table 7.5. DRASTIC Index (ID) in % and type of aquifer protection (MAPAQ, 1995).

Category	ID in %	Type of protection
Category 1	0–35	Certainly well protected aquifer
Category 2	35–75	Uncertain hydrogeological protection aquifer
Category 3	75–100	Certainly vulnerable aquifer

Québec (MAPAQ, 1995). It allows the conversion, as a percentage, of the DRASTIC index following the formula:

$$\text{DRASTIC index } (\%) = \times\ 100, \text{ avec ID} = \text{ordinary DRASTIC index.}$$

The DRASTIC index in % enables attribution of water supply hydrogeological protection type following as shown in Table 7.5.

7.4 RESULTS AND DISCUSSION

7.4.1 *Vulnerability maps*

The DRASTIC and GOD vulnerability maps to pollution (Figs. 7.4 and 7.5) indicate 4 classes of vulnerability: low to very high for DRASTIC and very low to high for GOD.

Both maps show a certain similarity in the coastal plain by assigning a high vulnerability to the aquifer. On the plateau, the index classes differ between the two methods (Table 7.6). The GOD method assigns very low to low vulnerability to the plateau aquifer (GOD index between 0.07 and 0.3) while the results of the DRASTIC method are low to moderate vulnerability ($85 \leq \text{ID} \leq 145$).

The results of physico-chemical analysis showed signs of pollution in the plateau aquifer with nitrate concentrations above $50\ mg \cdot \ell^{-1}$ and reaching $334\ mg \cdot \ell^{-1}$ in some wells (Figure 7.4). Therefore, the results of the DRASTIC method appear to reflect the known degree of pollution more accurately than those from GOD on the plateau. The GOD method under-estimated the aquifer vulnerability (Table 7.6) compared to the DRASTIC method. The evaluation of "total recharge", "hydraulic conductivity", "soil type" and "topography" in the DRASTIC method are of a crucial importance in the estimation of vulnerability to pollution in the study zone.

The DRASTIC vulnerability map shows 4 vulnerability classes: low to very high degree (Figure 7.4):

– The low vulnerability class represents 20.4% of the mapped areas and concerns the North and North-West of Porto-Novo plateau where the Ouando well field is located. In this sector, the depth to the water table exceeds 20 m and the unsaturated zone consists of clay and sandy clay.
– The medium vulnerability class is found in the Southern of Porto-Novo plateau and covers 21.7% of the study area. The southern part of the Ouando well field falls into this class. The groundwater depth is between 10 and 20 m and the unsaturated zone lithology is sandy clay and clayey sand.
– The high vulnerability class focuses on the wetlands of the coastal plain. These represent 24.3% of the mapped area, and the groundwater is shallow but the unsaturated zone is sandy clay formation and its thickness is approximately 0.5 to 3 m (Tastet, 1977).
– The very high vulnerability class comprises the sand in the coastal plain which covers 33.6% of the study area. The depth to the groundwater is between 1.2 and 7.5 m and the unsaturated zone consists of sand with gravels with rapid infiltration. The hydraulic conductivity of the aquifer is high.

Tables 7.5 and 7.6 show that the natural hydrogeological protection of the aquifers is mainly defined by the categories 2 and 3 (Table 7.7):

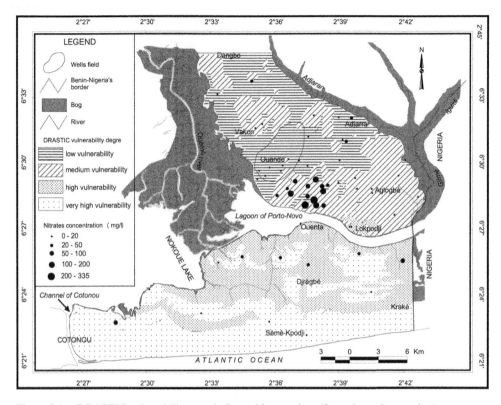

Figure 7.4.   DRASTIC vulnerability map in Porto-Novo region. (See colour plate section).

Category 2 covers 66.4% of the study area and characterizes almost the entire area of the Porto-Novo plateau and the swampy areas of the coastal plain. The groundwater natural protection is uncertain.

Category 3 concerns the coastal sand aquifer which represents 33.6% of mapped areas. The groundwater supply has no natural protection against pollution hazards.

The superficial aquifer in Porto-Novo region is mainly characterized by medium to very high vulnerability (80% of study area). Moreover, 99.95% of the mapped areas do not have any natural hydrogeological protection. It is clear from these results that this aquifer is naturally exposed to pollution and that there is need for measures to protect the quality of the water resources in this rural and semi-urban zone. The installation of public sanitation in the sector is essential. The depth generally considered for the construction of latrines is 2 m (CSI, 1997). However, the coastal plain is not appropriate for the building of latrines, while on the plateau, latrines must be wide and not deep. Chemical fertilizers must be used sparingly on the coastal plain. To prevent uncontrolled discharges of waste, the creation of refuse landfills and the organization of waste collection is needed. Extension of the drinking water distribution network in rural zones will restrict the installation of private wells. These wells are often poorly maintained and may become entry points of pollutants to the aquifer.

7.4.2   *Discussion*

The application of the **DRASTIC** and **GOD** methods produced vulnerability maps that provide information on areas of high risk potential to surface pollution hazards in Porto-Novo. Difficulties encountered in the application of the two methods include the evaluation of input parameters. The "total recharge of the groundwater" was estimated by empirical

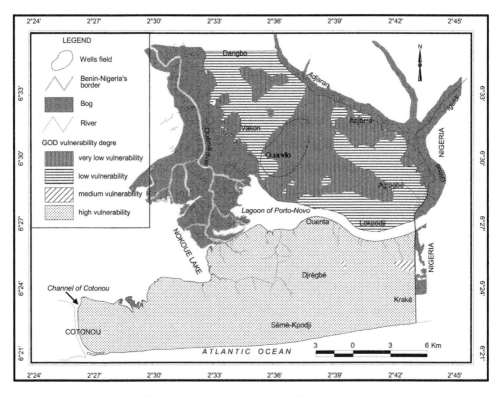

Figure 7.5.   GOD vulnerability map in Porto-Novo region. (See colour plate section).

Table 7.6.   Comparing class index for DRASTIC and GOD methods.

DRASTIC			GOD	
Class of index		Rate of vulnerability	Class of index	Rate of vulnerability
Ordinary ID	% of ID			
–	–	–	0.07–0.1	Very low
91–114	33.50–44.83	Low	0.1–0.3	Low
114–145	44.83–60.10	Moderate	0.3–0.5	Moderate
145–175	60.10–74.88	High	0.5–0.63	High
175–196	74.88–85.22	Very high	–	–

Table 7.7.   Comparing class index for DRASTIC and GOD methods.

	Type of protection	Area %
Category 1	Aquifer certainly well protected	0.1%
Category 2	Aquifer of uncertain hydro-geological protection	66.4%
Category 3	Aquifer certainly vulnerable	33.6%

formulae to determine actual evapotranspiration and runoff. This process involves manipulation of data and could lead to the multiplication of errors. But the recharge values obtained are of the same magnitude as those estimated by Boukari (1998), or 103 to 107 mm/year, using the Djonou method of hydrograph and piezometric levels fluctuation analysis. Maliki (1993) offers the same results (about 104.08 mm/year) using the water balance method.

Some parameters were estimated by interpolation in areas where data were lacking, notably recharge and hydraulic conductivity. The interpolation is reliable only within intervals defined by set data points (Jourda et al., 2006b) and can introduce errors.

In areas where data are lacking, the **DRASTIC** method represents the local situation most accurately, but it is inconclusive in the overlapping areas of medium vulnerability and areas with high levels of nitrate. The groundwater in the coastal plain is characterized by high to very high vulnerability and shows few signs of pollution by nitrate (Figure 7.4). Only two wells have revealed nitrate levels above 50 mg $\cdot$ l^{-1}. But, the aquifer of Porto-Novo plateau where the vulnerability is low and moderate reveals higher nitrate levels, exceeding the acceptable standard (50 mg/ℓ) and reaching 335 mg $\cdot$ l^{-1}. This contrast with the plateau and coastal plain reflects population density. The coastal area to the South of the Porto-Novo lagoon is entirely occupied by the village, but the population density in is low (30 to 500 inhabitants/km^2) (RGPH, 2002). On the plateau, the city of Porto-Novo, where the aquifer is heavily polluted (Figure 7.4), is densely populated (the population density is between 1000 to 7500 inhabitants/km^2) (RGPH, 2002) excepted in surrounding villages where the density is less than 500 inhabitants/km^2. This suggests that the spatial distribution of nitrate concentrations reflects population density. This factor needs to be included in the vulnerability evaluation parameters in Porto-Novo region. Jourda et al. (2006a) working on an aquifer near Abidjan in the Ivory Coast and Mohamed et al. (2003) on the Haouz aquifer in Morocco have made this same link.

7.5   CONCLUSION

After validation using the physico-chemical determinands, the **DRASTIC** method has proved more appropriate for the local conditions than the GOD method in the estimation of the groundwater vulnerability in Porto-Novo region. The **DRASTIC** map shows four (4) classes of vulnerability: low, medium, high and very high degree. Nearly 80% of mapped areas are characterized by medium, high and very high vulnerability and 99.95% do not have a natural hydrogeological protection. The groundwater in Porto-Novo region is naturally exposed to pollution and measures should be taken to protect the quality of water supply for use by the people in this rural zone. The spatial distribution of nitrate concentrations reflects population density, and this factor should be included in any future vulnerability assessment in the Porto-Novo region.

REFERENCES

Alassane, A. (2004). Etude hydrogéologique du Continental terminal et des formations de la plaine littorale dans la région de Porto-Novo : Identification des aquifères et vulnérabilité à la pollution de la nappe superficielle. Thèse de doctorat de 3ème cycle. Univ. C. A. Diop de Dakar. 145p + Annexes.

Alassane, A. (2000). Contribution à l'étude hydrochimique de l'aquifère du Continental Terminal du plateau de Sakété dans le bassin sédimentaire côtier du Bénin. Mémoire de DEA. Univ. C. A. Diop de Dakar. 81p + Annexes.

Aller, L., Benett, T., Lehr, J.H., et al. (1987). Drastic: a standardized system for evaluating ground water pollution potential using hydrogeologic settings. National Water Well Association; Rapport EPA-600/2-87-035; 622p.

Aubre, F., Isabel, D., Gelinas, P.J. (1990). Revue de littérature sur les méthodes d'évaluation des risques de contamination des eaux souterraines. Rapport GGL-90-21 présenté à Hydro-Québec, Vice-Présidence Environnement. Groupe de Recherche en géologie de l'ingénieur, Département de Géologie, Univ. Laval. 109p.

Boukari, M. (1998). Fonctionnement du système aquifère exploité pour l'approvisionnement en eau de la ville de Cotonou sur le littoral béninois. Impact du développement urbain sur la qualité des ressources. Thèse Doctorat ès-Science. Univ. C. A. Diop de Dakar. 278p + Annexes.

Boukari, M., Alassane, A., Azonsi, F., Zogo, D., Dovonou, L., Tossa, A. (2006). Groundwater pollution from urban development in Cotonou City, Bénin. Groundwater Pollution in Africa. Edited by Yongxin Xu and Brent Usher, Taylor & Francis/Balkema, Great-Britain, pp. 125–138.

Bouzid, M. (1971). Données hydrogéologiques sur le bassin sédimentaire côtier du Dahomey. Projet DAH3, FAO, Rome.

CSI-UNESCO (1997). Qualité de l'eau de la nappe phréatique à Yeumbeul, Sénégal. CSI info N° 3, 27p.

IGIP-GKW-GRAS (1983). Plan Directeur Alimentation en eau potable ville de Cotonou. Les ressources en eau. Ministère de l'Industrie des Mines et de l'Energie -SBEE. Vol. 2, Cotonou Bénin.

IGIP-GKW-GRAS (1989). Plans Directeurs et Etudes d'Ingénierie pour l'alimentation en eau potable et l'évacuation des eaux pluviales, des eaux usées et déchets solides. Ville de Cotonou. 90p. SBEE. Cotonou. Bénin.

Institut Geographique National-Paris (1963). Carte de l'Afrique de l'Ouest au 1/50000 (Type Outre-Mer), R. du Dahomey, Feuilles Porto-Novo 2c et 4a.

Jourda, J.P., Kouame, K.J., Saley, M.B., Kouame, K.F., Kouadio, B.H., Kouame, K. (2006a). A new cartographic approach to determine the groundwater vulnerability of the Abidjan Aquifer. Groundwater Pollution in Africa. Edited by Yongxin Xu and Brent Usher, Taylor & Francis/Balkema, Great-Britain, pp. 103–114.

Jourda, J.P., Saley, M., Djagoua, E.V., Kouame, K.J., Biemi, J.E.T., Razack, M. (2006b). Utilisation des données ETM + de Landsat et d'un SIG pour l'évaluation du potentiel en eau souterraine dans le milieu fissuré précambrien de la région de Korhogo (Nord de la Côte d'Ivoire): approche par analyse multicritère et test de validation. Revue de Télédétection, vol. 5, N° 4, pp. 339–357.

Longley, P.A., Goodchild, M.F., Maguire, D.J., Rhind, D.W. (2005). Geographic Information Systems and Science, 2nd Edition (Chapters 14 and 15). ISBN: 978-0-470-87001-3. 536p—Wiley edition, United States.

Maliki, R., (1993). Etude hydrogéologique du littoral béninois dans la région de Cotonou A.O). Thèse de Doctorat de 3ème cycle. Version Provisoire. UCAD; Dakar, Sénégal. 162p + Annexes.

MINISTERE DE L'AGRICULTURE, DES PECHERIES ET DE L'ALIMENTATION AU QUEBEC (1995). Guide d'application. Examen des projets de distribution au Québec d'eau embouteillée importée, 29p. Centre québécois d'inspection des aliments et de santé animale. http://www.mapaq.gouv.qc.ca/NR/rdonlyres/9B02EDDE-F148.4FC9-AB7E-C3D3F01B807B/0/no9.pdf

Mohamed, S., Rachid, M., Moumtaz, R. (2003). Utilisation des SIG pour la caractérisation de la vulnérabilité et de la sensibilité à la pollution des nappes d'eau souterraines. Application à la nappe du Haouz de Marrakech, Maroc, 25p.

Murat, V., Martel, R., Michaud, Y., Fagnan, N., Beaudoin, F., Therrien, R. (2000). Cartographie hydrogéologique régionale du piémont laurentien dans la MRC de Portneuf: comparaison des méthodes d'évaluation de la vulnérabilité intrinsèque. Commission géologique du Canada, Dossier public # 3664-d.

Nissaku Co. LTD, (1994). Projet pour l'exploitation des eaux souterraines du Bénin, phase III. Rapport final. 204p. + 54 photos + 3 vol. Annexes.

Oyede, L. M. (1991). Dynamique sédimentaire actuelle et messages enregistrés dans les séquences quaternaires et néogènes du domaine margino-littoral du Bénin (Afrique de l'Ouest). XIIème Congrès INQUA. Montréal., Résumé 237p.

RGPH (2002). Synthèse des résultats. Direction des études démographiques, INSAE, Cotonou, Bénin.

SERHAU-SA (2001). Plan directeur d'urbanisme de la ville de Porto-Novo, Rapport Projet FAC: Décentralisation et appui aux collectivités locales; 35p.

Tastet, J.P. (1977). Les formations sédimentaires quaternaires à actuelles du littoral du Togo et de la République Populaire du Bénin. Recherche française sur le Quaternaire—INQUA, 1977. Supplément au Bulletin AFEQ, 1977-1, N° 50, pp 155–165.

Volkoff, B. (1970). Carte pédologique de reconnaissance au 1/200000 du Dahomey, Feuille Porto-Novo, ORSTOM, 80p + Annexes.

Vrba, J., Zaporozec, A. (1994). Guidebook on mapping groundwater vulnerability. International Association of Hydrogeologists, vol. 16, 131p.

8

Vulnerability of dolomite aquifers in South Africa

K.T. Witthueser, R.C. Leyland & M. Holland
Department of Geology, University of Pretoria, Pretoria, South Africa

ABSTRACT: The karstified dolomites of the Chuniespoort Group are vital water resources for the expanding rural, urban and industrial complexes in Gauteng and Rustenburg, South Africa. Another example of the importance of the dolomites is their roles in sustaining the protected areas of the Blyde River Canyon and Sudwala Caves, Mpumalanga, South Africa, which are under threat from mining and forestry activities. The importance of the aquifer is in both cases not matched with any sound scientific based protection areas to prevent degradation of the resource from these potentially harmful activities. The present pressure on expansion of urban and agricultural development onto the dolomites is cause for major concern due to the enormous potential for pollution of the dolomitic aquifers underlying these areas as well as the renowned surface instability of these aquifers. The concept of vulnerability mapping is an important decision tool when expansion of urban, industrial or agricultural areas onto dolomitic land must be evaluated. For this reason the European COP vulnerability mapping method was adapted to suite the unique properties of South African karst terrains as well as the climatic conditions in semi-arid environments. The developed vulnerability mapping method is applied to the well known Cradle of Humankind World Heritage Site, Gauteng as well as to the Sudwala Caves/Blyde River Canyon dolomitic area in the Mpumalanga.

8.1   INTRODUCTION

Different interest groups define vulnerability in various ways to address their needs of evaluating potential impacts related to specific disasters or risks. Vulnerability is generally associated with the incapacities of a system or a population group to cope with an external event or resist the impact thereof (Villagrán 2006). External and internal sides to vulnerability, which relate to the exposure to external sudden shocks or continuous stresses and the defenselessness or incapacity to cope without damaging losses were proposed by Chambers (1989). Losses in this context can for example refer to human, economic, social or ecological losses, with the poorest groups of a society typically suffering the most. Vulnerability is therefore associated to the inverse of security. The consideration of losses is generally more closely related to risk, which addresses potential losses as a result of natural or man-made hazards and vulnerable conditions of the system or society. Newer approaches often include the coping capacity or deficiencies in preparedness of people or organizations in such risk assessments (Villagrán 2006). With regard to groundwater resources, vulnerability is usually used in the physical context of the susceptibility of an aquifer to droughts (quantitative aspect) or contamination (qualitative aspect). Groundwater drought vulnerability refers to the likelihood of well and borehole failures (drying up) during droughts as a result of increased stress on low-yielding groundwater sources and reduced recharge to the aquifer, which typically lags meteorological events (Calow et al., 1997).

   The National Research Council defined groundwater vulnerability to contamination as the likelihood for contaminants to reach a specified position in the groundwater system after introduction at a location above the uppermost aquifer (NRC, 1993). Representative municipal (e.g. sewer leakage), agricultural (e.g. animal wastes), industrial (e.g. pipeline leaks) or mining (e.g. acid mine drainage) sources of contaminants are considered (Sililo et al., 2001). In karst or dolomitic terrains, another dimension of "aquifer" vulnerability to be considered, is the vulnerability of the system to sinkhole formation with subsequent loss of the aquifer potential.

This chapter describes the adaption of a European vulnerability mapping method to conta-mination, to suit the unique properties of South African karst terrains as well as the climatic conditions in semi-arid environments. The developed vulnerability mapping method is applied to the well known Cradle of Humankind World Heritage Site (COHWHS), Gauteng as well as to the Sudwala Caves/Blyde River Canyon dolomitic area, Mpumalanga. Both areas of investigation are underlain by the karstified dolomites of the Chuniespoort Group (Figure 8.1).

These karstified dolomites are capable of sustaining high-yielding boreholes and are the only readily available water resources for many towns, rural areas and farms in parts of South Africa's Gauteng Province. In the Mpumalanga, these kartsified dolomites sustain the protected Blyde River Canyon and Sudwala Caves. The importance of the aquifer is in both cases not matched with any sound scientific based protection areas to prevent degradation

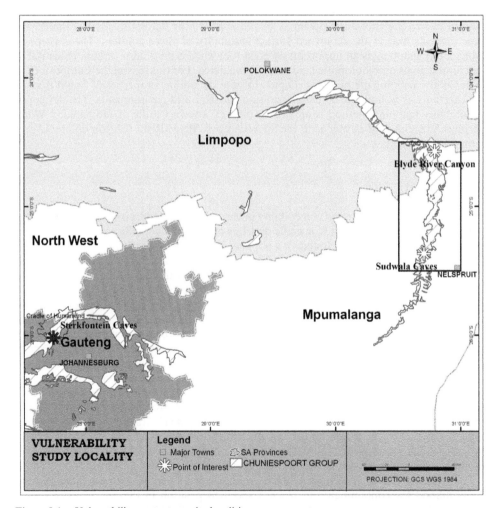

Figure 8.1.   Vulnerability assessment site localities.

of the resource from these potentially harmful activities. The present pressure on expansion of urban and agricultural development onto the dolomites is cause for major concern due to the enormous potential for pollution of the dolomitic aquifers underlying these areas as well as the renowned surface instability of these aquifers. While the vulnerability mapping of the COHWHS entailed detailed field work, including ground verification of dolines, infiltration tests and geological mapping, the mapping of the Sudwala Caves/Blyde River Canyon area represents a sole, though comprehensive, desktop vulnerability assessment using readily available data. In this regard the examples reflect also different levels of vulnerability assessment and applicable methodologies and data sources.

8.2   GEOLOGICAL DESCRIPTION

The main area of investigation and the focus of the activity are the Chuniespoort Dolomites, formed between 2650 and 2500 Ma. The regional geology and stratigraphy in both study areas show a wide diversity of rock types. The COHWHS mapping area covers an area of approximately 515 km^2 on the 1:250 000 scale geological map sheets 2626 West Rand and 2526 Rustenburg. It is predominantly underlain by strata of the Chuniespoort and Pretoria Groups of the Transvaal Supergroup with minor sections underlain by rocks of the Halfway House Granites, Ventersdorp Supergroup and Witwatersrand Supergroup (Figure 8.2).

The western boundary of the Sudwala focus area associated with the Chuniespoort dolomites is related to the Pretoria Group rocks, mainly the Rooihoogte, Timeball Hill and Hekpoort Formations. The eastern boundary is formed by the Black Reef Quartzite (Transvaal Supergroup), the Nelspruit Suite and intrusive basement rocks. The western boundary of Pilgrim's Rest/Blyde River focus (associated with the Chuniespoort dolomites) is also related to Pretoria Group rocks, i.e. the Timeball Hill and Hekpoort Formations. The eastern boundary is formed by the Black Reef Quartzite Formation, the Wolkberg Group and basement rocks (Figure 8.3).

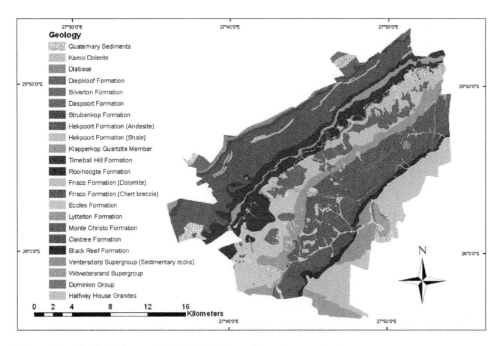

Figure 8.2.   Geological map of COHWHS. (See colour plate section).

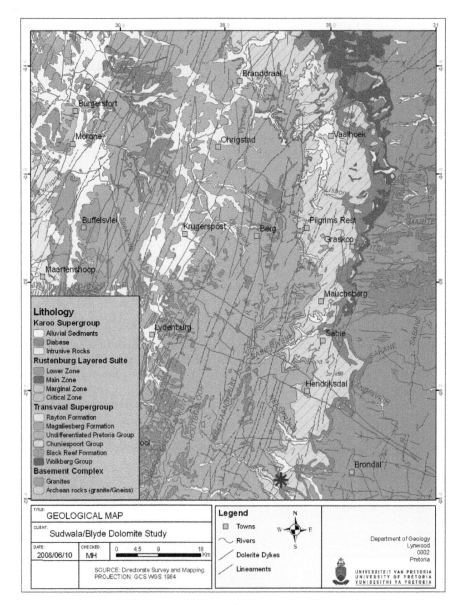

Figure 8.3.   Regional geological map of the Sudwala Caves/Blyde River Canyon area.

8.3   METHODOLOGY

The applied methodology of **Vu**lnerability mapping in **Ka**rst terrains (VUKA) is a modification of the COP (derived from the initials of the three vulnerability factors considered, surface **C**onditions, **O**verlying layers and **P**recipitation) aquifer vulnerability mapping method, developed by Vías et al. (2003) to suit South African karst terrains, lithologies and different climatic conditions. (Leyland et al., 2008). The COP method is an intrinsic, source vulnerability mapping method developed as part of the European COST (Co-operation in Science and Technology) Action 620, in which numerous approaches to "vulnerability and risk mapping for the protection of carbonate (karst) aquifers" were developed by various working groups in Europe. The COP method was selected because it is a detailed mapping method and considers the intensity of rainfall events in an area and the effects thereof on the aquifer vulnerability. The method is based on the determination of the protection offered by the unsaturated zone of the aquifer against a contaminant event, i.e. the capability of the unsaturated zone to filter or attenuate contamination by different processes. The protection provided by the overlying layers (the O factor) is modified by surface settings that control water flow towards areas of rapid infiltration (the C factor) and the characteristics of the transport agent (water), that transfers the contaminants through the unsaturated zone (the P factor). The three factors are multiplied to obtain a final vulnerability index, which is classified into five vulnerability classes, ranging from "very low" to "very high" and visualised. A flow diagram of the VUKA methodology including different sub-factor ratings is given in Figure 8.4 and described in detail.

8.3.1   *Overlying layers map (O-Map)*

The protection provided by the unsaturated zone (overlying layers) to filter or attenuate contamination is assessed by the type and thickness of the soil (OS sub-factor) and lithology (OL sub-factor) (Figure 8.4). The soil data can be derived by field mapping or, with a lower confidence, from the The South African Land Type data sets (spatial data and maps) and memoirs (ARC:ISCW, 2007), which assigns a soil type (depth and clay content) to each terrain unit, (based on slope gradient, length and shape), present in a land type. A GIS analysis of a digital elevation model (DEM) is required to determine the slope gradients and topographical units which are then used to derive the distribution of the different terrain units within each land type. Since most terrain units are defined by more than one soil series, a precautionary approach should be used and the lowest soil depth and lowest clay content given in the data should be assigned to the terrain unit. Based on the clay content and soil thickness an Os rating (soil sub factor) can be assigned to each terrain unit using the COP methodology (Figure 8.4).

The lithology sub-factor, specifying the thickness of lithologies above the water table, is derived from geological maps and a depth-to-water-table map based on field measurements (only done for COHWHS) and available databases (NGDB/DWAF data). Varying lithology on the down dip side of all plunging geological contacts were considered in the COHWHS by simple averaging.

The original COP lithology sub factors were modified to accommodate lithologies typically found in South African karst terrains (Figure 8.4). Confining conditions (cn sub-factor) were identified during a hydrocensus and/or derived from water levels versus geological setting and soil type in the COHWHS. Alternatively, a precautionary approach was applied for the Sudwala Caves/Blyde River Canyon area and all aquifers considered as being unconfined.

The final O Score is simply the sum of the OL and OS sub factors (Figure 8.4). The derived O maps for both areas of investigation clearly show the inferior (low or very low) protection of the aquifer by the dolomitic lithologies and associated soils, with the rating being for most parts determined by the underlying geology. Exceptions occur in areas with thick sediment cover or clay rich soils, which improve the aquifer protection to moderate, despite the underlying dolomites.

The outcropping shales of the Pretoria Group and Witwatersrand Supergroup, in the northwest and southeast parts of the COHWHS respectively, provide a very high or high protection of the aquifer, while the quartzite units within the Pretoria Group and Witwatersrand Supergroup offer only moderate protection (Figure 8.5). The basement complex rocks of the Halfway House Granite Dome offer moderate to high aquifer protection depending on nature (felsic or mafic) of the underlying rocks and consequent soil differences. Where soil properties differ across different topographic units, variations within areas underlain by the same lithologies are seen.

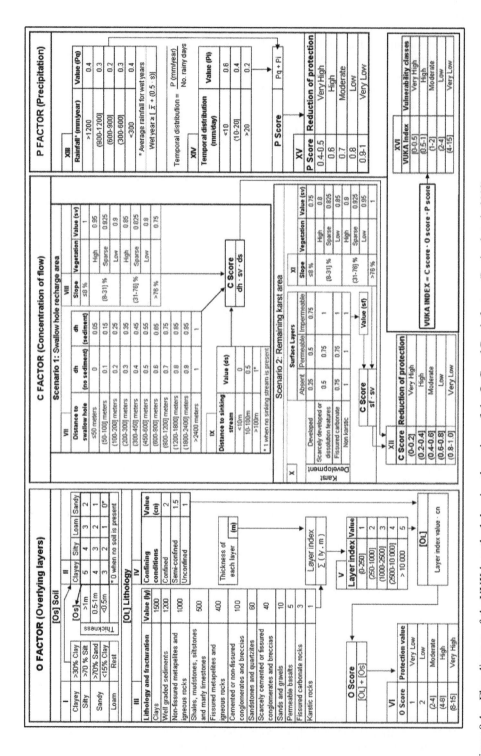

Figure 8.4. Flow diagram of the VUKA method showing the ratings of the O, C and P sub-factors (Leyland et al., 2008).

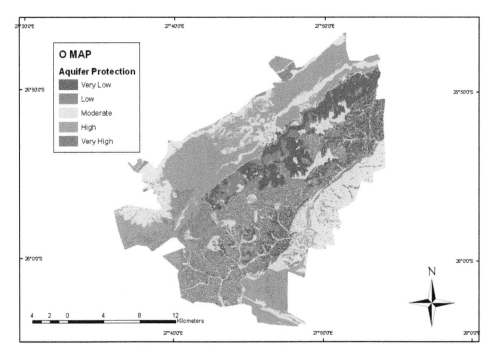

Figure 8.5.   Overlying (O) factor map for the COHWHS (Leyland et al., 2008). (See colour plate section).

The Sudwala Caves/Blyde River Canyon alluvium and dolomitic lithologies offer inferior protection of the aquifer, compared to the quartzites of the Pretoria group towards the east and the basement rocks towards the west. Considering that both the lithology and soil protection associated with these lithologies was low or very low. The dolomitic areas towards the northern parts of the map are mostly classified as very low or low aquifer protection zones, where thick sediment cover or clay rich soils cover occur, a moderate aquifer protection is assigned. This is clearly evident in the Sabie dolomitic area in the centre portion of the map (Figure 8.6).

8.3.2   *Concentration of flow map (C-Map)*

The determination of the C Factor (Figure 8.4) differentiates areas which fall within a swallow hole recharge area (2.4 km buffer around points of localised recharge, scenario 1) or in a remaining karst area (scenario 2). No flow concentration classification is done for non karstic areas beyond catchments of swallow holes. Concentrated recharge points (swallow holes), sinking streams as well as karst development of an area were identified during extensive field mapping (COHWHS) or with a lower confidence identified from aerial photographs and satellite photo imagery. No attempt was made for the Sudwala Caves/Blyde River Canyon area to delineate sinking streams from desktop interpretations, an obvious shortcoming of desktop assessments. To cater for the unique old karst features found in South African karst terrains, swallow holes within the COHWHS were classified into two categories based on the presence or absence of sediment within the feature as identified during field mapping. Those filled with a sediment layer were assigned values equal to half the values assigned to swallow holes in which no sediment was present (Figure 8.4). The desktop assessment of the Sudwala Caves/Blyde River Canyon area did not allow for such sub-division and following a precautionary approach, the absence of sediment layers was assumed for all swallow holes. The dh sub-factor ratings for the swallow hole buffer zones in the Sudwala Caves/Blyde River Canyon area was slightly modified to take the remaining karst area (scenario 2 sf sub-factors)

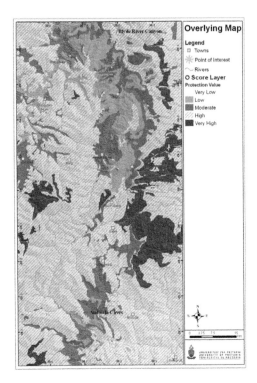

Figure 8.6.   Overlying (O) factor map for the Sudwala Caves/Blyde River Canyon area. (See colour plate section).

into account. This means that values outside the perimeter of a swallow hole cannot be lower than the area within the swallow hole perimeter. This improved approach allows for a smooth transition between scenario 1 and 2 flow concentration factors.

Field mapping (COHWHS) and aerial photograph interpretation (Sudwala Caves/Blyde River Canyon) was used to classify the karst development on land surface within all scenario 2 areas as either developed (high concentration of surface flow), scarcely developed or dissolution features (medium concentration of surface flow), fissured carbonate or non karstic (both low concentration of surface flow) and "surface features" sub factors assigned to these areas accordingly (Figure 8.4). The vegetation cover within both scenario 1 and 2 areas was defined as being either low, sparse or high based on aerial photograph interpretation and ground truthing (COHWHS only). A sparse vegetation class was added to the COP method to incorporate the large areas over which vegetation cover could neither be defined as low nor high, typically for the Highveld karst in South Africa. It is obvious that identification of sinkholes and vegetation classification by aerial photographs and satellite photo imagery is of lower confidence and efforts should be undertaken to update them with field data. Slopes within scenario 1 and 2 areas were determined using a digital elevation model and classified into four classes and a slope/vegetation sub-factor subsequently assigned to all areas (Figure 8.4). The product of the "surface features" and slope/vegetation sub factors for all scenario 2 areas and the product of the "distance to swallow hole" and slope/vegetation sub factors for all scenario 1 areas gives the final C Score (Figure 8.7).

The concentration of flow (C factor) maps (Figure 8.7 and Figure 8.8) clearly show the influence of the numerous swallow holes in the dolomitic outcrop on the reduction of aquifer protection in the already poorly protected karstic areas. The reduction in aquifer protection decreases with distance from these features, though the modification based on the slope and vegetation cover is visible as either an increase or decrease in protection reduction of the

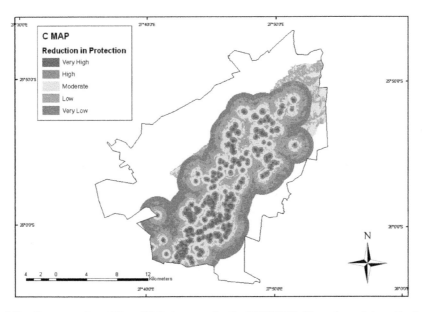

Figure 8.7.   Concentration of flow (C) factor map for the COHWHS. (See colour plate section).

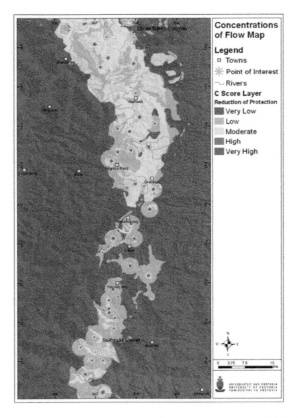

Figure 8.8.   Concentration of flow (C) factor map for the Sudwala Caves/Blyde River Canyon area. (See colour plate section).

zones around the swallow holes. Note the improved transition of C scores between scenario 1 and 2 areas for the Sudwala Caves/Blyde River Canyon area (Figure 8.8) in comparison to the COHWHS, where a sudden increase of reduction in aquifer protection is apparent (as ring buffers) in the north-eastern part (Figure 8.7).

8.3.3   *Precipitation map (P-Map)*

The P factor represents the degree to which the natural aquifer protection (O factor) is reduced due to the rainfall regime. The precipitation layer map is created by adding a rainfall amount (Pq) and a temporal rainfall distribution (Pi) sub-factor (Figure 8.4) to the mapping area subdivided using Thiessen polygons around meteorological stations, or more advanced de-clustering methods.

The Pq sub-factor is determined based on the average rainfall for wet years, which was defined for the sake of statistical consistency as a year in which the total rainfall exceeds or equals the long-term arithmetic average rainfall plus 0.5 times its standard deviation. While groundwater vulnerability increases initially with increasing recharge due to reduced transit times, recharge rates above a certain threshold (900 mm/year) are considered to dilute potential contaminants and subsequently reduce aquifer vulnerability (Figure 8.4).

The rainfall data for each of the influential stations in the areas of investigation were analysed to arrive at the average annual number of rain days per wet year. The ratio of average annual rainfall for wet years and average annual number of rain days per wet year characterises the temporal distribution/intensity of rainfall for each station (Pi sub-factor X). While intense storms promote Hortonian flow, the aquifer vulnerability will be lower in areas where an equal amount of rain is distributed over more rain days leading to slower infiltration. The illustration of the different P Scores within the mapping areas (P Factor map) shows how the different precipitation regimes within a mapping area reduce the aquifer protection offered by the overlying layers (Figure 8.9). Due to limited number of rainfall stations within the areas of

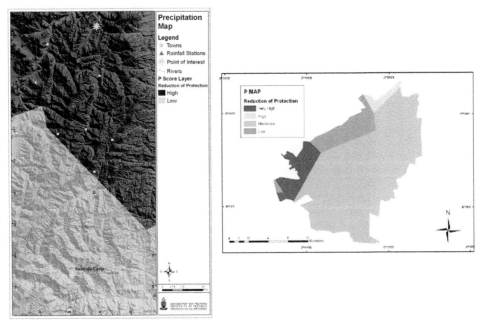

Figure 8.9.   Precipitation (P) factor maps for the Sudwala Caves/Blyde River Canyon area (left) and the COHWHS (right). (See colour plate section).

interest and the simple applied de-clustering method both maps show sharp differences in the reducing influence of precipitation on aquifer protection. It is obvious that a wider database and more advanced de-clustering methods would result in a more realistic representation of e.g. orographic effects on rainfall distribution and subsequent aquifer vulnerability.

8.4   VUKA INDEX MAP

The three layers (O, C and P maps) are overlain in a GIS to create the final aquifer vulnerability map. The final vulnerability or VUKA Index is calculated as the product of the three layers' scores and used to classify areas into one of five aquifer vulnerability classes, ranging from very low to very high.

   From the final aquifer vulnerability maps (Figure 8.10 and Figure 8.11) it is obvious that the vulnerability of the areas of investigation is clearly dictated by the lithology and the occurrence of swallow holes, i.e. the areas underlain by the Chuniespoort dolomites are the most vulnerable areas. As expected a strong correlation between lithology and aquifer vulnerability is seen in the non dolomitic areas due to the fact that the vulnerability of these areas is evaluated solely based on the lithology, soil and precipitation factors without consideration of flow concentration. If the precipitation shows little variation, so does the vulnerability for areas with non dolomitic rocks. Lithologies consisting of shales, andesites, granites and lavas are related to areas of low to very low vulnerability. Within karstic areas a larger variation in vulnerability is seen due to the additional consideration of the C factor.

8.5   SUMMARY

The presented intrinsic, resource vulnerability mapping method VUKA is an adaption of the European COP method to South African karst terrains, lithologies and climatic conditions. The method assesses the protection offered by the unsaturated zone of the aquifer and the driving agent (rainfall) against a contaminant event. It considers soil types, lithologies, depth to water table and confining conditions (O factor), settings that control water flow towards areas of rapid infiltration (C factor) and the characteristics of the transport agent

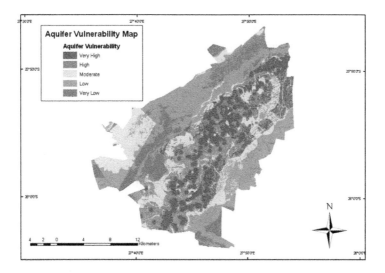

Figure 8.10.   Aquifer vulnerability map for the COHWHS. (See colour plate section).

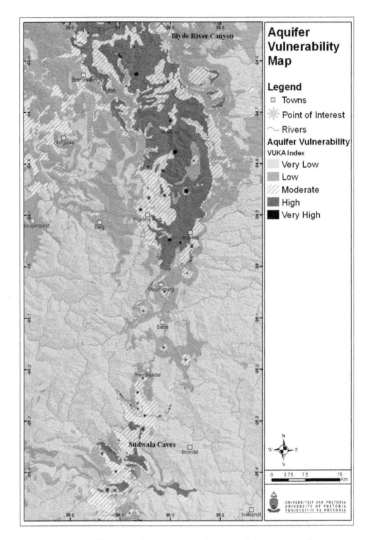

Figure 8.11.   Aquifer vulnerability map for the Sudwala Caves/Blyde River Canyon area. (See colour plate section).

(precipitation), that transfers the contaminants through the unsaturated zone (P factor). The method allows for rapid desktop vulnerability assessments based on readily available data but should preferably entail detailed field mapping for assessments of higher confidence level.

The final aquifer vulnerability map is an important planning tool for site developments on karst terrains (e.g. limiting development to areas with low vulnerability) as well as to guide further groundwater investigations in sensitive areas. Areas with moderate aquifer vulnerability usually require further investigations to arrive at suitable development types for the site (or engineering of the site) that will not result in a high risk of aquifer contamination.

REFERENCES

Agricultural Research Council—Institute for Soil, Climate and Water (ARC:ISCW). (2007). Land Types of South Africa, 1:250 000 map and Memoir series. ARC:ISCW. Pretoria, South Africa.

Calow, R.C., Robins, N.S., MacDonald, A.M., MacDonald, D.M.J., Gibbs, B.R., Orpen, W.R.G., Mtembezeka, P., Andrews, A.J. and Appiah, S.O. (1997). Groundwater Management in Drought-prone Areas of Africa. International Journal of Water Resources Development Vol. 13, No. 2, pp. 241–262.

Chambers, R. (1989). Vulnerability. Editorial introduction. IDS Bulletin Vol. 20, No. 2, pp. 1–7.

Leyland, R.C., Witthüser, K.T. and Van Rooy, J.L. (2008). Vulnerability mapping in Karst terrains, exemplified in the wider Cradle of Humankind World Heritage Site. WRC Report No. KV 208/08, Water Research Commission, Pretoria.

National Research Council [NRC] (1993). Groundwater vulnerability assessment, Contamination potential under conditions of uncertainty, Committee on Techniques for Assessing Ground Water Vulnerability, Water Science and Technology Board, Commission on Geosciences Environment and Resources, National Academy Press, Washington DC.

Sililo, O.T.N., Saayman, I.C. and Fey, M.V. (2001). Groundwater vulnerability to pollution in urban catchments. Report to the Water Research Commission, Project No. 1008/1/01, Water Research Commission, Pretoria.

Vías, J.M., Andreo, B., Perles, M.J., Carrasco, F., Vadillo, I. and Jiménez, P. (2003). The COP method. In Zwahlen, F. (ed) (2003): Vulnerability and risk mapping for the protection of carbonate (karst) aquifers, final report (COST action 620). European Commission, Brussels, pp. 163–171.

Villagrán De León, J.C. (2006). Vulnerability—A Conceptual and Methodological Review. SOURCE No. 4/2006, The Institute for Environment and Human Security. United Nations University, Bonn.

9

A low tech approach to evaluating vulnerability to pollution of basement aquifers in sub-Saharan aquifer

N.S. Robins
British Geological Survey, Maclean Building, Wallingford, Oxfordshire, UK

ABSTRACT: The application of quantitative approaches, such as DRASTIC, to assessing groundwater vulnerability at village scale in the weathered basement aquifer of sub-Saharan Africa is questionable. This is because the techniques are data intensive and were developed for regional scale evaluation. Deconstruction of the techniques allows the key influencing parameters to be identified. These parameters can be reviewed in terms of their likely influence on the weathered basement aquifer system and ranked as being of likely high, medium or low importance. Reassembly in the form of a scorecard, using only those parameters which can be judged subjectively in the field, produces a simple vulnerability assessment technique that can be applied at village and small town scale throughout the basement aquifer of sub-Saharan Africa. It is hoped that this tried and validated technique will provide a way forward for groundwater resource vulnerability assessment in the rural savannah lands underlain by the weathered basement aquifer system.

9.1   INTRODUCTION

Sub-Saharan Africa has been pre-occupied with water coverage statistics for a good few years. However, water supply is not the same as safe water supply and safe water supply coverage applies to a much smaller population of the sub-continent than overall water coverage statistics would imply. Safe water derives from a water supply that is sustainable at times of water stress and which remains potable both during a prolonged and intense rain season and progressively down 'the spiral of drought' (Calow et al., in press). In the arid and semi-arid savannah lands the only sustainable water supply is groundwater because most surface waters are ephemeral and even sand river supplies may dry up periodically. Groundwater is critically important to livelihoods in these areas as it is resistant to drought, responding with a delay time behind meteoric events, be they drought or flood. However, for this same reason groundwater requires longer to recover from an extreme event than surface water, albeit seasonal water courses, but groundwater nevertheless remains the key resource over much of the weathered basement aquifer of Sub-Saharan Africa.

Not all groundwater is safe water. The sight of domestic animals drinking from standing water around a village hand pump is commonplace. But commonsense dictates that animals and their faeces should be kept away from well heads and boreholes lest they contaminate the source. Commonsense also dictates that pit latrines and soakaways should be distant from water points for the same reason, and there are a number of simple guidelines to help address these issues (e.g. MacDonald et al., 2005).

Not so easy is the problem of the vulnerability of the resource—rather than the source, or water point—to pollution from anthropogenic activities on the ground surface. Groundwater vulnerability is variously defined, but perhaps most appropriately, as the hydraulic inaccessibility of the saturated zone to the penetration of pollutants, combined with the attenuation capacity of the strata above the water table (Foster, 1998).

Hydraulic inaccessibility depends largely on the physical properties of the substrate and its capacity to allow direct rainfall recharge to occur through it. In other words, it is the infiltration capacity of the soil cover and of the unsaturated or vadose zone. This capacity depends on the prevailing physical properties of the weathered zone, including the predominant grain size, porosity, soil cover and vegetation in addition to ground slope and aspect, and land use. Most importantly, it depends on whether there is a clay layer within the weathered zone. There are also a number of other features that may influence infiltration. For example, small scale ground topography is important because local surface ponding after rainfall creates a concentrated and prolonged zone of potential infiltration, and by pass-features such as cracks or fractures can offer a direct and rapid pathway from ground surface to the water table.

Attenuation is the physico-chemical retention or reaction of pollutants through biochemical degradation, sorption, filtration and or precipitation. A useful surrogate indicator for the chemical activity of the soil zone is the cation exchange capacity (CEC). However, CEC of soils is rarely measured and a useful secondary indicator is the amount of clay minerals that are perceived to be present in the soil (Griffiths et al., in press). Note that clay minerals are distinct from clay grade material and that rock flour does not contribute to soil CEC.

There are many techniques available to bring hydraulic inaccessibility and attenuation together to form an assessment of the vulnerability of a groundwater body. One of the more common ones is DRASTIC (Aller et al., 1987), but a variety of other methods are also available. All these methodologies are data intensive and require detailed knowledge of the soil cover and its properties and of the vadose zone and its physical and chemical properties. For the most part these data are not available for much of the basement aquifer system of Sub-Saharan Africa, and Robins et al. (2007) argue that few of the conventional approaches to vulnerability assessment used in the 'North' are readily transferable for application in the 'South'. Besides, at local scale, features such as laterite horizons and clay grade weathering products may provide some protection to the underlying regolith aquifer.

Some of the ideas from methodologies such as DRASTIC can be applied to form a robust but qualitative approach to vulnerability assessment. This approach needs to be readily applicable on the ground without the need for extensive data sets and expensive monitoring programmes, so that practitioners can make field judgements of groundwater vulnerability without recourse to intensive data gathering. New ideas can also be incorporated into the qualitative approach. One of these is weighting of the resource vulnerability according to its use, i.e. where there are people dependent on the resource for their livelihood then the resource is more valuable than locations where it is used only for marginal purposes, although even that is more valuable than groundwater that merely sustains a small dependent ecosystem.

This chapter describes the deconstruction of the accepted vulnerability assessment algorithms and their reassembly into a simple low-tech tick box methodology that can easily be applied in the field. The outcome of the approach is the ranking of shallow aquifers and component zones within such aquifers between highly vulnerable to weakly vulnerable pollution potential to groundwater, i.e. the converse to recharge potential. This is an easy to apply tool with which to protect groundwater resources from inappropriately located hazards, be they a pit latrine at one end of the scale to mine tailing dumps at the other. Because the assessment is measuring the converse of recharge potential it can also be used as a qualitative assessment of the available renewable resource.

9.2   THE WEATHERED BASEMENT AQUIFER

Recharge processes in the weathered basement aquifer are, as in all unconfined aquifers, dependent on the effective rainfall and its distribution with time. The difference with the basement aquifer is that it extends over a large area, i.e. is of regional scale, but it is extremely shallow. As a consequence, there is limited lateral transport of groundwater within it other than on a local scale, and natural discharge is invariably ephemeral, catchment scale, discharge to springs and seepages focussed on lower elevation ground such as valleys and other depressions. In essence, the component of the rainwater that infiltrates the ground and percolates to the water table is the sole source of groundwater in that vicinity. In some places ephemeral surface water flows and ponds lose to groundwater providing a secondary recharge source, but the scale is important as conditions can change over short distances (Adelana and MacDonald, 2008).

The typical weathering profile in the basement is a superficial lateritic layer at the top with some seasonal small scale lateral water flow. Laterite tends to be present largely in the tropical to sub-tropical regions of Africa and on the older weathering surfaces where it has not been removed by erosion (Davies and Robins, 2007). Below, may be a mottled clay layer that grades downward to a fine saprolite; a small quantity of water may be found within this layer which generally contains the water table and its zone of oscillation. The base of the fine regolith may occasionally be marked by a smectite clay horizon. The water table will fall to this level during prolonged drought. Below again is the coarse-grained saprolite, in which most groundwater is stored, the base of which commonly marks the weathering front. Limited weathering along decompression zones of horizontal fracturing may be observed within the upper parts of the underlying bedrock.

A key to understanding the performance of the weathered basement aquifer is that the thicker the regolith, or the deeper the weathering, the better developed will be the upper mottled clays. Although a thick saturated regolith is attractive from the point of view of storage, it may not be accompanied by free access via the overlying clay for potential infiltration and recharge. In addition, as seen in parts of Uganda, there is a risk of poor quality water in the thicker weathered zones in which the recharge potential is poor. Optimum resource potential, therefore, is a balance between depth of weathering and minimal development of the clay zone. In addition the occurrence of lateritic horizons near or at the surface may also inhibit recharge. But the converse to good recharge potential is good protection of the groundwater from surface pollutants, and for this the presence of a well developed clay and lateritic horizons are equally attractive.

Optimum development of the aquifer depends on locating areas of deepest weathering into which the groundwater will pond at times of water stress. These areas are likely to be the same areas that the mottled clay cover is best developed so providing a protected and sustainable source dependent on local scale lateral flow from areas where the weathering is thinner but the clay is absent or nearly so. As a consequence the evaluation of groundwater vulnerability in the aquifer can only be carried out sensibly at a local or village scale. That being said there are also indications that the age of the erosion surface dictates the depth of weathering on a regional scale, which in turn is reflected in the regional groundwater potential (Davies and Robins, 2007).

9.3   THE GROUNDWATER VULNERABILITY DRIVERS

Lessons can be learnt from previous vulnerability assessment campaigns and methodologies. Groundwater vulnerability maps were developed in the early 1990s for use in the UK as a planning tool and were heavily reliant on the soil zone as the key to vulnerability assessment. The maps do not consider travel time to the water table because depth to water was not then generally known on a regional basis (Robins et al., 1994). Furthermore, they were designed only for a conservative pollutant that would not degrade in the unsaturated zone (Palmer et al., 1995). The approach in Ireland, however, focussed more on the unsaturated zone (Daly and Warren, 1998) and in particular on porosity as an indicator of unsaturated permeability.

The standard assessment technique DRASTIC is a data intensive spreadsheet or GIS based assessment that relies on input from a range of datasets. These are Depth to water table, Recharge rate, Aquifer medium, Soil medium, Impact effect of the unsaturated zone and unsaturated hydraulic Conductivity. These data are rarely available in their entirety and invariably become subjective, besides the DRASTIC assessment only works at a larger scale than that of interest to a village community or township and disregards the human and social dimension of groundwater vulnerability. DRASTIC does indicate that the vulnerability of the basement aquifer system is low simply because recharge rates are low in strata such as weathered shales, greywackes and fractured crystalline rocks. However, a universal low does not help differentiate between land areas in which no polluting activity should take place and those where such activity would not significantly impact the groundwater body. DRASTIC, and other numerical weighting approaches to groundwater vulnerability assessment, will not, therefore, provide a useful assessment at village scale on the weathered basement aquifer.

It is, nevertheless, useful to deconstruct the standard methodologies and pull out the key controlling processes which might influence groundwater vulnerability in a basement type aquifer. Deconstruction needs to be sympathetic to the different environmental conditions in the African savannah with its shallow basement aquifer and the humid maritime lands of Western Europe where the aquifers tend to be small and compartmentalised and even the large scale generally thick continental aquifers of North America. Nevertheless, using DRASTIC and the various related methodologies as a guide, the most likely key physical processes affecting rainwater infiltration in basement strata in the semi-arid savannah lands appear to be:

- Degree and type of weathering.
- Land slope, shape and aspect.
- Vegetation and land use.
- Soil type and cover.
- Unsaturated zone properties.
- Depth to water table.

The processes promoting contaminant attenuation in the vadose zone are not quite as obvious and depend on the type of soil and rock and the types of contaminant. Attenuation is generally most active in the soil zone, where bacterial activity is greatest. The unsaturated zone, is nevertheless, of special importance as it represents a significant line of defence against pollution of groundwater. The key indicators in the unsaturated zone relate to the process of sorption, ion exchange, filtration and precipitation. Of these, ion exchange is the main overall process, the others being dependent on the nature of the pollutant as much as the nature of the unsaturated zone medium. A single value describing the potential for the medium to attract cations, the CEC, is the most useful parameter in assigning its attenuation potential. Thus CEC can be used as a meaningful surrogate for the overall attenuation processes that are likely to occur in the vadose zone.

CEC describes the process of attracting cations to a negatively charged surface—usually clay minerals. However, CEC values are commonly not available for the vadose zone and not universally available for the soil zone. It is useful, therefore, to use clay mineral content of the vadose zone as a surrogate for CEC. A second part of the attenuation process is controlled by the availability of carbon as a catalyst for adsorption and precipitation in the medium. Thus the two key indicators used to derive the attenuation potential of the soil and unsaturated zone of the aquifer are clay mineral and organic contents.

Finally, the socio-economic value of the resource needs to be factored into the assessment. Mato (2007) suggests a scheme where the lowest status of a groundwater body is in those places where it is not used as an alternative supply, and its highest status is where there is no piped supply and groundwater is the sole source.

9.4   SIMPLE FIELD APPROACH TO VULNERABILITY ASSESSMENT

Taking the component parts of a groundwater vulnerability assessment identified from existing methodologies it is a relatively easy step to create a tick box approach for field use that will provide a useable vulnerability score. In order to keep the score sheet simple, some of

Table 9.1.   Identification of focus parameters for the assessment of groundwater vulnerability in the weathered basement aquifer.

Parameter	Importance	Comment
Degree of weathering	Medium	e.g. age of weathering surface
Topography	High	
Vegetation and land use	Medium	
Soil type and cover	Low	Poor soils mean small influence
Unsaturated zone properties	Low	Generally not known
Depth to water table	High	Generally known
Clay zone present	High	
Laterite present	Medium	Can normally be seen in the field
Organic material present	Low	Generally absent in soils
Human dependence	Medium	
Polluting activities	Medium	e.g. mining, livestock, fuel dumps, etc.

Table 9.2.   Vulnerability scorecard—possible scores range from 4 to 30.

Parameter	Indicator	Score times bias	Score
Topography	Flat	0	
	With hollows	3×2	
Depth to water table	<5 m	3	
	5–10 m	2	
	>10 m	1×2	
Clay zone thickness	Absent	0	
	<2 m	2	
	>2 m	3×2	
Degree of weathering	Shallow <5 m	1	
	Thick >5 m	3×1	
Vegetation and land use	Sparse cover	2	
	Farmland	0	
	Livestock	1×1	
Laterite present	Absent	0	
	Patchy	1	
	Continuous	2×1	
Human dependence	High	2	
	Other sources	2	
	None	0×1	
Polluting activities	Mining	3	
	Fuel dumps	2	
	Livestock	1×1	

Score <12 Low vulnerability   Total
Score 13–22 Moderate vulnerability
Score >22 High vulnerability

the parameters can be downgraded as being least relevant to the assessment of the weathered basement aquifer. The poorly developed soils of much of the savannah lands, for example, mean that the role of these thin and sandy soils as a moderator of percolating pollutants from the surface is likely to be small. In addition the scorecard can be developed with an emphasis on those parameters that can be seen in the field rather than measured, for example, topography and slope, vegetation and land use. But the depth to the water table is a valuable parameter, as also is knowledge of clay development in the weathered zone, but in theory both these factors should be known in broad terms from well records or from the local drillers and well diggers.

A comprehensive list of parameters that influence groundwater vulnerability and their relevance to the basement weathered aquifer enables selection of a group of focus parameters (Table 9.1). This list suggests that the key focus parameters (high) are topography, depth to water and the presence or not of a significant clay horizon. The subordinate parameters (medium) are: degree of weathering including the fracture pattern within the less weathered bedrock, vegetation and land use, presence of laterite, livelihood dependence on the resource and the type of polluting activity. Those parameters classed as 'low' in Table 9.1 need no further consideration. The selection of the focus parameters is inevitably subjective but it does enable a reasonably well justified target set of parameters that are either known or can be seen in the field without the need for detailed measurement or monitoring.

Weighting of the parameters in terms of the focus and subordinate groups can simply be achieved by biasing the final score towards the focus parameters. The degree of bias can be determined by trial and error. Using a 100% bias, i.e. multiplying the focus parameter scores by two, reduces the assessment to a simple field score sheet (Table 9.2). The score can range from a minimum of 4 to a maximum possible of 30. This can be subdivided arbitrarily between high, moderate and low vulnerability to pollution as indicated on the scorecard.

9.5   SCORECARD VALIDITY

The validity of the scorecard, subjective and judgemental as it is, has been tested in various areas for which a detailed understanding of the prevailing conditions is available

Table 9.3.   Sample scorecard for Mangochi, southern Malawi.

Parameter	Indicator	Score times bias	Score
Topography	Flat	0	0
	With hollows	3×2	
Depth to water table	<5 m	3	
	5–10 m	2	4
	>10 m	1×2	
Clay zone thickness	Absent	0	0
	<2 m	2	
	>2 m	3×2	
Degree of weathering	Shallow <5 m	1	
	Thick >5 m	3×1	3
Vegetation and land use	Sparse cover	2	2
	Farmland	0	
	Livestock	1×1	
Laterite present	Absent	0	0
	Patchy	1	
	Continuous	2×1	
Human dependence	High	2	2
	Other sources	2	
	None	0×1	
Polluting activities	Mining	3	0
	Fuel dumps	2	
	Livestock	1	
	Few	0×1	
Score <12 Low vulnerability		Total	11
Score 13–22 Moderate vulnerability			
Score >22 High vulnerability			

and a perceived vulnerability rating can readily be arrived at. Using data derived from a comprehensive and well documented drilling programme at Mangochi in southern Malawi (Robins et al., 2003) a sample scorecard was derived (Table 9.3). The scorecard indicates a low vulnerability to pollution as would be expected. The score is approaching that of moderate vulnerability providing the assessment with a welcome degree of conservatism.

Similar tests for other data sets in the weathered basement aquifer provide valuable support for the methodology. However, it may be, as experience is gained with the scorecard, that the biasing factor, currently two, for the three focus parameters may require adjustment.

9.6   CONCLUSIONS

The valid application of quantified techniques such as DRASTIC for groundwater vulnerability assessment in the weathered basement aquifer of sub-Saharan aquifer is questionable. These techniques are data intensive as well as data selective, being developed for use in the 'North' where conditions are significantly different and where data coverage is greater.

A simple vulnerability assessment scorecard has been developed and is based on the parameters used in other assessments but requires only data that can be observed in the field. The scorecard can be used equally to assess those areas of vulnerability to polluting surface activities as well as the converse, to assess zones of greatest recharge potential. However, its main application is to help towards identifying areas where greatest care should be taken in protecting underground water resources in the weathered basement aquifer. These areas need to be recognised and safeguarded from polluting activities such as intensive cattle rearing, concentrations of people and pit latrines or even mine tailings.

The scorecard has been tested against various datasets and shown to be valid. It needs to be applied sensibly with the assessor mindful of the subjectivity of the technique. However, it is anticipated that this simple technique will provide a way forward for field assessment of groundwater vulnerability in rural and small town environments on the weathered basement aquifer.

REFERENCES

Adelana, S.M. & MacDonald, A.M. (Eds.) 2008. Applied Groundwater studies in Africa. International Association of Hydrogeologists, Selected Papers, 13, CRC Press/Balkema, London.

Aller, L., Bennett, T., Lehr, J.H., Petty, R.J. & Hackett, G. 1987. DRASTIC: a standardised system for evaluating groundwater pollution potential using hydrogeological settings. US-EPA Report 600/2-87-035.

Calow, R.C., MacDonald, A.M., Nicol, A.L. & Robins, N.S. In press. Groundwater and drought in Africa—the balance between water availability, access and demand. Ground Water.

Daly, D. & Warren, W.P. 1998. Mapping groundwater vulnerability: the Irish perspective. In: N.S. Robins (Ed.) Groundwater Pollution, Aquifer Recharge and Vulnerability. Geological Society Special Publications, 130, 179–190.

Davies, J. & Robins, N.S. 2007. Groundwater occurrence north of the Limpopo: are erosion surfaces the key? Geological Society of South Africa Groundwater Division Groundwater Conference, Bloemfontein (CD).

Foster, S.S.D. 1998. Groundwater recharge and pollution vulnerability of British aquifers: a critical overview. In: N.S. Robins (Ed.) Groundwater Pollution, Aquifer Recharge and Vulnerability. Geological Society Special Publications, 130, 7–22.

Griffiths, K.J., MacDonald, A.M., Robins, N.S., Merritt, J., Booth, S.J., Johnson, D. & McConvey, P. In press. Improving the characterisation of Quaternary Deposits for groundwater vulnerability assessments. Quarterly Journal of Engineering Geology & Hydrogeology.

MacDonald, A.M., Davies, J., Calow, R. & Chilton, J. 2005. Developing Groundwater, a Guide for Rural Water Supply. ITDG Publishing, Warwickshire, UK.

Mato, R.R.A.M. 2007. Modeling and mapping groundwater protection priorities using GIS: the case of the Dar Es Salaam city, Tanzania. In: Witkowski, A.J., Kowalczyk, A. & Vrba, J. (Eds.) Groundwater Vulnerability Assessment and Mapping. International Association of Hydrogeologists Selected Papers 11, 155–166.

Palmer, R.C., Holman, I.P., Robins, N.S. & Lewis, M.A. 1995. Guide to Groundwater Vulnerability Mapping in England and Wales. HMSO, London.

Robins, N.S., Adams, B., Foster, S. & Palmer, R.C. 1994. Groundwater vulnerability mapping: the British perspective. Hydrogéologie, 3, 35–42.

Robins, N.S., Davies, J., Hankin, P. & Sauer, D. 2003. Groundwater and data: an African experience. Waterlines, 21: 4, 19–21.

Robins, N.S., Chilton, P.J. & Cobbing, J.E. 2007. Adapting existing experience with aquifer vulnerability and groundwater protection in Africa. Journal of African Earth Sciences, 47, 30–38.

10

Preserving groundwater quality against microbiological contamination through differentiated aquifer management in Africa

Yasmin Rajkumar
Department of Water Affairs and Forestry, Durban, South Africa
Department of Earth Sciences, University of the Western Cape, South Africa

Yongxin Xu
Department of Earth Sciences, University of the Western Cape, South Africa

ABSTRACT: In Africa, groundwater specifically plays an increasingly important role as a source of potable water, more so in rural communities where the cost implication of routing surface water supplies may far exceed the budgets of local water service authorities. Protection of groundwater resources requires not only good planning and a concerted effort but also local information and knowledge of potential sources of contamination. The latter is essential in site-specific management and circumventing potential health threats that may result from the consumption or contact with contaminated groundwater.

Several African countries have reported microbiological contamination of their production aquifers to varying degrees. The effect of this microbial contamination is noticeable on the end users of the water source, with many African countries reporting cases of water borne diseases. Potential sources of microbial contaminants identified are burial sites, indiscriminately sited waste disposal facilities and on-site sanitation facilities, agricultural practices as well as animal kraals.

This chapter deals with issues of the microbiological pollution and advocates the principle of the differentiated protection strategy for groundwater protection on the African continent.

The key take home message is the extent of the vulnerability of our African aquifers and, that in order to safeguard groundwater quality various mitigation measures should be implemented.

10.1 INTRODUCTION

Globally, it is estimated that over 2 billion people rely on aquifers directly for drinking water. Within the African continent alone, it is approximated that over 435 million people are dependent on local aquifer systems for domestic water supply. These figures are indicative of the tremendously important role that groundwater resources play, not only on the African continent but globally, to provide people with a potable water source. It is thus imperative that groundwater is developed and managed in a responsible and coordinated manner to ensure sustainability of the resource, as well as maintaining the environmental balance of many sensitive groundwater dependent ecosystems (Morris et al., 2003; Pietersen, 2005; Shah et al., 2000). Groundwater offers precious opportunities for alleviating the misery of the poor, but it poses many, daunting challenges in preservation of the resource.

Within the African continent impingement of potable groundwater quality poses one the greatest threats to the sustained use of our aquifers. Bigger African cities have a complex mix of first world and third world conditions, subjecting the environment to a diverse range of both natural and anthropogenic activities that influence the quality of our groundwater. This in effect is exerting an ever increasing strain on potable supply.

The traditional lifestyle of many rural communities in Africa is undergoing a dynamic change, with families opting to move to urban areas in search of employment and better and more convenient lifestyles. This rapid influx of people from rural areas to urban city centres has resulted in an increased demand on water resources. Together with this an increase in contaminant load on the land surface can be noted as well. Population densities in urban areas are such that it results in a decrease in effective recharge area for the aquifer. This is resultant of an increase in intensive agricultural activities, changes in land uses as well as a general increases in hardened surfaces that promotes precipitation run-off instead of recharge to the aquifer. These activities require that water resource managers manage and protect groundwater resources 2 fold; both in quality and quantity aspects.

In many parts of sub-Saharan Africa, viral outbreaks, resulting from microbiologically unsafe drinking water and inadequate sanitation, is a daily occurrence. The rate of infectious diseases in developing countries is high and it is mostly the economically disadvantaged who are most vulnerable (Okeke et al., 2007). The World Health Organisation in 2000 estimated that approximately 4 billion cases of diarrhoea associated with a lack of access to clean water occurs annually in developing countries. Mortality figures from water related diseases are high, with figures in excess of 2.2 million deaths per year (Gleick, 2002). Gleick also projects that as many as 135 million people may die from water related diseases by 2020. Even if the Millennium Development Goals are reached, he projects a possible 76 million water related deaths by 2020.

This chapter addresses the concept of differentiated aquifer protection (source protection zoning around economically and socially important aquifers) with a focus on microbiological quality of groundwater resources within the African continent. Protection zoning entails the establishment of zones around specific groundwater sources. Normally subdivided into three zones of protection, this method seeks primarily to control land use activities, thus preventing or controlling pollution of groundwater resources. The second zone of this method specifically tackles microbiological contaminants, taking into account the concept of travel times and minimum safe distances to water supply boreholes. This distance ensures that the time taken for the horizontal travel of the contaminant is sufficient to allow physical and biochemical degradation of the contaminant. For microbiological pathogens a travel time of 30–50 days and a minimum distance ranging between 15–50 m (depending on aquifer conditions) has been proposed for South African conditions (Xu and Braune, 1995). The attenuation or elimination capacity of the sub-surface may in some cases reduce or completely eliminate the concentration of these contaminants via natural physical, chemical or biological processes. However the effectiveness of this natural process within the subsurface may vary greatly under the heterogeneity of the underlying geology within the African continent.

Typically, Africa is covered in geologically complex terrain. Over 78% of the continents aquifers are secondary in porosity and only 22% are of primary porosity. Groundwater plays an important role in provision of a potable water source in rural Africa. It is estimated that over 375 million people are dependent on the continents hard rock aquifers for their water supply.

10.2   MICROBIOLOGICAL TRANSPORT WITHIN SECONDARY AQUIFERS

The saturated zone in a large proportion (78%) of the aquifers in Africa has principally secondary porosity (Morris et al., 2003). Fractured rock aquifers show extreme spatial variability in their hydraulic conductivities thus making the determination of groundwater flow rates difficult and complex.

Generally pathogens within the sub-surface do not travel farther or faster than the water in which they are suspended if diffusion is negligible. Although there have been documented cases of microorganisms traveling faster than the average pore velocity. Bearing this generality in mind, the most logical manner in determining travel times of microbial pathogens within the sub-surface would be to determine the rate of groundwater flow. Analyzing groundwater flow data together with microbial size, lifespan, pH conditions of soils would give one a relatively good indication of the distance to be traveled by a contaminant, allowing it sufficient time for inactivation.

Fractured rock aquifers display different contaminant transport characteristics due to structure of water pathways as compared to unconsolidated sediments. One of the drawbacks of current protection strategies is that contaminants are assumed to be transported at the average linear velocity of groundwater flow (Taylor et al., 2004). In the African context this assumption seems incorrect as groundwater flow in fractured rock media may follow fracture networks and is largely not linear but highly variable due to the variability in hydraulic conductivities and connectivities.

It is important that the physiochemical characteristics of the microbe be taken into account when determining microbe transport in fractured rock media. The factors of cell size, structure, hydrophobicity, exopolymer production etc influences transport of the micro-organisms within the sub-surface even to the extreme where single cells within a monoclonal bacterial population will display variations in transport characteristics (McCarthy and McKay, 2004).

Many bacterial transport experiments are based on the use of indicator organisms as well as lab cultured microbial tracers and isotopic analyses (Leclerc et al., 2000). And whilst these type of lab experiments add much value to current knowledge on microbial subsurface transport it may lead researchers into the blanket assumption that microbes of similar sizes may be transported at the same rate, and hence reference bacterium models could be used to extrapolate the results to other microbes of similar sizes within similar geological media. However, Becker et al. (2003) were able to demonstrate that small differences in microbial cell size, morphology and motility can result in significant differences in transport times.

Recent research has focused not only on transport of individual organisms but into biocolloidal transport as well. Colloidal transport faces challenges in both the microscopic (interactions with air, water and solid interfaces) as well as macroscopic (preferential flow, spatial variability) levels (McCarthy and McKay, 2004).

Fractures in hard rock aquifers form conduits in which rapid transport of contaminants may occur even in media that is typically considered aquitards. These preferential pathways (fractures or fissures) enhance rapid transport of pathogens to deeper confined aquifers which were previously thought to be safe (Borchardt et al., 2007). The discovery of pathogens up to 60 mbgl, more so viable pathogens suggest extremely rapid movement to the subsurface. Pathogens will remain viable out of its natural host conditions for a limited period of time, before it will either die off, be inactivated or be subjected to natural predation.

In summary, the transport of pathogens in the sub-surface, both in the saturated and unsaturated zones is difficult to predict. One has to take into account the many factors around complexity of pathogen structure, episodic recharge and the air interface in the unsaturated zone, preferential flow in both zones, pathogen inactivation rates, sub-surface environmental changes all coupled with hydraulic variability in hard rock fractured aquifer media.

10.3   EXTENT OF MICROBIOLOGICAL IMPACTS ON SOME AFRICA AQUIFERS

Traditionally in Africa, not much emphasis has been placed on aquifer protection against pollution. Water resource managers now realise that there is an increasing need to assess their aquifers and put in mitigation measures such that it ensures sustainability of their resource.

It is important to mention upfront, that trying to quantify the extent of microbiological pollution within the African continent is difficult. This is mainly due to the fact that most African countries literally have very little microbiological monitoring data on their groundwater resources on a long term level from which one would be able to draw conclusions on trends.

From a UNEP-UNESCO led project (Xu and Usher, 2006), involving 11 African countries assessing their major supply aquifers for pollution vulnerability, several factors have been identified as potential reasons as to why African aquifers are under increasing stress. These reasons are repeatedly echoed within the African continent.

In Cameroon, from 1983–1994, a high incidence of water related diseases were noted in the country (8000 cases of cholera, 11500 cases of typhoid and 46400 cases of dysentery). The shallow groundwater in the densely populated areas of Yoaunde and Doula had high total and faecal coliform counts (in excess of 300 cfu/100 ml) (Mafany et al., 2006). These were thought to originate from indiscriminate disposal of domestic, industrial and hospital waste. Shallow hand dug unprotected wells also pose a risk of being polluted directly from surface water ingress via the normal indirect recharge paths. These results were predominant in low income areas where high contaminant loading occurs in a small area due to higher population densities.

Studies in the Ashanti region of Ghana showed that whilst on-site sanitation plays a major role as a potential source of microbiological contaminants, that the level of contamination noted is dependent on the distance between the abstraction point and pit latrines (Duah, 2006). As such water resource planners in Ghana use a generic 50 m rule of the minimum distance required between a water resource and point source of contamination.

In Zimbabwe, results of groundwater assessments at seven different sites in Harare, including industrial sites, a sewage treatment works, landfills and a cemetery supported the idea that rapid urbanization with limited and insufficient upgrading of infrastructure was a contributing factor to contamination. Shallow wells and water tables coupled with pit latrines resulted in high total and faecal coliform counts. The cemetery sites show high coliform counts in close vicinity of the graves. Results in the boreholes located furthest away from the grave sites indicate that there was no transport of the coliforms in the groundwater suggesting a filtering off, or death of viable coliforms as the groundwater flows within the subsurface. The horizontal distance between the grave site and furthest borehole is not given in the text cited (Love et al., 2006). However, the results obtained suggest that land use constraints as required in zoning can be technically justified, but will require refinement of exact delineation areas for high loading microbiological contaminant sources. The study highlights the need for integrated land-use planning and geotechnical mapping in order to minimize contamination and better protect groundwater resources as expansion of the metropolitan area occurs.

In Delmas, South Africa, an outbreak of typhoid in the local community was broadcasted on the national news. A combination of the bucket sanitation system, standpipes and dolomitic aquifers proved to be fatal for five people (Pienaar and Xu, 2007). This scenario permitted rapid movement of microbes to the groundwater via fractures and solute channels, especially in the non-serviced informal areas (Le Roux and du Preez, 2008).

As shown by the above mentioned outbreaks and studies, African aquifers are indeed under increasing strain. There are several factors that are repeatedly echoed regarding the increasing stress experienced on aquifers in the developing countries in terms of quality. Migration of rural communities to urban areas in search of better lifestyles has seen rapid population growths within urban areas, which are not designed in terms of service infrastructure to cater for these large numbers. Indiscriminate siting of waste disposal facilities and on-site sanitation poses as some of the primary microbiological threats to water resources. Also contributing to aquifer pollution is improper construction of boreholes and wells, in many cases with no protective measures to prevent ingress of contaminated surface water directly into the wells. Population density and socio-economic setting were also seen as contributing factors to pollution. Higher coliform counts were encountered in settlements within the lower income brackets. The greatest threat however, is the lack of management and lack of protection strategies to minimize and prevent future contamination.

Table 10.1. Indication of levels of microbiological contamination within the African continent.

Country	Total colifoms (min value) (cfu/100 ml)	Faecal coliforms (min value) (cfu/100 ml)	Faecal streptococci (min value) (cfu/100 ml)	General comments	Reference
Cameroon	>300	–	–	Reports on water born diseases: Typhoid, Cholera, Amoebic dysentry. Iron bacteria fouling	Mafany et al., 2006
Ghana	Yes but not widespread	–	–	High iron levels. Corrosion on ferrous products	Duah, 2006
Zimbabwe	>3000	>182	–	–	Love et al., 2006; Conboy and Goss, 2000
Mozambique	>39	–	49–956	Thermotolerant coliforms tested	Cronin et al., 2006 Godfrey et al., 2005
Benin	> lab count capacity	> lab count capacity	> lab count capacity	*Clostridium perfringens* also tested—counts > lab capacity	Boukari et al., 2006
Mali	No reported values	No reported values	No reported values	Presence of these microbes noted in groundwater	Orange and Palangie, 2006
Niger	>200	>10	>10	Reports of water born diseases	Ousmane et al., 2006
Senegal	–	No reported value	No reported value	Presence of parasitic cysts: *Entamoeba histolictica, Entamoeba Coli* and *Rabditis* also reported	Deme et al., 2006
Ethiopia	3–160	–	–	*E. Coli* present in concentrations from 1–160 cfu/100 ml	Alemayehu et al., 2006
Kenya	–	0–16 000	–	Reported water born diseases of diarrhea, malaria, worms, eye and skin infections. *E. Coli* present in concentrations from 0–10 000 counts/100 ml	Munga et al., 2006
Zambia	50–21 6000	25–11 400	–	Reports of cholera, dysentery and typhoid.	Nkhuwa, 2006
Abidjan	–	–	–	Reports of water born diseases: Cholera. Reported value of 909 cfu/100 ml of *E. Coli*	Jourda et al., 2006

Table 10.1 portrays the concentrations of some of the major microbial indicators within some of the African countries.

10.4 STATE OF MICROBIOLOGICAL MONITORING NETWORKS IN AFRICA

One can not overemphasis the need and importance of a dedicated monitoring network and the value of reliable data collection. In sub-Saharan Africa national monitoring programs for groundwater quality range from limited to non-existent. In these developing countries, focus is preferentially placed on provision of services rather than monitoring programs or enforcement of water quality standards. Existing monitoring networks are thus subjected to neglect, under funding and lack of focus. This results in a lack of continuous data to help in the management of the groundwater resources. In terms of microbiological contamination and monitoring networks, there is very limited focus and hence limited data available.

There is a need for protection strategies which may include land use restrictions to activities that will impact on the resource in a detrimental manner. But in order to assess which resource needs to be protected and to what extent protection is required the authorities once again need to have reliable monitoring data.

African countries that have implemented microbiological monitoring, albeit on a local, smaller scale are Benin, Mali, Niger, Senegal, Ethiopia, Kenya, Zambia, Cote d'Ivoire and South Africa. The countries of Nigeria, Cameroon, Ghana, Zimbabwe, Mozambique, Kenya, Mali and South Africa have all reported microbiological contamination to some extent in some of their municipal and rural boreholes (Xu and Usher, 2006).

Whilst not all African countries have implemented monitoring programs aimed specifically at pathogens, the high incidences of water related diseases is a clear indication that there exists to some extent, pollution of the groundwater resources. The Human Development Report (UNDP, 2006) reports that 5 billion cases of diarrhea in children are reported per annum in developing countries. It is also approximated that 1.8 million children die per annum due to illnesses related to unclean water supplies. In a study regarding fatality rates resulting from an outbreak of cholera in several southern African countries in 2001, Malawi, Zimbabwe, Swaziland, Zambia and Mozambique reported fatality rates greater than 20–50% (Chabalala and Mamo, 2001). The high rates of infections affects not only the health in both the short and long term of the individual but can be noted in decreased educational days for children and productive working days for adults and hence a loss of income.

Table 10.2 shows the extent of groundwater management and protection in terms of a microbiological perspective in Africa in relation to some of the more environmentally active and economically developed countries.

What is important to note from Table 10.2 is whilst Africa may experience the same type of contaminants which threaten groundwater quality as many developed countries, the difference in data collection mechanisms creates a huge data gap. Most African countries do not have dedicated monitoring networks. Those countries that do have dedicated monitoring networks rarely include microbiological contaminants as a standard constituent to be analysed. This lack of data translates to very few African countries developing their own set of water quality guidelines specific to their conditions as well as very little being done in terms of further protection mechanisms being implemented.

10.5 GROUNDWATER PROTECTION ZONING

Protection zoning is being recognised as a vital tool for the management and protection of groundwater resources (DWAF, 2000). This strategy offers three zones for differentiated protection. Each zone offers increasing strictness in land use constraints moving in from the

Table 10.2.   Protection mechanisms of African groundwater resources in comparison to first world countries.

Area	Contaminants	Monitoring networks	Water quality guidelines	Protection zoning	Vulnerability mapping
Africa	Chemical, microbiological and lesser extent organics except in industrialised areas	Dedicated microbiological networks in continent non existent except Egypt	WHO guidelines widely used. South African water quality, Botswana Bureau of standards 32, Ghanian water quality standards, Nigerian Federal Environmental Protection Agency standards, Kenyan Bureau of standards	Not implemented yet. SA working on policy to initiate zoning. Present in Egypt	Yes. Has been done for several African countries on a regional basis
USA	Chemical, organics, microbiological	Yes	Environmental Protection Agency Groundwater Rule	Wellhead protection zones implemented	Yes
Canada	Chemical, organics, microbiological	Yes	Canadian Drinking Water Quality guidelines		Yes
UK (England)	Chemical, organics, microbiological	Yes	Yes	Protection zones delineated countrywide	Yes
Germany	Chemical, organics, microbiological	Yes	German Drinking Water Regulations	Yes—Mainly agricultural settings	Yes

outer protection zone to the wellhead operational zone. By forming protective zones around important water sources, land uses that may have detrimental effects on groundwater can be controlled. Certain uses can be allowed under predetermined conditions, the intensity of development limited, and the locations where certain land uses are to be carried out, specified.

Whilst zoning has been widely implemented within many areas in the United States of America, according to the guidelines set out by the Environmental Protection Agency (EPA) (EPA, 1987) and within many of the countries of the European Union (Robins et al., 2007), this trend seems not to be successfully implemented in many developing countries. Developed nations like the USA and UK seem to have succeeded in implementing source protection zones due to having their institutional, legal, technical and socio frameworks in place (Robins et al., 2007). All these frameworks are not necessarily in place yet in most developing countries.

The lack of sufficient local and technical knowledge and capacity in contaminant travel characteristics within the specific fractured rock aquifers present on the African continent coupled with a lack of financial resources and insufficient political power and backing, maybe the key factors in hindering implementation of groundwater protection zones.

One also needs to assess the socio-economic impact of restricting activities around water resources. It is often the case that many of the very activities that sustain rural and poorer communities occur on land within the protection zone. In delineating the zones a balance needs to be maintained between resource protection and socio-economic development. Placing large portions of land under land constraints might lead to resistance from communities as the perceived loss of income generated from the land will supersede the importance of a safer water source. This resistance due to perceived loss of income could possibly be eliminated by monetary compensation for the loss of agricultural production from the zoned land.

It is encouraging to note though, that some African countries are becoming increasingly aware that land use activities around a water resource may impact negatively on the resource. A risk assessment survey carried out in Uganda on spring water quality showed that 51-80% of springs scored high risks to faecal contamination (Haruna et al., 2005). It was acknowledged that this risk was associated with activities in close proximity to the water resource e.g. pit latrines within 30m of the resource, inadequate protection measures and point source of contaminants. These results were comparable with earlier studies conducted in Uganda indicating an ongoing problem of microbiological contamination (Haruna et al., 2005). This scenario is evident in many areas of Africa where shallow, unprotected water resources are subject to continued microbiological pollution due to improper siting of sanitation facilities and poor construction of resource infrastructure and physical protection measures. It is in these types of scenarios where constraints to land based activities and zoning around the immediate area of the resource abstraction point would be extremely beneficial to communities and thus minimize contamination levels.

Currently South Africa is leading in sub-Saharan Africa in formulating and adopting policies on protection zoning. A three year research project had been initiated in 2005 by the national Department of Water Affairs and Forestry within the South African government to address the feasibility of introducing such a policy (DWAF, 2008). Having done the groundwork in protective legislation and classification of the country's aquifers, South Africa is now in a position to move into the differentiated protection strategy of zoning. The project is currently in the feasibility stage. The differentiated approach adopted by SA consists of 3 tiers (Xu and Reyders, 1995). The first tier addresses basic protection standards, guidelines such as safe location of latrines, etc to minimize impact of point sources in all aquifers country-wide. The second tier is the approach of planned protection of resources through classification, mapping and education to regulate point and diffuse sources at regional scale. The third and final tier is source protection zoning. This third tier forms an advanced stage of groundwater protection and is seen as a long-term objective. In order to implement zoning, detailed information regarding current land use, geohydrological characteristics and groundwater sources is required (Xu, and Braune, 1995). It would be interesting to follow how well the source protection zoning method can be translated to the developing world environment, given all the technical, social and legal difficulties faced ubiquitously by the African continent as a whole.

10.6   WAY FORWARD

Provision of safe drinking water to any community is of utmost importance. In rural Africa, the problems surrounding provision of water for basic human needs are numerous from supply to quality. Groundwater is the most logical choice in most rural areas, though one needs to take into cognisance all the potential problems as highlighted in this chapter, especially in terms of microbiological pollution. In order to provide a reliable service of clean safe water, African groundwater practitioners need to address various issues:

First and foremost, groundwater developers should ensure that groundwater resources have the appropriate sanitary seals and wellhead protection mechanisms as well as boreholes are sited hydraulically upgradient of point pollution sources. Deeper aquifers with well constructed boreholes (appropriate casing lengths and sanitary seals) should be investigated. As can be seen from the status on microbiological pollution of most African countries, many of the shallow easily accessible aquifers are already polluted and an alternative needs to be investigated.

Knowledge on the state of groundwater is minimal in most African countries. Complete surveys and hydro census needs to be conducted to establish baseline technical data on existing infrastructure (groundwater usage, well construction, pollution issues, water resource potential). Once conducted the data can be utilized to determine aquifer pollution vulnerability and subsurface contaminant loads. Pending these assessments protection measures can be prioritized and initiated.

Many countries need to increase focus on groundwater monitoring programmes for both quantitative and qualitative data, with more specification on microbiological qualitative monitoring to be carried out. This will help paint a clearer picture of the state of resources and help with planning of protection measures. Not enough microbial monitoring data are available allowing, characterization of microbial quality trends.

Leading on from this, once this information is obtained and assessed, appropriate management plans specific to each socially important and productive wellfields should to be drawn up. A well managed and routine system helps in identification of problems before they become detrimental.

The current geotechnical skills shortage within the African continent is another factor that needs to be urgently addressed. Africa's hard rock aquifers are complex in terms of understanding contaminant transport. In order to fully understand groundwater and contaminant movement in a dual porosity system one requires sophisticated and well planned experiments which are usually expensive to conduct and require highly technical skilled people.

Another issue that needs to be addressed is the extrapolation of lab based results to field tests in real systems, use of tracers and indicator organisms vs the true pathogenic organism. How well do these results correlate?

Also changing and improving of on-site sanitation technologies would greatly assist in minimizing risk to pollution of groundwater resources. Whilst the urine diversion system is the more expensive option as compared to pit latrines and VIP systems, it can be seen that in the longer term the environmental and health benefits far outweigh the initial higher cost implication (Von munch and Mayumbelo, 2007).

At a final and most advanced level, the application of land use constraints around water resources is recommended as well as creating awareness around water quality issues with local communities. This should take into account socio-economic impacts on communities, environmental, health and service delivery issues as well.

10.7 CONCLUSION

Growing populations, together with the increase in intensive agriculture and change in land use activities, it has become evident that our shallow groundwater resources are at risk of pollution. Whilst studies like that of the UN agencies assist in giving much required indication of the state of African groundwater resources and will help in defining aquifer differentiated protection strategies including the zoning approach, it is imperative that the gaps identified within the groundwater management sector in terms of technical issues be focused on and addressed in the near future. It is only once the foundation is firmly laid by addressing these issues can the management of groundwater resources be effected successfully, including the successful implementation of differentiated zones.

The challenge faced by many African managers now, is not only supply and demand issues to accommodate rapid influx of people into urban centres and meeting millennium development goals, but to start proactive aquifer management, such that the problem of pollution does not escalate to a scale where it is either insolvable or not worth solving. This involves a transition from a resource development mode to an active resource management mode (Shah et al., 2000).

REFERENCES

Alemayehu T, Legesse D, Ayenew T, Mohammed N and Waltenigus S (2006). Degree of groundwater vulnerability to pollution in Addis Ababa, Ethiopia. In Groundwater pollution in Africa—XU Y and USHER B eds. Taylor & Francis/Balkema. The Netherlands, pp. 203–213.

Becker MW, Metge DW, Collins SA, Shapiro AM and Harvey RW (2003). Bacterial transport experiments in fractured crystalline bedrock. Groundwater. 41: 682–689.

Borchardt MA, Bradbury KR, Gotkowitz MB, Cherry JA and Parker B (2007). Human Enteric Viruses in Groundwater from a Confined Bedrock Aquifer. Environ. Sci. & Technol. Web version.

Boukari M, Alsasane A, Azonsi, Dovonou FAL Tossa A and Zogo D (2006). Groundwater pollution from urban development in Cotonou City, Benin. In Groundwater pollution in Africa—XU Y and USHER B eds. Taylor & Francis/Balkema. The Netherlands, pp. 125–139.

Chabalala HP and Mamoo H (2001). Prevalence of water-borne diseases within the health facilities in Nakuru District, Kenya. Applied Epidemiology, University of Nairobi, Kenya.

Conboy MJ and Goss MJ (2000) Natural protection of groundwater against bacteria of fecal origin. Jrnl of Contam Hydro. 43: 1–24.

Cronin AA, Pedley S, Okotto-Okotto J, Oginga JO and Chenoweth J (2006). Degradation of groundwater resources under African cities: Technical and socio-economic challenges. In Groundwater pollution in Africa—XU Y and USHER B eds. Taylor & Francis/Balkema. The Netherlands, pp. 89–97.

Deme I, Tandia AA, Faye A, Malu R, Dia I, Dialllo MS and Sarr M (2006). Management of Nitrate pollution of groundwater in African cities: The case of Dakar, Senegal. In Groundwater pollution in Africa—XU Y and USHER B eds. Taylor & Francis/Balkema. The Netherlands, pp. 181–193.

DWAF (2008). Feasibility study towards the policy development on aquifer protection zoning by Xu Y, Nel JM, Rajkumar Y and Pienaar H. University of the Western Cape. Internal Report. ISBN 978-0-621-38235. Water Resources Information Programme, Department of Water Affairs and Forestry, Pretoria, South Africa.

DWAF (2000). Policy and Strategy for groundwater quality management in South Africa, (1 st Ed) DEPARTMENT OF WATER AFFAIRS AND FORESTRY, Pretoria.

Duah AA (2006). Groundwater contamination in Ghana. In Groundwater pollution in Africa—XU Y and USHER B eds. Taylor & Francis/Balkema. The Netherlands, pp. 57–65.

EPA (UNITED STATES ENVIRONMENTAL PROTECTION AGENCY) (1987) Guidelines for the delineation of wellhead protection areas. Office of Groundwater Protection, EPA, Washington DC, USA.

Gleick PH (2002). Dirty water: Estimated deaths from water related diseases 2000–2020. Pacific Institute Research Report, Pacific Institute for Studies in Development, Environment and Security. (www.pacinst.org)

Godfrey S, Timo F and Smth M (2005). Relationship between rainfall and microbiological contamination of shallow groundwater in Northern Mozambique. Water SA. 31: 609–614.

Haruna R, Ejobi F and Kbagambe EK (2005). The quality of water from protected springs in Katwe and Kisenyi parishes, Kampala city, Uganda. African Health Sciences. 5: 14–20.

Jourda JP, Kouame KJ, Saley MB, Kouadio BH, Oga YS and Deh S (2006). Contamination of the Abidjan aquifer be sewarage: An assessment of extent and strategies for protection. In Groundwater pollution in Africa—XU Y and USHER B eds. Taylor & Francis/Balkema. The Netherlands, pp. 291–301.

Leclerc H, Edberg S, Pierzo V and Delattre JM (2000). A REVIEW: Bacteriophages as indicators of enteric viruses and public health risk in groundwater. J. of Appl. Microbiol. 88: 5–21.

Le Roux W and Du Preez M (2008). Fate and transport of viruses in groundwater. Case study: Delmas, Natural Resources and the Environment, CSIR, Pretoria.

Love D, Zingoi E, Ravengai S, Owen R, Moyce W, Mangeya P, Meck M, Musiwa K, Amos A, Hoko Z, Hranova R, Gandidzanwa P, Magadzire F, Magadza C, Tekere M, Nyama Z, Wuta M and Love I (2006). Characterisation of diffuse pollution in shallow groundwater in Harare urban area, Zimbabwe. In Groundwater pollution in Africa—XU Y and USHER B eds. Taylor & Francis / Balkema. The Netherlands, pp. 65–77.

Mafany GT, Fantong WY and Nkeng GE (2006). Groundwater quality in Cameroon and its vulnerability to pollution. In Groundwater pollution in Africa—XU Y and USHER B eds. Taylor & Francis/Balkema. The Netherlands, pp. 47–57.

McCarthy JF and McKay LD (2004). Colloid transport in the sub-surface: Past, present and future challenges. Vadose Zone Journal 3: 326–337.

Morris BL, Lawrence ARL, Chilton PJC, Adams B, Calow RC and Klinck BA (2003). Groundwater and its Susceptibility to Degradation: A Global Assessment of the Problem and Options for Management. Early Warning and Assessment Report Series, RS. 03–3. United Nations Environment Programme, Nairobi, Kenya.

Munga D, Mwangi S, Ong'anda H, Kitheka JU, Mwaguni SM, Mdoe F, Barongo J, Massa HS and Opello G (2006). Vulnerability and pollution of groundwater in Kisauni, Mombsa, Kenya. In Groundwater pollution in Africa—XU Y and USHER B eds. Taylor & Francis/Balkema. The Netherlands, pp. 213–229.

Nkhuwa DCW (2006). Groundwater quality assessements in the John Laing and Missi areas of Lusaka. In Groundwater pollution in Africa—XU Y and USHER B eds. Taylor & Francis/Balkema. The Netherlands, pp. 239–253.

Okeke IN, Aboderin OA, Byarugaba DK, Ojo KK and Opintan JA (2007). Growing problem of multidrug resistant enteric pathogens in Africa. Emerging Infectious Diseases. 13: 1640–1646.

Orange D and Palangie A (2006). Assessment of water pollution and risks to surface and groundwater resources in Bamako, Mali. In Groundwater pollution in Africa—XU Y and USHER B eds. Taylor & Francis/Balkema. The Netherlands, pp. 139–147.

Ousane B, Soumaila A, Boubacar A, Garba Z, Daddy-Gao A and Margueron (2006). Groundwater contamination in the Niamey urban area, Niger. In Groundwater pollution in Africa—XU Y and USHER B eds. Taylor & Francis / Balkema. The Netherlands, pp. 169–181.

Pienaar H and XU Y (2007). Protecting whats underneath the tap. Water Wheel. 24–27, South Africa, ISSN 0258-2244.

Pietersen K (2005). Groundwater crucial to rural development. Water Wheel. 26–27, South Africa, ISSN 0258-2244.

Robins NS, Chilton PJ and Cobbing JE (2007). Adapting existing experience with aquifer vulnerability and groundwater protection for Africa. Journal of African Earth Sciences 47: 30–38.

Shah T, Molden D, Sakthivadivel R and Seckler D (2000). The global groundwater situation: Overview of opportunities and challenges. Colombo, Sri Lanka: International Water Management Institute.

Taylor R, Cronin A, Pedley S, Barker J and Atkinson T (2004). The implications of groundwater velocity variation on microbial transport and wellhead protection-review of field evidence. FEMS Microb. Eco. 49: 17–26.

UNDP (UNITED NATIONS DEVELOPMENT PROGRAM) (2006). Human development report 2006 Beyond Scarcity: Power, poverty and the global water crisis. ISBN 0-230-50058-7. New York, USA.

Von Munch E and Mayumbelo KMK (2007). Methodology to compare costs of sanitation options for low income peri urban areas in Lusaka, Zambia. Water SA. 33: 593–602.

Xu Y and Braune E (1995). A guideline for groundwater protection for the community water supply and sanitation programme (1 st Ed.), Dept of Water Affairs and Forestry. Pretoria.

Xu Y and Reyders AG (1995). A three-tier approach to protect groundwater resources in South Africa, ISSN 0378-4738, Water SA, Vol. 21 No. 3 July 1995.

Xu Y and Usher B (eds) (2006). Groundwater pollution in Africa. Taylor & Francis/Balkema. The Netherlands.

11

Fluoride in African groundwater: Occurrence and mitigation

Slavek Vasak
International Groundwater Resources Assessment Centre (IGRAC), Utrecht, The Netherlands

Jasper Griffioen
DELTARES, Utrecht, The Netherlands

Lourens Feenstra
TNO Built Environment and Geosciences, The Netherlands

ABSTRACT: Fluoride in groundwater has both natural and anthropogenic sources. Fluoride bearing minerals, volcanic gases and various industrial and agricultural activities can contribute to high concentrations. High intake of fluoride from drinking water is the main cause of fluorosis and may lead to many other health problems. Problems usually start with intake of water containing more that 1.5 mg/ℓ of fluoride (WHO guideline). Many water supplies in Africa are contaminated by much higher concentrations of fluoride. Alternative water sources, improvement of the nutritional status of populations at risk and appropriate defluoridation methods are potential options for mitigation of high fluoride effects. Regarding defluoridation, there is not a universal method which is appropriate under all social, financial, environmental and technical conditions.

This chapter assesses the probability of occurrence of excessive fluoride concentrations in African groundwater. The assessment combines available information from reported cases with knowledge on geochemical behavior of fluoride in different geographical settings, defined by geological and climatic conditions. A better understanding of the distribution pattern can provide valuable information for design of new water supplies, particularly those located in Precambrian Basement regions or volcanic areas of the African Rift system. For the benefit of existing water supplies already affected by high fluoride concentrations, an overview of available defluoridation techniques is presented in terms of efficiency, required technology and costs. Simple decision trees are used to provide guidelines for selection of an appropriate removal method. Such guidelines are easily understandable to a "problem owner" seeking practical solutions for improvement of his/her groundwater resource.

11.1 INTRODUCTION

11.1.1 *Health effects of ingested fluorine*

Fluoride (F-), the free ion of the chemical element fluorine, forms strong bonds with calcium and phosphate as so-called biological apatite in bones and teeth (Dorozkhin and Epple, 2002). Depending on the quantities entering body, intake of fluoride has both a positive and a negative health effect. Low quantities of fluoride (0.2 mg/day) are essential for building bones and teeth. Slightly higher quantities of fluoride (up to 2.0 mg/day) stimulate growing teeth and formation of stronger enamel. At higher quantities, however, health effects can be negative. Quantities exceeding one gram a day may be lethal (Backer Dirks, 1992). Death results from respiratory paralysis or cardic failure (Nochimson, 2008). Fluorosis is a disease directly related to the high intake of fluoride that cannot be cured. Preventive measures are therefore necessary. Besides causing various grades of fluorosis, excessive fluoride may affect

Table 11.1.   Possible effect of high fluoride concetrations in drinking water.

Concentration (mg/ℓ)	Possible effects
>1.5	Dental fluorosis
>3	Skeletal fluorosis
>10	Crippling skeletal fluorosis

Source: WHO, 2004.

Table 11.2.   Standards for fluoride in drinking water in some African countries.

Country	mg/ℓ	Reference
Morocco	0.7	WHO, 2006b
South Africa	0.75	DWAF, 1996
Egypt	0.8	WHO, 2006b
Kenya	1.5	Gikunju et al., 2002
Ethiopia	3	Gossa, 2006
Tanzania	8	Edmunds and Smedley, 2005

the neurological and digestive systems and reduce immunity (Meenakshi and Maheshwari, 2006). A comprehensive overview of the effects of fluoride on health is given by the US National Research Council (NRC, 2006). Table 11.1 summarizes possible effects of high fluoride concentrations in drinking water.

11.1.2   *Fluoride content as a water quality parameter*

Drinking water is the most important source of fluoride. Because of its negative health effects at high concentrations, drinking water limits have been established in many countries. The World Health Organization guideline value for fluoride in drinking water is 1.5 mg/ℓ (WHO, 2006a).

The health effects are strongly related to physiographic conditions. Many groundwater supplies with elevated fluoride levels in Africa are located in arid and semi-arid zones characterized by high temperatures. In such zones, water consumption is high and negative health effects can occur even at concentrations below 1.5 mg/ℓ. High altitudes can increase fluoride retention within the body (Manji et al., 1986)

Socio-economic situations are often taken into account when developing national drinking water standards. National standards for fluoride in drinking water for some African countries are listed in Table 11.2. Note that in Tanzania, the national standard is 8 mg/ℓ, reflecting the difficulties with compliance in a developing country with regionally high fluoride concentrations and problems with water scarcity (Edmunds and Smedley, 2005).

11.2   SOURCES OF FLUORIDE IN GROUNDWATER

11.2.1   *Natural sources*

Rainfall containing marine aerosols has fluoride but concentrations are low. Hydrogen fluoride enriched volcanic gases and fluoride bearing minerals are by far the most important natural sources of fluoride in groundwater. According to Strunz (1970), there are about 150 different minerals in which fluorine is found. Fluorite (CaF_2) is the most important fluoride mineral. In some other minerals (amphiboles, micas and apatite), fluoride can replace the hydroxyl ion, where fluorapatite is an end-member for continuous mixing between OH- and F- in apatite.

In Africa, groundwater is exploited from aquifers situated in many geological environments. In igneous rocks (including volcanic rocks), fluoride is mostly bound in micas and amphiboles, but some volcanic rocks such as basalt have fluoride bound in apatite.

Carbonate sedimentary rocks usually have low fluoride contents, primarily in fluorite. Clastic sediments with clay fractions rich in micas and illites have much higher fluoride contents. Phosphate beds having fluorapatite and volcanic ash layers can also be important sources of fluoride. In metamorphic rocks, the highest contents (>5000 ppm) of fluoride are found in rocks that were formed by contact metamorphism (Allman and Koritnig, 1974).

11.2.2 *Anthropogenic sources*

Anthropogenic sources of fluoride are less important than the natural sources.

Industrial chemicals, such as hydrofluoric acid and phosphate fertilizers, are the main anthropogenic sources of fluoride. Processing of phosphatic raw materials, use of clays in ceramic industries and burning of coal also release fluoride into the environment (WHO, 2002). Contributions from phosphate fertilizers and industrial sludge can considerably increase fluoride contents in agricultural soils (Edmunds and Smedley, 2005).

11.3   FORMATION OF FLUORIDE-RICH GROUNDWATER

The fluoride concentration in groundwater is primarily controlled by the leachability of fluoride from rock, being favoured by a large surface area per unit weight. At pH > 7, adsorption of OH- over F- is favoured on hydroxides of Al and Fe (Omueti and Jones, 1977). Secondly, the Ca concentration is important, because the activity of F- is restricted by the fluorite solubility. Low Ca concentrations can be caused by:

1. absence in a leachable state,
2. cation-exchange of Na and Mg for Ca or
3. calcite ($CaCO_3$) precipitation with alkalinity > hardness.

A high pH can also lead to low Ca concentration due to limitation by calcite solubility.

Conditions favourable for high fluoride concetrations are found in alkaline or peralkaline rocks, which have naturally high F- contents. In these rocks, volcanic glass with high surface area can produce F-rich groundwater during percolation. The fluoride concentration can be increased by hydrothermal activity, because (1) fluorite solubility increases with increasing temperature for all solutions and (2) fluoride may be added by dissolution of HF gas. Acidic hydrothermal waters can have high fluoride concentrations. The fluoride concentration is no more restricted by fluorite solubility, but by F- availability with complex forming of fluoride with Fe, Al, B, Si and H.

Another way in which the fluoride concentration can increase is by evaporative concentration. If alkalinity > hardness, the Ca concentration decreases by calcite precipitation whereas the HCO_3 concentration increases. The fluoride concentration will then increase with the alkalinity until fluorite solubility is reached. The likelihood of evaporative concentration is largest in arid and semi-arid regions, where evaporative concentration factors above ten are not uncommon. Closed basins are typically regions where evaporative concentration is high, but the condition that alkalinity > hardness is not found in all closed basins across the world.

11.4   DETERMINATION OF FLUORIDE

Fluoride in water is tasteless, odourless and colourless. It is a "minor" anion in groundwater, occurring in relatively low concentrations when compared to "major" anions such as chloride or bicarbonate. Consequently, determination of fluoride demands sensitive methods. Standard equipped laboratories can perform fluoride analysis on a routine basis. However, many rural water supplies are located in regions that have inadequate water quality testing infrastructure or have logistical problems for transportation of water samples to the laboratory. In such cases, field test kits might provide the first indications for fluoride presence in groundwater. Most of the field kits use the principles of titration. A specific amount of reagent is added to a water sample and the change of color is identified visually or measured by means of a portable colorimeter. Purchase of a pocket colorimeter and required fluoride reagents will usually require an investment of several hundreds of dollars. Several Indian institutions claim development of cheap (below US$ 50) fluoride test kits (IWP, 2008). Shriram Institute for Industrial Research (2005) has evaluated the performance of some of the fluoride field test kits. Out of the 15 tested kits, only four were labeled as "effective". It must be stressed that fluoride field kits provide only qualitative or semi-quantitative estimate of the total fluoride concentrations. Accurate fluoride analysis should always be carried out in a certified laboratory.

11.5   FLUORIDE OCCURRENCE IN AFRICAN GROUNDWATER

11.5.1   *Occurrence based on documented cases*

Information on the occurrence of fluoride in African groundwater is limited by the availability of chemical analyses. In many African countries, nation-wide monitoring networks of groundwater are lacking (Jousma and Roelofsen, 2004).

Within the framework of IGRAC's Global Groundwater Information System (GGIS), a global inventory of occurrence of fluoride was carried out based on publicly available information from the internet, publications, technical reports and maps. A simple approach, based on qualitative estimates of the number of locations or zones with fluoride concentrations in excess of 1.5 mg/ℓ, was used to provide country-wise coverage. Twenty one African countries have some records of high fluoride concentrations in groundwater. These countries are shown in Figure 11.1.

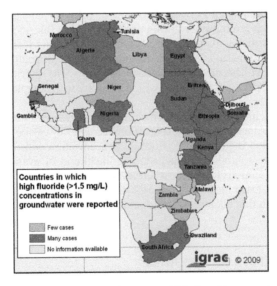

Figure 11.1.   African countries with reported cases on high fluoride concentrations in groundwater (IGRAC, 2009).

Edmunds and Smedley (2005) evaluated the available data for 9 African countries and linked the occurrence of high fluoride to various geological environments (Table 11.3 and Figure 11.2).

Georeferenced fluoride data for Africa are scarce. Only South Africa and Ethiopia have produced maps showing countrywide distributions of fluoride in groundwater (Ncube and Schutte, 2005; Gossa, 2006). For many countries, information on fluoride occurrence is based on samples collected during various (short-term) regional and/or local projects. The Swiss Institute of Aquatic Science and Technology EAWAG (Amini et al., 2008) produced a global map with measured fluoride concentrations in some regions of Africa, including Morocco, Eritrea, Ethiopia, Ghana, Kenya, Senegal, South Africa and Tanzania. The areas with data are shown in Figure 11.3.

As can be seen by comparing Figures 11.2 and 11.3, no publicly accessible georeferenced data are available for some areas, described in literature as having high-fluoride

Table 11.3.   African countries with documented cases of high fluoride groundwaters in various geological environments.

Country	Geological environment			
	Geothermal	Crystalline	Volcanic	Sedimentary
Ethiopia			■	
Ghana		■		
Kenya	■			
Libya				■
Senegal		■		
South Africa				■
Sudan			■	■
Tanzania	■			
Tunisia	■			■

Source: Edmunds and Smedley, 2005.

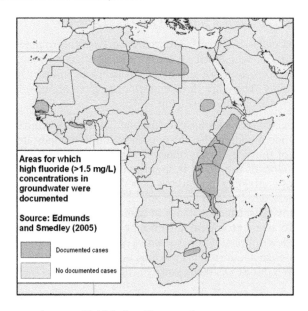

Figure 11.2.   Documented areas with high-fluoride groundwaters.
Source: Edmunds and Smedley, 2005.

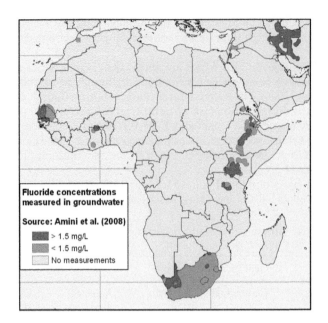

Figure 11.3.   Distribution of fluoride measurements in Africa.
Source: Amini et al., 2008.

groundwater (e.g. North Africa). For Bushveld complex in the Northeastern part of South Africa only the spots with exact coordinates are indicated in Figure 11.3. On the other hand, Figure 11.3 includes new data for the Northwestern part of South Africa. EAWAG (Amini et al., 2008) obtained these data from South African Department of Water Affairs and Forestry through personal communication.

11.5.2   *Can we predict distribution pattern of fluoride in Africa?*

Because documented cases of fluoride occurrence in African groundwater are scarce, the spatial distribution cannot be mapped solely on the basis of georeferenced fluoride analyses. We need other information available at the pan-African scale in order to bridge the gaps in fluoride data availability. There are three important sources of such "proxy" information relevant to the occurrence of fluoride (Figure 11.4):

- Geological information provides insight on potential fluoride sources. From the hydrochemistry of fluoride, we know that the concentration in groundwater is strongly related to the weathering of fluoride-bearing minerals or the presence of volcanic gases. In Africa, fluoride-bearing minerals are found in granites and gneisses of Precambrian Basement areas, in sediments containing argillaceous deposits and phosphatic layers or in hyperalkaline volcanic rocks developed in the rift zones. The African rift zones are characterized by geothermal activity which contributes to high fluoride concentrations (Edmunds and Smedley, 2005). To assess geological influence on fluoride occurrence in groundwater we need lithological descriptions of the main rock formations, information on the tectonic history and knowledge about (recent) volcanic activity. For example, the geological map of Africa (CGMW/UNESCO, 2000) shown in Figure 11.4 shows the location of volcanic rocks.
- Climate plays a major role in groundwater recharge and evaporation process. Evaporation increases the concentrations of ions, including fluoride, in water in the unsaturated zone.

Proxy information on fluoride occurrence in groundwater

Geology   Climate

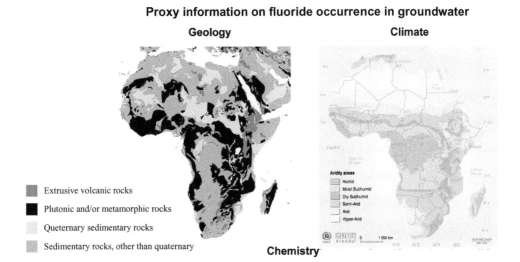

■ Extrusive volcanic rocks

■ Plutonic and/or metamorphic rocks

Queternary sedimentary rocks

Sedimentary rocks, other than quaternary

Chemistry

Expected fluoride if →	T	EC	pH	Na	Ca	Mg	HCO₃
Low	Low	Low/High	Low	Low	High	High	Low
High	High	Low/High	High	High	Low	Low	High

Figure 11.4.   Main sources of "proxy" information on fluoride. (See colour plate section).
Source: CGMW/UNESCO, 2000; UNEP/GRID-Arendal, 2002.

Table 11.4.   Chemical analyses of groundwater from Kenya (Griffioen and Kohnen, 1987, Vasak, 1992).

Borehole number	Location district	T °C	EC mS/cm	pH –	Na mg/ℓ	Ca mg/ℓ	Mg mg/ℓ	Cl mg/ℓ	HCO₃ mg/ℓ	F mg/ℓ
BH 5540	West Pokot	23	0.49	7.0	8	37	26	9	180	0.5
BH 5211	Elgeyo M.	24	0.30	6.9	68	7	3	2	217	0.8
BH 2406	Machakos	26	4.65	6.9	526	352	253	770	572	1.7
BH 2023	Laikipia	24	0.37	8.0	37	22	14	10	247	1.9
BH 3780	Machakos	24	0.64	7.8	105	32	7	16	326	2.5
BH 932	Laikipia	22	0.82	7.7	155	19	8	66	371	5.7
BH 120	Baringo	24	0.50	7.5	116	6	1	13	248	6.2
BH 3868	Baringo	35	2.80	8.5	880	2	1	108	1990	20.0

This process can produce elevated fluoride in groundwater upon recharge. Dissolution of evaporative salt deposits may also contribute to high fluoride concentrations in pore water. In humid regions, the overall dilution effect of abundant rainfall usually results in lower concentrations of most chemical compounds dissolved in groundwater, including fluoride. To asses climatic influences on fluoride we need an aridity index according to precipitation/potential evaporation ratio. Figure 11.4 shows a map of aridity zones in Africa (UNEP/GRID-Arendal, 2002).

- The chemical type of groundwater can provide a geochemical framework for fluoride occurrence in the absence of direct fluoride analysis. Major constituents, such as sodium, calcium, magnesium, chloride and bicarbonate, are commonly included in routine chemical analyses of groundwater. Examples of chemical analyses of groundwater collected in five districts in Kenya are shown in Table 11.4.

High-fluoride groundwaters are mainly associated with a sodium-bicarbonate water type. In such waters, calcium and magnesium concentrations are (relatively) low, even if the total dissolved solid contents (in Table 11.4 represented by EC) is high. Chemical analyses of major ions are therefore important indicators for presence of fluoride in groundwater. Groundwater analysis showing low calcium and magnesium and high bicarbonate indicate potential for a high fluoride concentration. In samples containing hot water (e.g. borehole 3868 in Table 11.4), the high fluoride concentration is attributed to the magmatic gas flux of CO_2 and HF in the active hydrothermal area.

11.5.3   *Occurrence based on probability*

IGRAC has developed a simple approach to map the probability of high fluoride groundwater on a hydrogeochemical basis (Griffioen et al., 2005). Two global data sets, available in GIS format, are used to recognize respectively basic lithological units and climatic zones. From the global geological map (CGMW/UNESCO, 2000) the distribution of main endogenous rocks, extrusive volcanic rocks,, consolidated and unconsolidated sediments, and geological age can be distinguished. From the climatic map (Milich, 1997; CRU, 2004) six classes of precipitation/potential evaporation ratio (hyper-arid to humid) are obtained. A distinction was made between: (a) countries for which occurrence of fluoride rich groundwater was documented and (b) countries for which a probability was assumed. When a classified unit continues from a documented country to a non-documented one, the unit was continuously classified. Three classes were distinguished in countries where fluoride-rich groundwater is documented and a fourth class is added for other countries (Table 11.5).

The literature references are provided by Brunt et al. (2004). The probability of fluoride occurrence in Africa is shown in Figure 11.5.

Global probability maps based on proxy information cannot replace detailed distribution maps based on georeferenced measurements. However, the probability approach described above helps to assess potential risks in unmonitored regions. Another approach has been developed by EAWAG. They used statistical modelling to predict the probability of fluoride concentrations in groundwater on a global scale. The results of modelling and a world map are described by Amini et al. (2008).

Table 11.5.   Classification of probability of high-fluoride groundwater (Griffioen et al., 2005).

Probability	Hydrogeology	Climate	Confirmed by references
High probability/ documented	Geol. formation with F-rich groundwater	Hyper-arid/arid	yes
Medium probability/ documented	Geol. formation with F-rich groundwater	Semiarid/dry-subhumid	yes
	Potential F-rich & known fluoride problem country (or adjoined)	Hyper-to semiarid	no
Low probability/ documented	Geol. formation with F-rich groundwater	Moist-subhumid/humid	yes
	Potential F-rich environment & known fluoride problem country (or adjoined)	Dry subhumid to humid	no
Assumed risk	Potential F-rich environment & *no* known fluoride—problem country	Hyper-arid/dry-subhumid	no

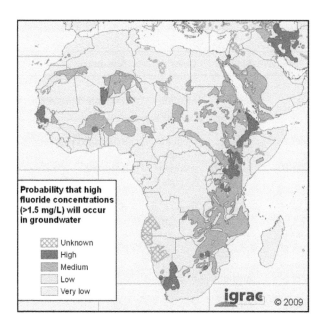

Figure 11.5.   Probability of occurrence of excessive fluoride concentrations in groundwater. (See colour plate section).
Source: Brunt et al., 2004, updated with data from Figure 11.3, Amini et al., 2008.

11.6   REMOVAL OF FLUORIDE FROM DRINKING WATER SUPPLIES

A review of fluoride removal methods, published by Heidweiller (1992), states that most defluoridation methods require a certain level of technological skills, centralized distribution systems or expensive chemicals. In African rural water supplies, low-cost and low-technology removal methods are usually required. Feenstra et al. (2007) presented an updated overview and evaluation of fluoride removal methods with a focus on domestic and community water supplies.

11.6.1   *Removal processes*

Technologies used for the removal of fluoride from drinking water are based on three processes. A precipitation process uses the addition of chemicals, usually calcium and aluminium salts, to precipitate fluoride out of solution. Precipitation chemicals must be added daily and a certain amount of sludge is produced. An adsorption process involves passage of water through a contact bed. Fluoride is removed by ion exchange or surface chemical reaction with the solid bed matrix. After a period of operation, the reaction column must be refilled or regenerated. Adsorbents used for fluoride removal include activated alumina, carbon, bone charcoal and synthetic ion exchange resins. Finally, there is a membrane filtration process that includes reverse osmosis and electrodialysis. Reverse osmosis generally removes any molecular compounds smaller in size than water molecules.

11.6.2   *Fluoride removal methods*

Table 11.6 compares the removal efficiency, operation skills and relative costs of common fluoride removal methods based on the different removal processes. The main advantages and disadvantages are also listed.

Many of the methods based on precipitation or adsorption and ion exchange may be suitable for both community and household water supplies.

Table 11.6.   Fluoride removal methods (based on Heidweiller, 1992; Pickard and Bari, 2004 and BGS, 2003).

Technique	Efficiency	Skills	Costs	Advantages	Disadvantages
Precipitation					
Alum (aluminum sulfate)	>90%	low	medium-high	established process	sludge; acid water; residual Al
Lime	>90%	low	medium-high	established process	sludge; alkaline water
Alum + Lime (Nalgonda)	70–90%	low	medium-high	low-tech; established process	sludge; high dose; residual Al
Gypsum + fluorite	low	medium	low-medium	simple	residual $CaSO_4$
Calcium chloride	>90%	–	–	–	–
Adsorption & ion exchange					
Activated carbon	>90%	medium	high	–	many interferences; change of pH
Plant carbon	>90%	medium	low-medium	availability	soaking in KOH
Zeolites	>90%	medium	high	–	low capacity
Defluoron 2	>90%	medium	medium	–	disposal of chemicals
Clay pots	60–70%	low	low	availability	low capacity; slow
Activated alumina	85–95%	low	medium	effectiveness; low energy; sludge non-hazardous	pH adjustment; chemical and sludge handling
Bone	low	low	low	availability	taste; acceptance
Bone char	60–70%	low	low	availability; capacity	acceptance
Filtration					
Electrodialysis	85–95%	medium	very high	removal other contaminants; public perception	water loss; energy consumption
Reverse osmosis	85–95%	medium	very high	removal other contaminants; public perception	water loss; energy consumption

There are several new technologies being developed and tested (Feenstra et al., 2007). The Solar Dew Collector System (Solar Dew, 2008) and WaterPyramid (Aqua-Aero, 2008) are already suitable for a small-scale use at relatively low cost.

11.6.3   *Critical assumptions for application*

Defluoridation of drinking water is technically feasible at point-of-use (at the tap) and for small communities of users (e.g. wellhead application). Point-of-use systems can produce sufficient quantities of treated water for the drinking and cooking requirements of several people. Numerous plumbed-in, small distillation units are marketed that have been tested and can produce 10 litres per day or much larger volumes. Many certified low pressure reverse osmosis units are available with rated capacities in the range of 30–100 litres per day. Point-of-use defluoridation using activated alumina anion exchange is capable of removing fluoride from small volumes of water, but international performance standards have not been developed to date. However none of the methods listed in Table 11.6 has been implemented successfully at a large scale in Africa. All available defluoridation methods do have disadvantages, such as:

1. High Cost-Tech; i.e. either the price and/or the technology is high, demanding imported spare parts, continuous power supply, expensive chemicals, skilled operation or regeneration, etc. Reverse osmosis, ion exchange and activated alumina have this disadvantage.
2. Limited efficiency; i.e. the method does not sufficiently remove fluoride, even when appropriate dosages are used. For example, in the Nalgonda technique the residual concentration is often higher than 1 mg/ℓ, unless the raw water concentration itself is relatively low.
3. Unobserved breakthrough; i.e. the fluoride concentration in the treated water may rise gradually or suddenly, typically when a medium in a treatment column is exhausted or when the flow is out of control. Bone charcoal and other column filters necessitate frequent monitoring of the fluoride residual.
4. Limited capacity; for example, large amounts of calcined clay have to be used in order to obtain appropriate removals.
5. Deteriorated water quality; the water quality may also deteriorate due to bacterial growth, poorly prepared medium (bone charcoal) or due to medium escaping from the treatment container, e.g. ion exchange material, alumina, Nalgonda sludge, etc.
6. Taboo limitations; in particular, the bone charcoal method is culturally not acceptable to some religions.

Based on available information, a matrix is composed to give an idea about the applicability of the methods for some given situations. The matrix is shown in Table 11.7. The colours in the matrix correspond with theh appropriateness of the method for the given situation:

- Green colour means that the method is very suitable
- Orange colour means average suitability
- Red colour means that the method is unattractive or not applicable for the given situation.

To help water users by choosing the most appropriate method for their situation a process selection tree is shown in Figure 11.6.

High removal efficiency refers to reduction of original fluoride by 90%. High capacity refers to production of more than 100 litre water/day at a community scale or 10–50 litre water/day at a household scale.

Activated alumina is the most favourable technique in case high removal efficiency and high capacity is needed. For situations where moderate removal efficiency is sufficient, the

Table 11.7.   Matrix of fluoride removal methods (Feenstra et al., 2007).

Methods	Domestic + low costs	Community + low costs	Domestic + high F removal	Community + high F removal	Domestic + brackish water	Community + brackish water
Activated Alumina						
Ion exchange						
Reverse osmosis						
Electrodialysis						
Nalgonda process						
Contact precipitation						
Bone Charcoal						
Calcined Clay						
WaterPyramid/ Solar Dew						

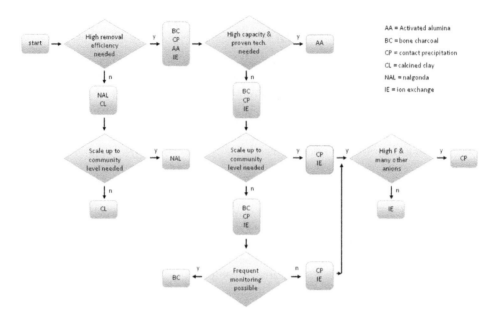

Figure 11.6. Decision tree for fluoride removal techniques applicable in developing countries (Feenstra et al., 2007).

Nalgonda technique is preferable. In other situations (high removal efficiency, small scale) contact precipitation or ion exchange (only high F-ions) is advisable. Bone charcoal can also be used if frequent monitoring is possible.

11.7   ALTERNATIVE MITIGATION

In case it is impossible to apply one of the fluoride removal methods (due to social, financial and/or technical various constraints), alternative solutions should be found. Improved nutrition of the local population in contaminated areas or a search for fluoride-free water sources can help mitigate the negative health effects of fluoride.

11.7.1   *Improved nutrition*

Epidemiological studies in endemic regions of India (Jaffery et al., 1998) and China (Chen et al., 1997) suggest that a diet rich in calcium, magnesium and vitamin C might mitigate the effect of fluoride toxicity. Milk is an important source of calcium. Fruits and vegetables or, if necessary, vitamin supplements should provide the necessary daily requirement of 40 milligrams of vitamin C. However, due to logistical and/or financial constraints it is often not possible to optimize the diet of African rural populations in order to mitigate the effects of fluorosis.

11.7.2   *Fluoride-free water sources*

Finding an uncontaminated water source is, in theory, the best solution when dealing with the negative effects of hazardous compounds found in groundwater. In practice, however, alternative water sources are non-existent or too difficult to develop.

Rainwater, surface water and groundwater from another part of an aquifer system are potential water sources that might replace a groundwater supply contaminated with fluoride.

Fluoride concentrations in rainfall are low and surface water usually contains much less fluoride than groundwater. On the other hand, the availability of both rainfall and surface water in large parts of Africa show large variation in time and space. Storage tanks and dams are necessary to bridge dry spells and chemicals are required to protect these sources from bacterial pollution. This requires substantial financial investments and there is no guarantee for a sufficient volume of water on a long term basis.

Optimizing the existing groundwater supply also requires financial investments in terms of investigation and construction, but these investments can help to find fluoride-free groundwater. Transfer of a rural groundwater supply to another location is not feasible due to various technical and social constraints. But in cases where fluoride concentrations in groundwater vary with depth, redesign of supply can be considered.

For example, Ethiopian volcanic aquifers contain groundwater having high fluoride concentrations (e.g. Tekle-Haimanot et al., 2006). Groundwater found in the overlying alluvium has diminished concentrations as result of increased calcium concentrations following likely reactions with carbonate minerals in the sediments (Ashley and Burley, 1994). Boreholes tapping the fault zones in Kenyan rift system have a high yield but are often contaminated with fluoride from gaseous emanations (Vasak, 1992). Groundwater in alluvial aquifers in this area is also mostly fluoride-free. Increases in fluoride concentrations with depth have also been reported from Ghana (Apambire et al., 1997) where shallow hand-dug wells produce groundwater having lower fluoride concentrations than deeper boreholes drilled in fractured basement granite.

What is always needed is a proper understanding of the local hydrogeology. Hydrogeological knowledge is obtained through review of existing (hydro)geological maps and documents. If such information is not available, remote sensing study and field investigations are required. There are many guidelines describing the principles of basic groundwater investigations. An inventory of existing guidelines and protocols for groundwater assessment and monitoring (Jousma and Roelofsen, 2003) is available on the IGRAC web-site (www.igrac.nl). In 2004, UNESCO-IHP published an international guide for hydrogeological investigations (Kovalevsky et al., 2004).

11.8   IMPLICATION FOR RURAL WATER SUPPLIES

Groundwater is the major source for many rural water supplies in Africa, but many supplies are contaminated with naturally occurring fluoride. Numerous cases of endemic fluorosis have been reported in Senegal, Ghana, Tanzania, Kenya, Ethiopia and South Africa.

The regional occurrence of fluoride can be estimated using proxy information such as geological provenance. A large part of the rural population lives in areas underlain by rocks that might release fluoride to groundwater. MacDonald and Davis (2000) estimated that in sub-Saharan Africa more than 60% of the rural population lives in areas underlain by basement (40% of the land area of sub-Saharan Africa) or volcanic rocks (6%). Many African countries have arid and semi-arid climates; groundwater in such zones is vulnerable to high fluoride concentrations due to evaporation.

While the health effects and causes of high fluoride concentrations in African groundwater are clear, immediate solutions for an improvement of water quality in affected water supplies are not.

We cannot predict fluoride contamination at the well-field scale in most parts of Africa, since fluoride levels are variable. Studies in Sri Lanka, carried out in geological environments similar to Africa, indicate that one well may contain high levels while a nearby well contains little. Senewirante et al. (1973) reported a difference of 8 mg/ℓ of fluoride from wells situated just 149 m apart.

In contrast to anthropogenic contamination of groundwater (e.g. organics pollutants), fluoride contamination cannot be avoided by proper protection of production wells. Because the occurrence of fluoride is primarily controlled by the local geology and climate, there are no preventive measures, under the given areal limits of a water supply, to avoid contamination.

There is not a universal removal method which is appropriate under all social, financial, economic, environmental and technical conditions.

The great challenge for geochemists, hydrogeologists, and sanitation engineers is to refine knowledge about the geochemical behavior of fluoride in the major geological environments, the pathways of fluoride from rock to drinking water and efficient, low-cost remediation techniques. It is of great importance that all information and knowledge on the occurrence of fluoride and its remediation is shared among African professionals and stakeholders.

It must be stressed that fluoride contamination might be just one of many problems for a particular rural water supply. Low yields, malfunctioning pumping equipment or acute life threatening pollution might put fluoride contamination low on the priority list of water management issues. In arid and semi-arid parts of Africa, there always is a conflict of interests between water quantity and quality. Sustainable groundwater management will always require looking at larger managerial issues, including optimum use of alternative water supplies. It is obvious that the availability of water and the distance to the sources are important factors which should be taken into account when looking for alternative water sources. Like other water resources management issues, the management of excessive fluoride in groundwater must be undertaken through stakeholder participation.

Elements of efficient and sustainable information sharing related to water were outlined during the Organisation for Economic Co-operation and Development (OECD) workshop on International Science and Technology Cooperation for Sustainable Developments in South Africa (Vasak and Van der Gun, 2007). Dissemination of information is important for raising public awareness on groundwater. The provider of information should anticipate the stakeholder's perception about water issues. Sharing information can save time in the search for effective measures. All parties involved can benefit from analogies and do not need to duplicate the whole process of developing concepts, identifying problems and formulating solutions. The same applies also to the reduction of costs if "lessons learned" are included in the sharing process.

Finally, a high concentration of information and knowledge can initiate important breakthroughs in knowledge and perception, leading to new views and better groundwater management approaches. Modern technology opens many possibilities for effective and efficient information sharing. Information sharing can be seen as a shortcut to proper understanding of issues in sustainable water resources management.

In this context the following recommendations are formulated:

- develop a continental database of fluoride concentrations in water
- develop a continental meta-information system for documents and maps related to fluoride
- include fluoride measurements in regular groundwater monitoring programs
- develop an on-line discussion forum on fluoride
- initiate regional programmes for appropriate removal technologies in which the laboratory experiments will be tested under field conditions
- translate the "professional" information into low threshold practical approaches applicable under given physical and socio-economical conditions
- disseminate these solution approaches to local population in affected areas by regular public awareness campaigns on "safe" water.

REFERENCES

Allman, R. and Koritnig, S. 1974. Fluorine. In 'Handbook of Geochemistry', K.H. Wedepohl, ed. Springer Verlag, Berlin, Heidelberg.

Amini, M., Mueller, K. Abbaspour, K.C., Rosenberg, T., Afyuni, M., Møller, K.N., Sarr, M. and Johnson, A. 2008. Statistical Modeling of Global Geogenic Fluoride Contamination in Groundwater. Environ. Sci. Technol., vol. 42, pp. 3662–3668.

Apambire, W.B., Boyle, D.R. and Michel, F.A. 1997. Geochemistry, Genesis, and Health Implications of Fluoriferous Groundwater in the Upper Regions of Ghana. Environ.Geol., vol. 33/1, pp. 13–24.

Aqua-Areo WaterSystems, 2008. WaterPyramid. http://www.aaws.nl/home.htm

Arancibia, J.A., Rullo, A., Olivieri, A.C., Di Nezio, S., Pistones, A.L. and Band, B.S. 2004. Fast spectro-photometric determination of fluoride in ground waters by flow injection using partial least-squares calibration. Analytica Chimica Acta, vol. 512/1, pp. 157–163.

Ashley, P.P. and Burley, M.J. 1994. Controls on the occurrence of fluoride in groundwater in the Rift Valley of Ethiopia. In 'Groundwater Quality', H. Nash and G.J.H. McCall, eds. Chapman and Hall, pp. 45–54.

Backer Dirks, O. 1992. Biokinetics of fluoride in relation to dental and skeletal fluorosis. In 'Endemic fluorosis in developing countries: causes, effects and possible solutions', J.E. Frencken, ed. Publication 91.082, NIPG-TNO, Leiden, The Netherlands, pp. 20–30.

British Geological Survey (BGS), 2003. Water Quality Fact Sheet: Fluoride. WaterAid International site. http://www.wateraid.org/documents/plugin_documents/fluoride1.pdf

Brunt, R., Vasak, L. and Griffioen, J. 2004. Fluoride in groundwater: Probability of occurrence of excessive concentration on global scale. Report nr. SP 2004-2, International Groundwater Resources Assessment Centre (IGRAC).

CGMW/UNESCO, 2000. Geological Map of the World at 1:25.000.000. Commission for the Geological Map of the World, UNESCO, September 2000 (second edition).

Chen, Y.X., Lin, M.Q., Xiao, Y.D., Gan, W.M. and Chen, C. 1997. Nutrition survey in dental fluorosis-affected areas. Fluoride, vol. 30/2, pp. 77–80. Research Report 77. http://www.fluoride-journal.com/97-30-2/302-77.htm

Climate Research Unit (CRU), 2004. Data sets. University of East Anglia, Norwich. http://www.cru.uea.ac.uk/

Dorozkhin, S.V. and Epple, M. 2002. Biological and medical significance of calcium phosphates. Angew. Chem. Int. Ed., vol. 41, pp. 3130–3146.

DWAF, 1996. South African Water Quality Guidelines, vol. 1: Domestic Water Use, 2nd edition (1996). Department of Water Affairs and Forestry, Pretoria.

Edmunds, M. and Smedley, P. 2005. Fluoride in natural waters. In 'Essentials of Medical Geology, Impacts of Natural Environment on Public Health', O. Selinus et al, eds. Elsevier Academic Press, London, pp. 301–329.

Feenstra, L., Vasak, L. and Griffioen, J. 2007. Fluoride in groundwater: Overview and evaluation of removal methods. Report nr. SP 2007-1. International Groundwater Resources Assessment Centre (IGRAC).

Gikunju, J.K., Simlyu, K.W., Gathuru, P.B., Kyule, M. and Kanja, L.W. 2002. River water fluoride in Kenya. Fluoride, vol. 35/3, pp. 193–196. http://www.fluoride-journal.com/02-35-3/353-193.pdf

Gossa, T. 2006. Fluoride Contamination and Treatment in the Ethiopian Rift Valley. Presentation at the WWF 4, Mexico. Ministry of Water Resources Addis Ababa. http://www.cepis.ops-oms.org/bvsacg/e/foro4/19%20marzo/Strategies/Fluoride.pdf

Griffioen, J., Brunt, R., Vasak, S. and Van der Gun, J. 2005. A global inventory of groundwater quality: First results. In 'Bringing Groundwater Quality Research to the Watershed Scale', N.R. Thomson, ed. Proceedings of GQ2004, the 4th International Groundwater Quality Conference, Waterloo, Canada. IAHS Publ. 297, pp. 3–10.

Griffioen, J. and Kohnen, M. 1987. Hydrogeochemistry of Fluorine in Four Different Hydrogeological Environments in Kenya. M.Sc thesis, Free University Amsterdam and State University Utrecht.

Heidweiller, V.M.L. 1992. Fluoride removal methods. In' Endemic fluorosis in developing countries, causes, effects and possible solutions', J.E. Frencken, ed. Publication number 91.082, NIPG-TNO, Leiden, The Netherlands, pp. 51–85.

International Groundwater Resources Assessment Centre (IGRAC), 2009. Global Groundwater Information System; Global Overview. www.igrac.nl

IWP, 2008. Water testing kits for field use. India Water Portal. http://www.indiawaterportal.org/data/kits/waterqualitykits.doc

Jaffery, et al., 1998. Impact of Nutrition on Fluorosis (A desk review). Industrial Toxicology Research Centre, Lucknow/UNICEF New Delhi. http://www.itrcenvis.nic.in/publications/fluorosis.pdf

Jousma, G. and Roelofsen, F.J. 2003. Inventory of existing guidelines and protocols for groundwater assessment and monitoring. Report nr. GP 2003-1, International Groundwater Resources Assessment Centre (IGRAC).

Jousma, G. and Roelofsen, F.J. 2004. World-wide inventory on groundwater monitoring. Report nr. GP 2004-1, International Groundwater Resources Assessment Centre (IGRAC).

Kovalevsky, V.S., Kruseman, G.P. and Rushton, K.R. 2004. Groundwater studies. An international guide for hydrogeological investigations. IHP-VI, Series on Groundwater No. 3. UNESCO/TNO, 430p.

MacDonald, A.M. and Davies, J. 2000. A brief review of groundwater for rural water supply in sub-Saharan Africa. Technical report WC/00/33. Overseas Geology Series. British Geological Survey. http://www.bgs.ac.uk/hydrogeology/ruralwater/manual.html

Manji, F., Boelum, V., and Fejerskov, O. 1986. Fluoride, Altitude and Dental Fluorosis, Caries Res, vol. 20, pp. 473–480.

Milich, L. 1997. Deserts of the world. http://ag.arizona.edu/~lmilich/desert1.html

National Research Council, 2006. Fluoride in drinking water. A scientific review of EPA's standards. BEST. The National Academic Press Washington DC. http://books.nap.edu/openbook.php?record_id=11571&page=R1

Ncube, E.J. and Schutte, C.F. 2005. The occurrence of fluoride in South African groundwater: A water quality and health problem. Water Sa, vol. 31/1, pp. 25–40.

Meenakshi and Maheshwari, R.C. 2006. Fluoride in drinking water and its removal. J. of Hazardeous Materials, vol. 137/1, pp. 456–463.

Nochimson, G. 2008. Toxicity, Fluoride. E-medicine. http://emedicine.medscape.com/article/814774-overview)

Omuet, J.A.I. and Jones, R.L. 1977. Fluoride adsorption by Illinois soils. J. Soil Sci., vol. 28, pp. 564–572.

Pauwels, H. and Ahmed, S. 2007. Fluoride in groundwater: origin and health impacts. Geosciences no. 5. March 2007. BRGM.

Pickard, B., and Bari, M. 2004. Feasibility of Water Treatment Technologies for Arsenic and Fluoride Removal from Groundwater. AWWA Water Quality Technology Conference, San Antonio, Texas.

Rakshit, P.K., 2004. Studies on estimation of fluoride and defluoridation of drinking water. Master Thesis, Department of Chemical Engineering, Indian Institute of Science, Bangalore. http://www.rainwaterclub.org/docs/Fluoride.pdf

Senewiratne, B., Senewirante, K., Hettiarachchi and Thambipillai, S. 1973. Assessment of the fluoride content of water in wells selected randomly and after examining schoolchildren for dental fluorosis. Bull WHO, vol. 49/4, pp. 419–422.

Shriram, 2005. Evaluation of Water Quality Field Tests Kits. Shriram Institute for Industrial Research/UNICEF India.

Solar Dew, 2008. The Solar Dew Collector System. http://www.solardew.com/

Strunz, H. 1970. Mineralogische Tabellen. 5 Auflage, Leipzig, Geest & Parting.

Tekle-Haimanot, R., Melaku, Z., Kloos, H., Reimann, C., Fantaye, W., Zerihun, L. and Bjorvatn, K. 2006. The geographic distribution of fluoride in surface and groundwater in Ethiopia with an emphasis on the Rift Valley. Sci. Total Environ., vol. 367, pp. 182–190.

UNEP/GRID-Arendal, 2002. Aridity Zones. UNEP/GRID-Arendal Maps and Graphics Library, 2002. http://maps.grida.no/go/graphic/aridity_zones

Vasak, S. 1992. Primary sources of fluoride. In 'Endemic fluorosis in developing countries, causes, effects and possible solutions', J.E. Frencken, ed. Publication number 91.082, NIPG-TNO, Leiden, The Netherlands, pp. 1–9.

Vasak, S. and Van der Gun, J. 2007. Sharing Information and Knowledge about Water: Groundwater Examples. In Integrating Science and Technology into Development Policies: An International Perspective. Proceedings OECD-SA workshop on International Science and Technology Cooperation for Sustainable Development, Pilanesberg, South Africa 21–22 November 2005, pp. 183–191.

Weaver, M.C.J., Cave, L. and Talma, A.S. 2007. Groundwater Sampling: A Comprehensive guide for sampling method, WRC Report No TT 303/07, 2007, 168p.

WHO, 2002. Environmental Health Criteria 227—Fluorides. World Health Organisation. http://www.inchem.org/documents/ehc/ehc/ehc227.htm

WHO, 2004. Fluoride in Drinking-water. Background document for development of WHO Guidelines for Drinking-water Quality. WHO/SPE/WSH/03.04.96. http://www.who.int/water_sanitation_health/dwq/chemicals/en/fluoride.pdf

WHO, 2006a. Guidelines for Drinking-water Quality; Third Edition. WHO, Geneva. http://www.who.int/water_sanitation_health/dwq/gdwq3rev/en/

WHO, 2006b. A compendium of drinking-water quality standards in the Eastern Mediterranean Region. Regional Office for Eastern Mediterranean. CEHA. WHO-EM/CEH/143/E.

12

Practical methods to reduce Iron in groundwater with a case study

Max Karen
Earth Science Systems, Botswana

Lytone Kanowa
Ministry of Local Government and Housing, Zambia

Jim Anscombe
GITEC Consult, Zambia

ABSTRACT: Iron rich groundwater poses a serious problem in many parts of the world. In rural water supply projects in Africa this problem can affect large numbers of people who rely on handpumps for a safe, reliable water supply. Groundwater drawn by handpumps is often rejected based on the taste and appearance of the water when in fact the water is much safer than traditional sources. The problem was encountered during the North Western Province Rural Water Supply Project in Zambia. The problem was approached in two ways; the first was to try to find a geophysical signature next to existing boreholes with high iron using Vertical Electrical Soundings (VES); the surveys suggested that there was a link between conductive layers at between 30–45 mbgl and high iron. The other method to mitigate the problem was to build a pilot Iron filter based on designs from around the world which were adapted to local conditions, this filter proved successful in reducing Iron to acceptable levels however the flow rates from the filter were found to decrease over time to unacceptable levels. During the construction phase of the project, less than 5% of the boreholes drilled encountered iron rich groundwater, however areas were found where high levels of Iron could not be avoided. Modifications have now been made to the filter, which have led to improved flow rates. The filter is now being constructed as an integral part of the civil works in the North West of Zambia.

12.1 INTRODUCTION

Iron rich groundwater poses a serious problem in many parts of the world because it leads to the rejection of functional boreholes. In rural water supply projects in Africa this problem can affect large numbers of people who rely on groundwater drawn from handpumps for a safe, reliable water supply. The problem of iron in groundwater is not due to toxicity, the reason iron is such a major problem is that it causes water discolouration and can impart an unpleasant taste to water as well as causing staining of food and laundry; this causes people to return to unprotected sources of water.

12.2 PROJECT AREA

In Zambia the problem of Iron in groundwater had been reported across the country (Chenov, 1978). However the source of the Iron and a way of dealing with the problem had not been identified. In 2005, the North Western Province Rural Water Supply Project found during a

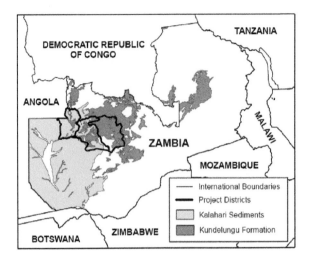

Figure 12.1.   Project area location and distribution of Kundelungu formation.

water point survey that 38% of all the boreholes in the project area had levels of iron above 1 mg/ℓ; the levels of Iron were measured in the field using dedicated Iron testing equipment. The implications for the project were that potentially a significant proportion of the 350 planned boreholes could be rejected.

The project districts are underlain in most populated areas by the Kundelungu Formation and in the West, Kalahari sediments (Figure 12.1). The Kundelungu is an ancient (800Ma) series of shales, mudstones and siltstones. This formation occurs over large areas of Zambia and is also present beneath the Kalahari sediments in Western Zambia (MacDonald, 1990). The main source of the Iron in groundwater in Zambia, based on the Water Point survey (Karen, 2007) and reports of problems with iron in other areas (Chenov, 1978), such as Luapula Province, is from boreholes drilled into the Kundelungu Formation.

The levels of iron varied throughout the project area (Karen, 2005), with boreholes within 200 m of each other showing large variations in Iron content. Any analysis of iron content in comparison to depth was not possible due to a lack of data, however it was possible to disprove a theory that was held at the time that the iron content was due to the steel in the construction of the borehole; this was done by finding a borehole which had PVC casing and screen and an iron problem.

In the Western part of the project area, specifically in Kabompo District, levels of dissolved iron in the Manyinga River were high enough to be measured, this explains how the river got its name, since Manyinga in the local dialect means blood. This suggested that there was a strong possibility that areas would be encountered where high levels of iron could not be avoided. Based on the evidence of the spatial distribution of iron it was decided that in order to supply potable water to 350 communities across the project area, with acceptable levels of iron, the problem had to be confronted and a practical way to reduce iron had to be found.

12.3   METHODOLOGY

In an attempt to mitigate the problem, the source of the iron was investigated using ground geophysical surveys; the main instrument used was the Geotron G41 Resistivity meter. The project used an in house geophysical programme which enabled the project hydrogeologist to revisit boreholes where high levels of Iron had been identified during the water point survey;

at these locations Vertical Electrical Soundings (VES) were carried out, using a Schlumberger array to an AB of 240 m. Boreholes with low Iron levels were also surveyed to balance the sample.

The results of the VES next to boreholes with high levels of iron revealed that each location had conductive layers of below 30 ohm-m, between 30–45 mbgl. The low resistance of these layers indicated clay layers or mudstones; this was also taken to indicate a link between high levels of Iron and low permeability.

The results of the investigation were then used during the in house geophysical surveys that were conducted at each community. The approach used was to combine Electromagnetic traversing, using the Apex Parametrics MaxMin, with Resistivity soundings using a Geotron G41. The main objective of the geophysical surveys was to find areas of deep weathering and/or fractures and therefore areas of higher permeability. The results from the drilling programme showed that the number of boreholes drilled with levels of Iron above 1 mg/ℓ were less that 6%, which showed a large contrast with the existing boreholes with iron levels above 1 mg/ℓ in 38% of boreholes measured (Karen, 2005).

The lower levels of Iron could also have been attributed to boreholes being constructed deeper than the pre project boreholes, which were often terminated once sufficient water was encountered. This can be explained by the weathering profile since the first water was often encountered at the base of the weathered zone. Project boreholes were not terminated at this level; in most cases boreholes were continued into bedrock where fractures were encountered due to the geophysical siting process. The yield was found in most cases to significantly increase below the base of the weathered zone. The larger yields encountered at deeper levels indicate that there is a link between high permeability areas and low Iron for these sites.

12.4   IRON REMOVAL

The geophysical approach has led to lower numbers of boreholes having unacceptable levels of Iron, however areas were found where Iron was found throughout the local area, the worst area being along the banks of the Manyinga River, this had been anticipated during the initial stages of the project and the methods used to remove Iron from water were researched (SKAT, 2001; Tyrrel, 2001; Smith, 1993). The research indicated that there were many methods used around the world to reduce Iron, which vary from complex to basic, however all the methods used the same main principles.

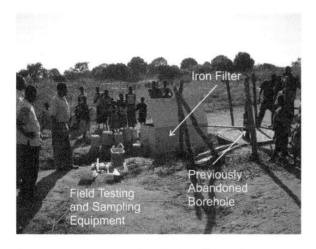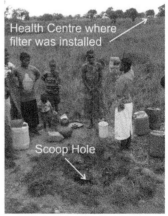

Figure 12.2.   Images showing pilot iron filter and unprotected nearby source of water.

Based on this research a pilot Iron filter was designed and constructed which was designed to be both robust and simple. The pilot filter was installed at four locations throughout the project area. Figure 12.2 shows a borehole at a rural health centre that had been abandoned, this was solely due to the Iron content, as the flow rate was measured at 0.25 ℓ/s. The first image shows the filter in use with a queue of people waiting to collect water. The water from the filter at this location was particularly valued since the only other source of water was from scoop hole (Figure 12.3) located in the local dambo; the water in the scoop hole was visibly contaminated, this was the only source of drinking water for the local community and the health centre which is visible in the background on the top of the hill.

12.5   RESULTS

The levels of Iron were measured at the Inlet and the outlet of the filters using a Hanna instruments iron photometer, which measured the levels of iron using a dedicated reagent added to a sample of water which was then measured by the photometer. In the case of the borehole with the highest levels of Iron, such as Chilemba Basic School (Karen, 2005), the levels of Iron were measured at 10 mg/ℓ at the inlet; these fell below 0.3 mg/ℓ measured from a sample from the outlet. To verify the results and to make certain that no other toxic elements were present the acidified water samples were sent to the British Geological Survey, the laboratory results confirmed the decrease in Iron (See Table 12.1).

The image below shows a schematic of the Iron Filter showing the three main processes used to lower the levels of Iron. After water enters from the handpump it drops onto a splash plate, which increases aeration, the water then flows into a settlement chamber where it is forced to flow under a baffle into a second chamber. The water is then forced to flow over a weir into a chamber filled with sand, where filtration takes place. Finally the water was forced to flow under a third baffle and flow through larger aggregates before reaching the outlet. The filter was constructed from steel and the interior was painted with water tank paint.

Table 12.1.   Iron concentration changes from pilot filters.

Borehole name	District	Iron filter	Fe (mg/ℓ)
Mafuliwanjamba RHC	Kabompo	Inlet	6.91
Mafuliwanjamba RHC	Kabompo	Outlet	0.21
Chilemba BS I	Mufumbwe	Inlet	10.60
Chilemba BS I	Mufumbwe	Outlet	0.24
Kivuku	Kasempa	Inlet	4.41
Kivuku	Kasempa	Outlet	0.01

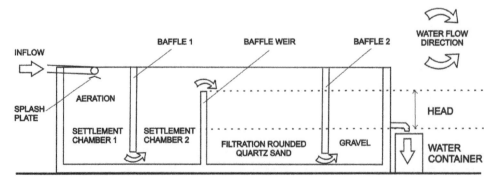

Figure 12.3.   Schematic side view of pilot Iron filter.

The pilot filters were monitored both in terms of the levels of Iron and the flow rates; the monitoring indicated that the main problem with the filters was the flow rate. The filter initially showed flow rates of about 60% of the flow rate of the handpump, however in the case of Chilemba this was found to decrease rapidly. In order for the filter to be successful it was decided that flow rates below 0.15 ℓ/s would be unacceptable; however in some cases flow rates were measured at 0.05 ℓ/s, in these cases the filter continued to be used for drinking water only and was still much appreciated by the communities.

In order to identify the cause of the low flow rate a detailed study (Karen, 2007) was carried out looking at water usage, different types of filter media and variations in head. Based on this study the problem of low flow rates were identified as:

1. The sand filter media was not adequately washed and contained fine material, some of the fines were identified as mica, the mica appeared to combine with the iron bacteria to form a slurry which significantly impeded flow.
2. The type of sand was found to be crucial for the filtration process, the ideal size and material was found to the rounded quartz grains between 1–2 mm.
3. The head between the weir and the outlet was insufficient to push water through the sand filter when it became partially blocked by the fine material and Iron bacteria.
4. While basic maintenance was easily accomplished, thorough cleaning of the sand and gravel was time consuming, the interval between major maintenance was also found to decrease rapidly with increased concentration of Iron.

The issue of maintenance is an issue that is crucial to any rural water supply project. Without trained people to maintain the handpump and a source of spares it is now well understood that the borehole is likely to fail, the other issue is that there must be a sense of ownership and responsibility for the handpump. During the NW Province rural water supply project the sensitisation and training of the community was an integral part of the project; it was therefore decided to extend the training to the filter so that the handpump caretaker understood the filter and was able with the help of the community to carry out necessary maintenance.

12.6 IRON FILTER CONSTRUCTION

The other main issue was how to incorporate the filter into the civil works construction, with the main datum point being the outlet height being high enough for a standard container to easily fit underneath with a 2 cm margin of safety. The other main consideration was that the head must be increased from the 15 cm in the original pilot filter to at least 30 cm. Increasing the head meant that the inflow height would have to change. The easiest solution to the problem was to increase the height of the pump by putting in a spacer.

The Figure 12.4 shows a schematic of the filter where the head has been increased. The increase in the height of the pump was compensated for by installing steps into the existing civil works design, therefore minimising changes to the construction process. A further increase in the head was made possible by placing the filter in line with the drain, this allowed the base of the filter to be placed on top of the drain while the water container can be placed below this level. These modifications allowed the head to be increased from 15 cm to 35 cm.

Other modifications from the pilot filter included putting in a bypass so that water could be allowed to flow directly from the pump; this modification was done for two reasons:

- It was found that the easiest way to clean the sand was to give each community a sieve made from metal mesh, this sieve was then placed directly under the pump bypass allowing water to flow over the sand, the water passing through the sand could then be observed below the sieve, thus when any red tint to the water disappeared the sand was deemed to be clear.
- Some of the boreholes with reported Iron problems were later measured and found to have levels of Iron that varied according to the season, in a number of cases the issue of

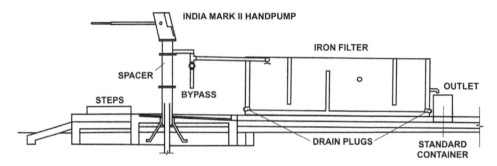

Figure 12.4.   Cross section of modified civil works and Iron filter.

Iron only became a problem during and just after the rainy season, the bypass provides a way to allow the user group to use the handpump without the filter, this was done on the condition that the community drain and clean the filter so avoid standing water and algal growth.

The material for the construction was determined initially by cost, the ideal material for the filter would have been a stainless steel box which was then built into a brick or concrete surround, however due to the massive increases in the cost of steel in August of 2007 this was difficult to justify. The filter was therefore made from bricks and cement with a plaster coating.

The Iron filter was designed to be as simple as possible, however certain elements of the construction required high precision, in particular the height of the outlet, the height of the baffle weir, the lid and obviously it must be waterproof, without these being accurately constructed the filter would not be last for the design period of 25 years.

The construction of the Iron filter began in September 2008, the modification to the head resulted in a large increase in flow rate, this was also due to the size of the sand filter being increased and the correct sand media being used. The flow rate at the outlet after the modifications was 85% of the flow rate of the handpump.

Early on in the construction process it became apparent that the teams that were able to build decent civil works were not necessarily able to construct the Iron filters, due to the level of detail and precision being significantly higher. Based on the experience of the construction of ten filters in the last quarter of 2008 it is recommended that the original plan is reverted too where the filter box in constructed from stainless steel which is them emplaced in a cement or brick surround.

Stainless steel was used in most pump parts which were constructed in India, the use of this material is due to the long term sustainability of this material, therefore the use of the same material for the filter, which is an integral and essential part of the water point is logical. If the filters are made on a larger scale and a single design in incorporated in the SADC region then costs can be dramatically reduced, Iron filters are another example of how project knowledge and experience gets lost over time and the reinvention of the wheel occurs, finding a solution, that is simple as robust only needs to be done once, it is hoped that this filter is the solution to this problem.

12.7   CONCLUSION

The project has shown that there are practical ways to mitigate the problem of high iron in groundwater. The geophysical approach, which used VES's, indicates a link between highly conductive layers and high levels of iron in boreholes, the results of the drilling programme where conductive layers were avoided show a large reduction in boreholes which have a problem with iron.

The second way used to mitigate the iron problem was to reduce the levels of iron at the borehole using an iron filter; after pilot filters had been constructed and tested over time the main factors that were crucial for the filter were identified and incorporated into a modified filter which was installed as part of the civil works at boreholes where iron could not be avoided.

Iron filters are an example of how project knowledge and experience can be lost over time and the reinvention of the wheel occurs, finding a solution, that is simple as robust only needs to be done once, it is hoped that this filter is the solution to this problem.

REFERENCES

Chenov, C.D., 1978. "Groundwater Resources inventory of Zambia", National Council for Scientific Research, UNESCO/NORAD Water Resources Research Project.

Karen, M.P., 2005. "Hydrogeological Assessment of North West Province, Zambia". Unpublished Report, GITEC Consult/Ministry of Local Government, Zambia.

Karen, M.P., 2005. "Project Area Water Point Survey", Unpublished Report, GITEC Consult/Ministry of Local Government, Zambia.

Karen, M.P., 2007. "Iron Removal in the North Western Province Rural Water Supply Project, Unpublished Report, GITEC Consult/Ministry of Local Government, Zambia.

MacDonald and Partners Ltd., 1990. Hydrogeological Map of Zambia.

Sarikaya, H.Z., 1990. "Contact aeration for Iron removal", Wat. Res. 24(3), pp. 329–331.

SKAT/SDC: Vol. 1 of the working papers on Water Supply & Environmental Sanitation (2001): Iron and Manganese Removal.

Smith, G., Gaber., Hattab, I., & Halim, H.A., 1993. "A study into the removal of Iron from ferruginous groundwater using limestone bed filtration", Wat.Sci.Tech. 27(9), pp. 23–28.

Tyrrel, S.F. and Carter, R.C. 2001. Iron in groundwater: prevention or cure? Waterlines, 20, 2, 16–18.

Tyrrel, S.F., Gardner, S.J., Howsam, P., & Carter, R.C. 1998. Biological removal of iron from well-handpump water supplies. Waterlines 16(4), 29–31.

13

Investigation of borehole failures—experience from Botswana

B.F. Alemaw & T.R. Chaoka
Department of Geology, University of Botswana, Gaborone, Botswana

ABSTRACT: Monitoring of both groundwater quality and quantity to support rural water supplies in Africa is almost non-existent. However, monitoring is the only way to identify changes that may later adversely affect borehole performance. Besides, baseline and historical production data are essential for planning and designing remedial measures such as identifying appropriate water treatment mechanisms should water quality deteriorate as a result of increasing pumping water levels. This case study describes groundwater monitoring in eight boreholes in the Gaotlhobogwe wellfield in south east Botswana during the period 1995–2003. Continually declining borehole yields were observed. Declining specific capacity can be correlated with change in some chemical determinants, including Mn, Mg, Fe, NO_3, HCO_3, CO_3 and SO_4. Sound interpretation of the geological and hydrogeological data can pinpoint the causes of groundwater decline. The study highlights the benefit of groundwater monitoring in water supply systems as the basis of an overall groundwater conservation strategy. Groundwater monitoring does have a significant cost, but it ensures long term and sustainable groundwater supplies for the rural communities of Africa.

13.1 INTRODUCTION

Monitoring the impact of a water management policy or management option provides a measure of just how well the policy meets expectations. A groundwater monitoring plan requires identification of suitable performance indicators both for the groundwater resource and its operational system and the plan needs to identify how frequently and accurately the indicators need to be measured over time and space in order for it to be effective. The major challenge in monitoring is selecting the parameters to be measured to optimise information recovered and so aid future decision making.

Monitoring of borehole yield and groundwater quality is a normal part of the groundwater management systems for the various wellfields in Botswana. One such wellfield is that at Gaotlhobogwe in south eastern Botswana (Figure 13.1).

The main objectives for the monitoring programme are:

a. to determine the trend and significance of groundwater level decline in the Gaotlhobogwe wellfield;
b. to determine the major causes of water well failure and declining yield in the well field; and
c. to identify the contributing variables for the decline of the water table in the various boreholes as well as the wellfield as a whole.

Previous work by Chaoka et al. (2006) highlights the importance of groundwater monitoring in support of national groundwater protection programmes with particular regard to rural water supply systems.

A linear trend model was first used to determine the significance of the underlying declining or increasing trend in groundwater levels. This technique has been successfully applied to study variability of numerous types of environmental variables including air and groundwater

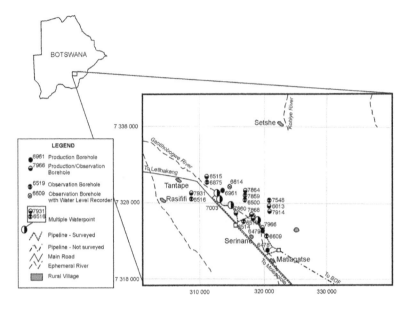

Figure 13.1.   Distribution of boreholes and location map in UTM coordinates of the study area.

temperatures (Gazemi, 2004), air temperature (Comani, 1987), precipitation (Rodhe & Virji, 1976; Maheras & Kolyva-Mahera, 1990), lake levels (Marchand et al., 1988), annual river flows (Alemaw et al., 2001; Alemaw 2002; Giakoumakis & Baloutsos, 1997; Kite, 1993; Negum & Atha, 1992; and Srikanthan, et al., 1983).

An attempt has also been made to explore the factors that cause aquifer depletion and borehole failure that relate to:

a. increase in water abstraction;
b. decline in recharge and/or rainfall; and
c. decline in specific capacity.

This investigation looks at the possible processes causing groundwater level decline. Statistical correlation analysis has been applied between the groundwater level and selected hydrogeological variables recorded over the period 1995–2003 in sixteen boreholes ain the Gaotlhobogwe wellfield. The groundwater yield determinants that have been considered are: hydrogeological variables (i.e. water abstraction, recharge, rainfall, specific capacity), physio-chemical variables (i.e. Mg, Fe, Ca, Na, NO_3, HCO_3, CO_3, SO_4, pH, Ec and TDS) and micro-biological indicators such as the presence of iron and sulphur reducing bacteria.

Statistical correlation analysis has a wide range of applications. Moore et al. (1998) used it to correlate well yield with geologic features in New Hampshire, New England. It has also been used successfully to study factors that affect well yield in the fractured bedrock aquifer of New Hampshire (Moore et al., 2002) and in relating well yields to site characteristics in the same area (Moore et al., 2001).

13.2   HYDROGEOLOGY OF THE STUDY AREA

The Gaotlhobogwe wellfield is located NW of Gaborone, the capital of Botswana (Figure 13.1). The wellfield is situated alongside the ephemeral Gaotlhobogwe River which runs sub-parallel to the Letlhakeng-Gaborone tarred road. The wellfield covers an area of approximately 160 km^2 and supplies water to the villages of Molepolole, Thamaga and the

Thebephatshwa Air Base (DWA, 2000). Earlier work in this area shows that most of the geological and lithological successions penetrated by the boreholes are of the Karoo Supergroup largely overlain by the Kalahari Group.

Gaotlhobogwe is located on the southeastern fringe of the Central Kalahari Karoo Basin in which the uppermost units are absent. This is where the lower Karoo wedges out into pre-Karoo basement and also where an active groundwater system exists (DWA, 1995). At a regional level the groundwater gradient is from the south east to the north west and tends more to the north as flow approaches the Zoetfontein Fault that is located 25 km to the northwest of the study area (DWA, 1997). The gradient is steeper in the south east where it approaches the perceived recharge zone. The aquifer units exhibit confined conditions even in the absence of clear confining layers (DWA, 1997). This is due to the heterogeneous character of the units. Water strikes are generally shallow (50–60 m below ground level) below lower ground and 80–95 m below ground level beneath higher ground. The water quality is potable in most of the wells. Salinity increases down gradient.

13.3   METHODOLOGY AND DATA UESD

13.3.1   *Methodology*

The following techniques were applied to investigate the declining groundwater levels in the wellfield:

a. Trend analysis, using visualization followed by statistical trend analysis, used to quantify the water level decline over a number of years and its significance in the Gaotlhobogwe wellfield.
b. Analysis of potential factors for aquifer depletion in the wellfield which can be related to (i) possible increase in water abstraction; (ii) possible decline in recharge; (iii) decline in specific capacity; (iv) imbalance between yield and pumping rate; and (v) possible groundwater quality deterioration.
c. Correlation analysis to identify the hydrogeo-chemical determinants of groundwater level decline. Seasonal and annual changes in the chemical properties of the groundwater are tracked. The results from the chemical analyses as well the biological parameters were linked to water level decline and the correlation between them and the water level analyzed.

For an aquifer to maintain hydrologic balance and operate under sustainable conditions, the total output including abstraction must equal recharge plus or minus any change in storage. If the total output is more than the total input then the aquifer is said to be mined and is bound to depletion. This can cause a decline in the well yield. This problem persists in most of the wellfields in Botswana (DWA, 2000). Aquifer depletion can relate to the following factors: (1) increase in water abstraction; (2) decline in recharge or rainfall; and (3) decline in specific capacity.

13.3.2   *Data used*

The study is based on physical hydrogeological and groundwater quality data collected from ten observation wells and six pumping wells as part of the wellfield monitoring network. The various datasets used and the potential factors contributing to groundwater level decline are:

i. Increase in water abstraction: increase in borehole water abstraction can contribute significantly to the decline of groundwater levels. Annual water abstraction data for the various boreholes in the wellfield were obtained from the Department of Water Affairs (DWA). The annual water abstraction increases in proportion to demographic growth and water demand has increased between1998 and 2003.
ii. Potential decline of rainfall and groundwater recharge: decline in recharge will affect groundwater levels. Annual recharge values were determined for the Gaotlhobogwe

Table 13.1.   The extent of the decline in specific capacity for the pumping wells in the Gaotlhobogwe wellfield.

| Borehole ID | Specific capacity | | | |
	Start (1995) (m³/hr/day)	Recent (2003) (m³/hr/day)	Absolute decline (m³/hr/day)	Decline in specific capacity (1995–2003) (%)
7860	1.562	0.94	0.622	40
7864	1.637	0.622	1.015	62
7914	4.143	2.461	1.682	41
7931	1.408	0.812	0.596	42

Data source: DWA, 2003.

Table 13.2.   Well yield and pumping rate comparisons in the various boreholes.

BH ID	Estimated yield of the well (m³/hr)	Actual yield (m³/hr)	Current pumping rate (m³/hr)	Prevailing status
7858	60	33	29.1	Under-pumped
7860	30	20	25.9	Over-pumped
7864	35	15	14.4	Under-pumped
7914	75	40	39.5	Under-pumped
7931	50	35	34.4	Under-pumped
7966	80	72	71.7	Under-pumped

wellfield from the annual rainfall data using a model developed for the Karoo aquifer systems by Sami (2003). The model was based on chloride mass balance and an integrated surface and subsurface model. Groundwater recharge follows the rainfall pattern, but there is no obvious trend over the period of observation.

iii. Possible decline of specific capacity: a decline in specific capacity has been reported in the production boreholes in the wellfield and this is reflected by the declining groundwater levels. DWA (2003) studied decline at four boreholes and the results are shown in Table 13.1.

iv. Exceedence of design pumping rate: comparison has also been made between the actual yield of the production boreholes and their pumping rate over time. Table 13.2 summarizes these data which show that the boreholes are operated below their rated capacity as established from test pumping interpretation.

13.4   RESULTS AND DISCUSSION

13.4.1   *Evidence of possible recent decline in groundwater yield*

Results from the statistical significance of the trends obtained in the observation boreholes is summarized in Table 13.3. The slope of the linear trend model is declining (negative slope), and all the negative slopes were found to be statistically significant in all of the boreholes. The linear trend model seems to be an appropriate model to represent much of the variability in the historical groundwater level fluctuations as there is a high coefficient of determination (R^2) value except in Borehole 6613.

A plot of the annual variation of the observed water level fitted with a trend line and the corresponding residual diagram of errors after the linear model fit for BH 6615 are shown in Figure 13.2. The errors are more or less randomly distributed without any possible

Table 13.3. Slope and the statistical significance of the declining trend in water levels borehole.

Borehole ID	Location Latitude (Decimal degrees)	Longitude (Decimal degrees)	Slope of linear trend model [–]	Coefficient of Determination, R^2 (%)
6500	–24.1547	25.32667	–0.76	92
6514	–24.1325	25.17583	–0.47	74
6515	–24.1403	25.15611	–0.86	96
6516	–24.1492	25.15972	–0.89	89
6517	–24.1631	25.18444	–0.55	82
6609	–24.1911	25.23111	–0.23	64
6613	–24.1722	25.21694	–1.55	33
6614	–24.1325	25.17583	–0.47	91

unexplained variance in the observed water levels and not otherwise accounted for by the assumed linear model.

The visual plots and the results from the linear trend modelling reveal a significant decline in the groundwater table. The reported 40–60% decline in the specific capacity of the pumping wells can partly be explained by the significant drop in the groundwater level.

13.4.2 Correlation analysis

Pearson correlation analysis was conducted between groundwater level in m below groundwater level (m bgl) and the various hydro-geochemical parameters, which include: (1) abstraction, (2) specific capacity, (3), recharge, (4) rainfall, (5) pH, (6) EC, (7) TDS, (8) CO_3, (9) HCO_3, (10) SO_4, (11) Ca, (12) Mg, (13) Fe, and (14) Mn.

The correlation coefficients (r) and their statistical significance for four pumping wells are presented in Table 13.4. The confidence intervals for the prevailing correlations between the groundwater level and the hydro-geochemical variables in the wellfield are also listed. The data were first normalized for quantitative assessment of the correlation between the variables. The values of r range between –1 and 1, where values close to –1 and 1 indicate a strong negative and positive correlation, respectively. The larger confidence intervals correspond to low values of significance of the error unaccounted for by the correlation between the variables. Values of r that fall close to 0 signify weak correlation in which significant variation is not accounted for by the correlation.

All of the variables except iron, manganese, carbonate and TDS possess a weak positive association or no association at all with water level. Ca shows a weak positive association in Boreholes 7860, 7864 and in Borehole 7914, but a weak negative association with water level in Borehole 7931. The association between the water level and the hydrogeological variables is weak and negative showing that when there is high rainfall there is high recharge and the decline in water level is reduced. It can also be seen that there is a weak positive association between the water level and the abstraction indicating that an increase in abstraction leads to an increase in the water level drop.

13.4.3 Reliability analysis of groundwater monitoring data

The quantified association between water level and the various hydrogeological and geochemical variables was determined from the linear correlation coefficient, r (also called the product-moment correlation coefficient, or Pearson's r). The number of sampled records from the monitoring of the groundwater is short, i.e. from 1995–2003 with some missing data. In order to assess whether a correlation is significant, the test statistic, t, has been

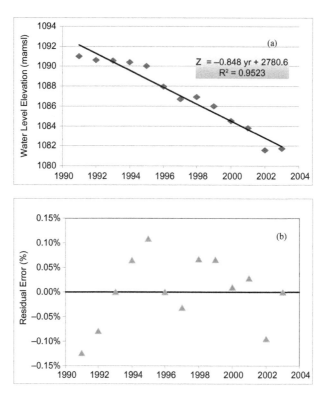

Figure 13.2.   (a) Water level fluctuations in observation well BH6615 with a fitted linear trend line and (b) Residual error (standardized) diagram for BH6615 in the Gaotlhobogwe wellfield after fitting linear trend model.

applied to identify the confidence level of the correlations between groundwater level and the hydrogeochemical variables. The test statistic, t, is defined by Hirsch et al. (1993) as:

$$t = r\sqrt{\frac{N-2}{1-r^2}} \tag{1}$$

which is distributed in the null hypothesis case (of no correlation) like Student's -distribution with $v = N - 2$ degrees of freedom. The confidence level, γ for $t(r, v)$ is then computed as given by Press et al. (1992) as follows:

$$\gamma(t,v) = \int_{-t(r,v)}^{t(r,v)} f(t,v)\, dt \tag{2}$$

By combining equations (1) and (2), the confidence interval, $\beta = (1 - \gamma)100\%$ is derived (see Figure 13.3 for the sample size of 6, which is the minimum number of concurrent data-sets attributed to the groundwater levels that were recorded for the 1995–2003. A confidence level γ indicates that $(1 - \gamma)100\%$ of the variability is attributable to random variability. Fortran routines provided in Numerical Recipes (Press et al., 1992) are used to compute the confidence intervals. Before applying the test, the data were normalized. It was also possible to demonstrate the likely inaccuracy in the t-test as a result of the small size of sample for a sample size of 6, 10 and 15. The correlation coefficient increases as the number of observed or monitored data sets increase which demonstrates the value of sustained and long-term monitoring data in groundwater monitoring programmes.

Table 13.4. Pearson correlation coefficients (r) and confidence interval (γ) between groundwater level and various hydro-geochemical variables.

Boreholes variables	BH 7864		BH 7860		BH 7931		BH 7914	
	(r)	(γ)	(r)	(γ)	(r)	(γ)	(r)	(γ)
Water level	1.000	–	1.000	–	1.000	–	1.000	–
Sp. capacity	0.004	0.503	−0.050	0.538	−0.329	0.787	−0.412	0.791
Abstraction	−0.093	0.569	0.019	0.514	0.510	0.902	0.641	0.915
Recharge	0.135	0.600	−0.368	0.764	−0.333	0.790	−0.466	0.824
Rainfall	−0.052	0.539	−0.538	0.865	−0.395	0.833	−0.357	0.756
pH	0.078	0.559	0.114	0.585	−0.699	0.973	0.668	0.927
EC	0.186	0.638	0.182	0.635	−0.005	0.504	−0.148	0.610
TDS	0.536	0.864	0.746	0.956	−0.436	0.860	−0.170	0.627
Ca	0.311	0.725	0.406	0.788	−0.379	0.823	0.457	0.819
Mg	0.453	0.817	0.266	0.695	−0.427	0.854	0.231	0.670
Fe	−0.129	0.597	−0.767	0.963	−0.293	0.759	−0.495	0.841
Mn	0.009	0.507	0.009	0.507	0.079	0.574	−0.618	0.905
CO_3	−0.878	0.989	0.988	1.000	−0.419	0.849	−0.049	0.536
HCO_3	0.061	0.546	0.607	0.900	−0.456	0.872	0.874	0.989
SO_4	0.047	0.535	0.114	0.585	−0.484	0.888	−0.368	0.763

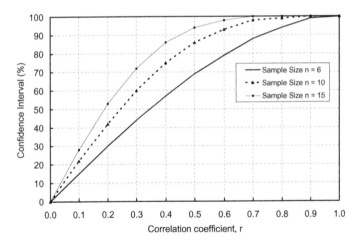

Figure 13.3. The effect of number of observations on the confidence interval of the value of correlation coefficients between monitored water level and hydrogeochemical variables.

13.5 SUMMARY, CONCLUSION AND RECOMMENDATION

The study revealed few conclusions regarding the assessment of the potential causes for the failure and decline of yield of the Gaotlhobogwe wellfield. Evidence from the statistical trends analysis and the visual presentation of time series graphs show that the water table in the wells is declining, and a consistent decline in the water level averaging 1.6 m per annum is recorded. The coefficient of determination of a fitted declining trend model for the ground-water level reaches the range of 64% to 96%.

The results from statistical correlation analysis reveal that the correlation of groundwa-ter level with recharge and specific capacity is negative and very weak. Among the hydro-chemical variables, a weak positive association exists between Ca and groundwater level. CO_3,

TDS, Fe and Mn are the only components which show a strong association with water level. This association is positive in some wells and negative in others. Calcite was identified as the substance responsible for clogging some boreholes in the wellfield. In particular in Borehole 7931, located in the western end of the low lying area of the wellfield, clogging caused by precipitation of calcite is inhibiting yield. This feature was identified using XRD analysis and thermodynamic data on calcite.

It can be concluded from the statistical correlation analysis that the decline in groundwater level is more strongly related to hydro-geochemical parameters than it is by the hydrogeological variables such as recharge and specific capacity. The important physical parameters governing the state of the various cations and anions are pH, Ec and TDS. Of these the most important is pH as it governs the solubility of minerals into the water system and their precipitation out of solution.

Calcium shows a strong positive association with the water level in all the boreholes. This may be because it is capable of forming precipitates with anions like CO_3 which can block the groundwater flow to wells. Well performance is affected by a number of factors. Chief among these is the aquifer geochemistry, which in turn influences the chemical make-up of the groundwater. Other contributory factors include the way the wells are designed, constructed and operated. Groundwater monitoring data for the rural supply wellfields in Botswana are extremely valuable. The data can be used to demonstrate the status of the wellfield and the groundwater resource and identify causal relationships. There are a number of obvious ways of improving the national monitoring network, including increased coverage and baseline monitoring in wellfields where no significant monitoring has yet been initiated.

The monitoring effort should not just be limited to physical and geochemical parameters. Microbiological monitoring should be considered for further evaluation of the impact from microorganisms on groundwater and wellfield (see Cullimore, 2007).

The Botswana Groundwater Quality Protection Programme, initiated in the early 1990s, should be expanded by further re-evaluating its effectiveness both potential and otherwise. The programme employs two approaches (Chaoka et al., 2004): (1) delineation of wellhead protection zones, and (2) groundwater quality vulnerability assessment. This approach needs to be rolled out to all aquifers and monitoring intensified to improve the groundwater quality monitoring network in order to underpin groundwater management decisions.

The tradeoff between the cost of monitoring and the optimum number of monitored variables and their frequency of measurement should be re-evaluated periodically to ensure a sustained groundwater monitoring plan which provides prudent benefits towards the sustainable groundwater management of all rural water supply systems.

ACKNOWLEDGMENTS

The authors would like to thank the Departments of Water Affairs and Geological Surveys of the Government of Botswana for providing assistance and hydrogeological data. The University of Botswana is also acknowledged for support for an MSc student who contributed to this study.

REFERENCES

Alemaw, B., Fanta, B., Zaake, B.T. and Kachroo, R.K. (2001). A study of variability of annual river flow of the southern African region. *Journal of Hydrological Sciences.* Vol. 46, No. 4, Aug. 2001. pp. 513–523.

Alemaw, B.F. (2002). Discussion on 'A study of variability of annual river flow of the southern African region'. *Journal of Hydrological Sciences –Journal–des Sci. Hydrologiques,* 47(6) December 2002, pp. 983–989.

Chaoka, T.R., Alemaw, B.F., Molwalelfhe, L. and Moreomongwe, O.M. (2006). Investigating the causes of water-well failure in the Gaotlhobogwe wellfield in southeast Botswana. *J. Appl. Sci. Environ. Mgt.* 10(3), 59–65.

Chaoka, T.R. Shemang, E.M. Alemaw, B.F. and Totolo, O. (2004). Impediments to the effective implementation of a groundwater quality Protection strategy in Botswana. In eds: D. Stephenson, E.M. Shemang, T.R. Chaoka. Water Resources of Arid Regions (Proceedings in the International Conference on Water Resources of Arid and Semi-arid Regions of Africa). Taylor and Francis Group plc, London, UK pp. 551–558.

Comani, S. (1987). The historical temperature series of Bologna, Italy from 1716 to 1774. *Climate change*, 11(3): 375–390.

Cullimore, D.R. (2007) Practical Manual of Groundwater Microbiology, 2nd ed. SRC press/ Taylor & Francis Group. 400p.

DWA (1995). Water supply System and Immediate Sanitation measures for the Villages of Kanye and Molepolole, Gaotlhobogwe Valley Area, Phase 1, Final Report, Wellfield Consulting Services (Pty) Ltd. Gaborone. 53p.

DWA (1997). Water supply System and Immediate Sanitation measures for the Villages of Kanye and Molepolole, Gaotlhobogwe Valley Area, Groundwater Resource Evaluation, Phase 3. Final Report, Wellfield Consulting Services (Pty) Ltd. Gaborone. 130p.

DWA (2003). Water supply system and immediate sanitation measures for the villages of Kanye and Molepolole. Gaotlhobogwe wellfield valley Phase 3 Final Report.

DWA (2000). Groundwater Monitoring. Vol. 1. Main Report. Geotechnical Consulting Services.

Gazemi, G.A. (2004). Temporal changes in the physical properties and chemical composition of the municipal water supply of Shahrood, northeastern Iran. *Journal of Hydrology*, Vol. 12, number 6. pp. 723–734.

Giakoumakis, S.G. and Baloutsos, G. (1997). Investigation of trend in hydrological time series of the Evinos River basin. *Hydrological Sciences Journal*, 42(1), 81–88.

Hirsch, R.M., Helsel, D.R., Cohn, T.A. and Gilroy, E.J. (1993) Statistical analysis of hydrologic data. *Handbook of Hydrology*, D. R. Maidment, Ed., McGraw-Hill, 17.1–17.55.

Kite, G. (1993). Analyzing hydro-metrological time series for evidence of climate change. *Journal of Hydrology*, 24, 135–150.

Maheras, P. and Koyva-Mahera, F. (1990). Temporal and spatial characteristics of annual precipitation in the twentieth century. *Journal of Climatology*. 10, 495–504.

Marchand, D., Sanderson, M., Cavadias, G. and Varianos, N. (1988). Climate change and Great Lake levels-the impact on shipping. *Climate Change* 12(2), 107–134.

Moore, R.B., Clark, S.F., Ferguson, E.W., Marcoux, G.J. and Degnan, J.R. (1998). New Hampshire Bed-rock assessment: Correlating Yield from 18,000 wells with geologic features, *in* North East Focus Groundwater Conference, Burlington, Vt., October 20–21, 1998, Proceedings: Burlington, Vt., National Groundwater water association. 18p.

Moore, R.B., Schwartz, G.E., Clark, S.F. Jr., Walsh, G.J., Degnan, J.R. and Mack, T.J. (2001). Factors Related to Well Yield in the Fractured Bed-Rock Aquifer of New Hampshire. United States Geological Survey (USGS), paper 1660.

Moore, R.B., Schwartz, G.E., Clark, S.F., Walsh, G.J. and Degnan, J.R. (2002). Relating well yield to site characteristics in fractured bed-rock of New Hampshire: Abstracts with programs. Geological Society of America, *in* Geological Society of America, 2001 Annual Meeting, November 2001.

Negum, A.N. and Atha, B.B. (1992). A power spectrum analysis for the Nile's natural flow time series, *in Climatic fluctuations and water management*, Butterworth, Worster, Britain.

Press, W.H., Flannery, B.P., Teukolsky, S.A. and Vettering, W.T. (1992) *Numerical Recipes: The art of scientific computing.* Cambridge University Press. New York. ISBN 0-521-43064-X.

Rodhe, H. and Virji, H. (1976). Trends and periodicities in East African rainfall, *Monthly Weather Review* Vol. 104, 307–315.

Sami, K. (2003). A comparison of recharge estimation in a Karoo aquifers from a Chloride Mass Balance in groundwater and an integrated surface and a subsurface model In: Groundwater Recharge estimation in South Africa Xu, Y., and Beekman, H.E., eds. UNESCO-IHP Series No. 64, 207p.

Srikanthan, R., McMahon, T.A. and Insh, J.L. (1983). Time series analysis of annual flows of Australian streams, *Journal of Hydrology.* 66, 213–226.

14

Cost-effective boreholes in sub-Saharan Africa

K. Danert
SKAT, St Gallen, Switzerland

R.C. Carter
Water Aid, London, UK

D. Adekile
Water Surveys and Resources Development Limited, Kaduna, Nigeria

A. MacDonald
British Geological Survey, Edinburgh, UK

ABSTRACT: A common assertion is that the cost of water well drilling in sub-Saharan Africa is too high and that construction quality is regularly compromised. Over the last 20 years, several studies regarding this have been undertaken, covering more than ten countries in the region. Although drilling costs in sub-Saharan Africa are generally higher than in India, there are valid reasons for this. However, changes to borehole designs, procurement and contract management practices, well clustering for economies of scale, siting and supervision practices as well as support to and professionalization of the private sector can all serve to bring drilling costs down, and improve construction quality. This chapter provides an overview of how drilling costs can be calculated. It pulls together the key issues that affect drilling costs and prices into a conceptual framework. The framework is subsequently used to compare policies and practices for the countries where information is readily available. The chapter thus intends to raise awareness and improve the analytical capacity of implementers and decision-makers regarding measures that could be adopted to improve the cost-effectiveness of borehole drilling in their particular context.

14.1 INTRODUCTION

It has been estimated that about 35,000 boreholes per year need to be drilled in sub-Saharan Africa to meet the MDGs for domestic water supply. This is based on Joint Monitoring Programme (JMP 2004) data: 412 million people served in 2004: MDG of 701 million people served in 2015 and full coverage of 1625 million served. Assumptions: 50% of people will be served with a hand dug well, treated surface water or spring; 37.5% of people will be served with a handpump (300 people per pump) and 12.5% with a mechanised borehole (2,000 people per system). Assumes 3% of existing boreholes are re-drilled annually. If one considers full coverage by 2050, and water for irrigation as well as industrial supply, at least 50,000 boreholes per year are required.

Government, private enterprises, NGOs and donors have all raised concerns about the high costs, variable construction quality and the inadequate volume of boreholes drilled in sub-Saharan Africa (e.g. in Kenya, about 250 wells are drilled annually, compared to the required 650 to meet targets (Doyen, 2003); in Tanzania, the investment plan provides for 1,600 boreholes annually, requiring a doubling in capacity (Baumann et al., 2005)).

Concerns regarding the disparity between the relatively low costs of handpumps and the high costs of drilled wells were raised at the UNDP-World Bank International Handpump

Workshop in 1992 (Doyen, 2003). Cost savings on conventional drilling of as little as 10% would have a significant impact on extending access to improved water supplies. Use of manual drilling where feasible could also extend access at very low cost. However, it is essential that cost-savings do not adversely jeopardise quality and that water well infrastructure

Table 14.1. Estimated and actual drilling cost and prices.

Country, year (ref)	Cost/Price per: well	meter	Description
Burkina Faso 2006 (ANTEA 2007)		$152	Average cost of drilling and installation of casing and screen (PVC) but not the pump, as established by study of drilling costs.
Chad 2005 (Practica, 2005)	$12,000–15,000		Range of machine drilled well prices paid by different agencies.
Ethiopia, 2005 (Carter, 2006)	$37,800	$252	Estimated price for a 200 mm diameter, steel cased borehole to 150 m. No pump or supervision (based on analysis of inputs).
Kenya, 1996 (Doyen, 2003)	$8,400	$120	Price estimated for 70 m well in specific programme (includes drilling, testing but not siting, supervision or failure).
Malawi, 2001 (Mthunzi, 2004)	$2,730	–	Estimated average well cost including capital, recurrent, personnel & materials; assuming 45 wells per year with small rig by NGO.
Niger, 2005 (Danert, 2005)	$10,000	$160	Estimated price on a bill of quantities, 60 m depth, 700 km from capital city, excluding supervision and pump installation.
Mozambique, 2006 (WE Consult, 2006)		$151	Average drilling price according to the report. Includes siting, pump installation and VAT.
Nigeria, 2006 (Adekile, 2007)	$11,700	$195	Federal Ministry of Water Resources 2006 borehole price. PVC lined, 60 m depth fitted with handpump.
Nigeria, 2008 (Adekile et al., 2008a)	$6,000	$120	Estimated price for a 110 mm diameter, PVC lined borehole to 50 m depth without pump or supervision (based on analysis of inputs).
Nigeria, 2008 (Adekile et al., 2008a)	$2,140	–	Hand drilled, 110 mm, PVC lined.
Senegal, 2006 (ANTEA, 2007)	–	$500	Average cost of drilling and installation of casing and screen (stainless steel) but not the pump, as established by study.
Tanzania, 2004 (Baumann, 2005)	$6,000	–	Budget for borehole with a handpump, as in the National Rural Water Supply and Sanitation Programme (2004), Main Report V 1.
Uganda, 2007 (MWE, 2007)	$8,700	–	Average price of private sector drilled deep boreholes (with handpumps) paid for by district local go.vernments in F/Y 2006/7.
Nigeria, 2008 (Adekile et al., 2009)	$500,000	$2,500	Contract price for 200 m borehole in River State

can be sustained over the long term. Further, in order to attract private investment, water well drilling must be a viable business venture.

In order to improve the health of the borehole drilling sector, decision-makers and implementers should be able to easily identify issues that reduce efficiency and compromise quality. This chapter sets out a conceptual framework for cost-effective boreholes and analyses policies and practices from several countries including Burkina Faso, Chad, Ethiopia, Kenya, Nigeria, Niger, Malawi, Mali, Mauritania, Senegal, Tanzania and Uganda.

14.2   ASSERTIONS, INFORMATION AND EVIDENCE OF HIGH DRILLING PRICES

It is essential to distinguish between borehole 'price' and borehole 'cost'. 'Price' refers to the amount paid by the Government or a particular project for a successfully completed borehole, whereas 'cost' is borne by the drilling enterprise. The difference between the two is the sum of overheads, taxes, profit and a margin for risk (e.g. dry holes, payment delays, insecurity and breakdown).

Accurate information on drilling prices or costs in sub-Saharan Africa is not easy to access (ANTEA, 2007). Systematic analysis is a challenge because there is poor, fragmented and non-standardized record keeping of water supply projects and programmes in sub-Saharan Africa as well as lack of transparency. Table 14.1 provides examples of estimated and actual borehole costs and prices, ranging from $2,000 to $500,000 ($120 to $1,271 per meter).

Some wells are shallow, and installed with a handpump while others are deep and use a motorised pump. There is variation in how they are calculated (e.g. what aspects are included or left out such as siting or supervision). The type and size of casing varies, as does the geology, distance travelled and equipment. This means that one cannot make simplistic comparisons of borehole prices between counties, or within the same country.

Carter et al. (2006), WE Consult (2006), ANTEA (2007) and Adekile and Olabode (2008a) emphasise that every borehole is unique. Estimates of borehole prices are vital for budgeting but averages hide more than they reveal. International and in-country benchmarking would be useful to consider value-for-money in service delivery but simple league tables of national drilling costs are not very useful for driving down the price of boreholes and ensure construction quality. In order to make useful comparisons, a standard accounting framework is needed, as well a methodology for modelling the effects of key variables on overall cost.

14.3   CONCEPTUAL FRAMEWORK

It is essential to understand and address the parameters that affect cost-effectiveness in context (ANTEA, 2007; Carter et al., 2006). Borehole costs and quality are primarily influenced by six core factors and thirteen elements as set out in the conceptual framework in Figure 14.1 and discussed below. Note that this framework builds on previous work by Wurzel (2001), Smith (2003), Ball (2004), Carter (2006), Carter et al. (2006), Danert (2008) and Adekile and Olabode (2008b).

14.3.1   *Borehole costs and quality*

The recommended way to analyse borehole cost is to examine each of the following components:

- Mobilisation—all costs involved in transporting equipment to site and back to base (Box 1).
- Drilling—allows for the per-hour (converted to per-meter) costs of equipment depreciation, labour consumption of fuel, lubricants and drill fluids and replacement of drilling tools. Affected by depth; diameter; drilling and standby time (Box 2).

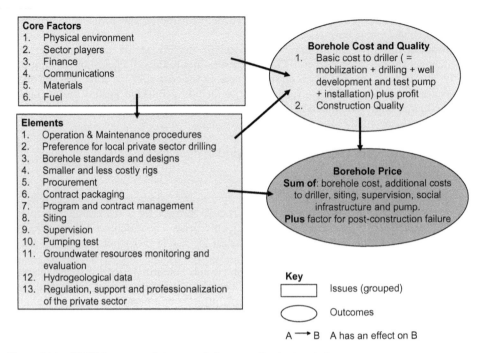

Figure 14.1.   RWSN conceptual framework for cost-effective boreholes.

Box 1.   Mobilisation cost component.

The table below shows the mobilisation (and demobilisation) cost for a hypothetical project, 100 km from the contractor's base. Two examples are given: (a) equipment purchased at US$ 170,000 and (b) equipment purchased at US$ 85,000.

Mobilisation	Calculation method	Amount (a)	Amount (b)
Capital equipment depreciation	Cost of rig, freight insurance, loan charges converted to daily cost based on a 10 year lifespan and 60% utilisation	$78	$39
Vehicles	Rental market rate/real running cost of: (a) 2 trucks and 1 pickup; (b) 1 truck and 1 pickup.	$297	$186
Fuel and lubricants	For a travel distance of 100 km (20 l of fuel/vehicle at $0.58/l)	$46	$35
Human resources	Salaries and per diems on daily basis for hydrogeologist, driller, assistant driller, 2 labourers, security person and 1 driver (rig a)/2 drivers (rig b).	$149	$141
Sub-total mobilisation		*$570*	*$401*
Sub-total demobilisation	Estimated at 80% of mobilisation cost	*$456*	*$321*
Total		**$1,026**	**$721**

Clearly, it is considerably more economical to spread the mobilisation cost over 10, 20 or 50 wells (i.e. a clustered contract with wells relatively close in distance) than to pay this amount for each individual well drilled.

Box 2.　Drilling cost component (adapted from Rowles, 1995).

Item		Explanation	Cost (US$)	
Capital equipment		Cost of rig, freight insurance, loan charges	$170,000	

Fixed costs	Lifetime (Hours)			Cost per hour
Depreciation	20,000	Capital cost divided by lifetime (i.e. 10 years at 60% utilisation) US$/h		$8.50
Maintenance	–	Maintenance (5% of depreciation) US$/h		$0.43
Labour	–	US$/h		$17.00
Fuel and Lubricants	–	US$/h		$10.00
Mud/foam	–	US$/h		$13.00
Sub-Total		*Sum of above US$/h*		*$48.93*
Cost per meter		*Convert to US$/m by dividing by drilling speed*		

Variable costs	Lifetime (Meters)	Explanation	Replacement cost	Costs per meter
Drilling string	20,000	Convert to US$/m by dividing replacement cost by lifetime	$15,254	$0.76
Hammer	3,000	Convert to US$/m by dividing replacement cost by lifetime	$8,136	$2.71
Hammer bit	300	Convert to US$/m by dividing replacement cost by lifetime	$1,186	$3.95
Drag bit	300	Convert to US$/m by dividing replacement cost by lifetime	$508	$1.69
Sub-Total Rock		*Sum of drill string and drag bit US$/m*		
Sub-Total Overburden		*Sum of drill string, hammer and hammer bit US$/m*		

Example

Formation	Depth (m)
Overburden depth (m)	20
Rock depth (m)	30
Total depth (m)	50
Drilling speed (m/h)	3

Calculation		Cost
Fixed costs	= 50 m × (48.93/3)	$815
Variable cost: overburden	= 20 m × (0.76 + 1.69)	$49
Variable costs: rock	= 30 m × (0.76 + 2.7 + 3.95)	$227
Total Cost—Drilling		**$1,091**
Drilling cost per m		**$22**

- Installation—includes the supply and installation of plain casing and screen, gravel pack, sanitary seal and well-head construction.
- Well development refers to the cleaning of the borehole after construction and test pumping is the post-construction assessment of borehole and aquifer performance.

The time taken to undertake these activities affects the basic drilling costs. The average execution time for a borehole in Burkina Faso is 2 days and 45 days in Senegal (ANTEA, 2007). Note that while savings on say casing can have a considerable effect on the installation cost, the proportion saved on the total cost depends on how much the installation component affects the total construction cost.

Construction quality refers to the degree to which the borehole is straight; the quality of well development and gravel packing; the casing/screen quality including its installation; the permeable backfill material and placement; the quality of the sanitary seal and head works. From the user perspective turbid water, low flow rates and seasonal functionality are a poor quality of service.

14.3.2   Borehole price

Figure 14.1 differentiates between borehole cost and borehole price. The borehole price includes:

- Borehole cost (as described in section 14.3.1)
- Additional costs to the driller (e.g. taxes, overheads and kickbacks) plus profit. An astute driller will assess the requirements for a particular tender, consider the risks involved and load particular items in the Bill of Quantities accordingly (Carter et al., 2006).
- Pump costs. These vary and may or may not be included in quoted borehole prices.
- Siting costs. These can be borne by the programme, driller or consultant. In the case of the latter, the costs are more visible. In cases where supervision is undertaken by programme or Government staff, the costs are often concealed.
- Supervision costs are can be borne by the programme, or consultants (as for siting).
- Costs of Social Infrastructure refers to community mobilisation and training. These costs are sometimes hidden within programme expenditure.

14.3.3   The core factors

The core factors indicated in Figure 14.1 are independent variables that cannot easily be influenced but have a bearing on the cost of boreholes. They are summarised in Table 14.2. It is important to understand them and be realistic about the extent to which they can be changed in a given time frame, if at all.

14.3.4   Key elements

Given that the core factors change very little, if at all, in order to improve borehole cost effectiveness it is essential that proper attention is paid to the 13 elements given in Figure 14.1. The following list outlines the basic principles that should be adhered to with respect to these 13 elements. These principles are drawn from on-going work to develop a code of practice for cost-effective boreholes.

A list of Key Elements for Cost-Effective Boreholes:

1. ***Operation and maintenance (O&M) procedures*** to ensure the sustainability of pumped groundwater sources for the expected lifetime of the facility should be established, adhered to and monitored.
2. ***Who drills?*** The preferred option is that local private sector enterprises undertake construction of water wells and pump installation. This should encourage in-county capacity to grow and foster competition.

Table 14.2.   Core factors that affect drilling costs.

Physical Environment (geology, hydrogeology, climate)	Water well construction in different formations has different requirements in terms of equipment, casing and depth requirements. If plentiful groundwater is available at shallow depths, it can be cheaper to drill than for deep groundwater. If formations are soft, and groundwater is within the first 15 to 20 m, low cost hand drilled wells may be feasible. Rainfall and recharge affect groundwater availability and sustainability. Although the physical environment cannot be changed, the understanding of it by sector professionals and practitioners can be improved.
Sector players and sector structure	There are numerous sector reforms throughout the region comprising a shift to public sector coordination, regulation, and policy formation with private sector implementation (e.g. Uganda, Ethiopia, Malawi and Ghana). These provide more opportunities for private drilling. There is a growing interest in the Sector Wide Approach (SWAp), although forms vary, which can make the market for borehole drilling more coherent and transparent. Where major structural changes are taking place roles and responsibilities are in a state of flux and the introduction of cost-saving measures is not always easy. In most countries there are numerous discrete water supply projects being implemented with different objectives, standards and conditionalities. A coherent legal and regulatory framework is key for cost-effective borehole provision but takes time to develop and be enforced.
Communication networks	Road networks are often poor, particularly in remote rural areas. This can render large parts of the country inaccessible for the rainy seasons (e.g. South Sudan), which impacts on equipment down time and amortisation costs. Telecommunications is changing rapidly. Mobile phone technology can have a huge impact on decision-making of junior field staff, and thus impact on waiting times considerably.
Finance	Coherent finance for investment in water supply development and maintenance over several years provides continuity of work and thus encourages investment by private drilling enterprises in equipment and human resources.
Materials and equipment	The cost and availability of materials such as casing, gravel pack, cement and drilling fluids and equipment such as drilling rigs and spares varies widely. Some countries are fortunate that casings are manufactured in-country while others have to import. Landlocked countries tend to be at a disadvantage as the raised cost of transport renders everything more expensive. This has to be carefully considered when making any comparisons.
Fuel	Fuel prices vary widely, not only over time (as we have seen over the last two years with fluctuating prices for a barrel of oil), but also between and within countries. Some countries (e.g. Egypt) have for many years operated a policy whereby oil prices are highly subsidized. In other countries, fuel taxation is a key contributor to Government revenues. In some areas, fuel is not readily available and mainly sold, at a higher price from plastic containers by the roadside. Kano, Nigeria is a case in point.

3. ***Standards and design***: Boreholes should be designed and constructed so that they are fit for their intended purpose in terms of diameter, depth, casing and screen.
4. ***Drilling equipment***: Smaller and less costly rigs should be utilized to provide boreholes that are fit for their designed purpose. Manual drilling should be brought into the mainstream of water supply programmes, with appropriate quality control.
5. ***Procurement***: Systematic, transparent and timely processes of advertising, pre-qualification, tendering, evaluation and award need to be established and followed.
6. ***Contract Packaging***: Contracts should be packaged for multiple boreholes in close proximity and for boreholes with similar geology.

7. **Programme and contract management**. It is essential that drilling programmes have sufficient skills to design and manage the programmes or bring in expertise. Payment for works must be timely.
8. Appropriate *siting* practices should be utilized.
9. High quality, timely **construction supervision** should be emphasized.
10. **Test pumping requirements** should be matched to borehole purpose while taking into account the importance of data to improve the understanding of hydrogeology and water resources.
11. Rigorous evaluation of **groundwater resources** should be undertaken and information made available.
12. **Hydrogeological data** collection and storage should be undertaken.
13. **Regulation and private sector professionalism**: A strong public sector is needed to oversee and regulate the private sector. The private sector needs better access credit and should professionalise.

14.4 ANALYSIS OF THE THIRTEEN ELEMENTS OF COST-EFFECTIVE BOREHOLES

This chapter sets out each of the key elements of CEB and provides examples of policies and practices from within the continent.

14.4.1 *Operation and maintenance procedures*

Operation and maintenance (O&M) procedures to ensure the sustainability of pumped groundwater sources should be established, adhered to and monitored.

Post-construction failure increases actual borehole costs significantly. A 50% failure rate effectively doubles the well price. Unfortunately broken down handpumps and abandoned boreholes are a frequent site across the continent. An estimated 30% to 50% of installed facilities in Nigeria are broken down at any one time (Adekile and Olabode, 2008b). Comparable figures for Malawi and Uganda are 30% and 20% respectively. Reliable and comprehensive data in this regard is lacking, but recent water point mapping work (e.g. Malawi, Angola) is capturing more information. Alas, update mechanisms are often weak.

Unless initial construction quality is high, water is of an acceptable quality and long term operation and maintenance procedures are established and adhered to, cost-effective borehole provision will never be realised. Drilling programmes often neglect the much-needed community sensitisation and mobilisation aspects. Water users rarely contribute more than a small proportion of the capital cost towards construction of the borehole, if anything, and ownership tends to be unclear. The development of and support to social infrastructure is often neglected and spare parts are frequently not available. These, combined with lack of follow-up support (e.g. to retrain committees and mechanics; ensure spares are available) contribute to poor operation and maintenance and thus broken down sources. Uganda has developed an operation and maintenance framework and a similar initiative started in 2008 in Malawi. Standardisation of hand pumps to two or three types has been undertaken in several countries in order to simplify maintenance procedures and reduce the different types of spares required.

14.4.2 *Who drills water wells?*

The preferred option is that local private sector enterprises undertake construction of water wells and pump installation. This should encourage in-county capacity to grow and foster competition.

In order to keep drilling costs down, a rig should be used for about 220 days per year (60% of the time) and be subject to regular maintenance and repair. This equates to drilling 20,000 hours over a ten year period. Unfortunately such high usage is rarely achieved by State-owned equipment. There are numerous rigs lying idle in Government yards, broken down or rarely used. There is a growing consensus (by the World Bank, several bilateral donors and African Governments) that private sector drilling tends to be more efficient and effective than direct implementation by the State. Governments and donor support agencies are encouraged to provide support so that the private sector can be built up rather than supporting the purchase of State-owned drilling rigs. Moves towards more private sector drilling vary widely in sub-Saharan Africa (Table 14.3) and is usually part of wider policy reforms. Note that it may be desirable for Government to retain at least some minimum drilling capacity to deal with emergency situations. There is a grey area with respect to NGO drilling wells directly. Clearly, this reduces the market for private drilling. Competition between NGOs and the private sector is clearly unfair if NGOs are able to cross-subsidise, or benefit from tax exemptions.

14.4.3 *Borehole standards and designs*

Wells should be designed so that they are fit for their intended purpose (Carter et al., 2006). This means that the diameter, depth, lining and backfill materials, screen open area and other design features should be well-matched to need (expressed as water demand, longevity, hydraulic efficiency and cost). Differentiating between different magnitudes of abstraction requirements is particularly important. Unfortunately, this is not always the case, as shown by the examples below.

Doyen (2003) points out that often, wells drilled for rural handpump are being constructed to give high yields, and are forced to conform to higher standards necessary. Well yields of 0.25 ℓ/s are adequate for hand pump wells.

Handpump boreholes diameter requirements and the small diameter submersible pumps that are now on the market mean that 4″ (102 mm) internal diameter boreholes are usually sufficient. However, diameter requirements vary considerably between countries:

- Tanzania—the internal diameter for deep and shallow wells are specified at 150 mm and 117 mm respectively.
- Mozambique—4″ casing is installed, but there are no official standards (We Consult, 2006).
- In Uganda 4–5″ casing is specified (MWE, 2007a).

Table 14.3.   Organisations undertaking water well drilling in Sub-Saharan Africa.

	Who drills?
Ethiopia	State enterprises are often the first choice for the Regional Bureau; private sector comprised 23 contractors with 64 rigs in late 2005; eight NGOs had 11 rigs in late 2005 (Carter, 2006).
Malawi	Drilling is undertaken by the State (mainly at regional level) as well as by 10 to 20 private drilling companies.
Nigeria	Nearly all drilling done by private sector although some Government agencies (e.g. Kano State RUWASSA) also construct water wells in-house. There are hundreds of private well drillers in Nigeria (Adekile and Olabode, 2008b).
Tanzania	The Drilling and Dam Construction Agency (DDCA) employs many well-trained drillers and hydrogeologists. It covers about 60% of the drilling market. DDCA staff skills are underutilised while private sector consultants are still lacking (Baumann et al., 2005). Private enterprises have drilled an estimated 9,000 private boreholes in Dar Es Salaam.
Uganda	All drilling done by private contractors.

- Six inch casing is used in Ethiopia, although drilling diameters are often 10″ or 12″ (Carter, 2006).
- Malawi specifies the installation of 110 mm casing (Mthunzi, 2004).
- In Burkina Faso and Senegal, final drilling diameters are 8″ and 12″ respectively (ANTEA, 2007).
- In Nigeria, there are five different borehole designs depending on the geology and aquifer depth in different parts of the country (Adekile and Olabode, 2008b).
- In Kenya, well diameters for boreholes with handpumps are 152 mm (Doyen, 2003).

In countries where boreholes are drilled into stable basement formation, it is possible to make savings by casing the collapsing formation only, grouting at the joint to the hard formation only and not casing the hole drilled into the basement, as is the standard in Uganda (MWE, 2007a). In Tanzania, all boreholes are fully cased and gravel packed, although Baumann et al. (2005) state that the specifications are not very precise. In Nigeria, boreholes are lined to the full depth. Concerns about silting of partially cased boreholes have been raised. A study in Malawi (Mthunzi, 2004) of 60 partially cased and 23 fully cased boreholes found that 73% of the partially cased boreholes had no depth reduction over 4–6 years and that 5% of boreholes showed an increase exceeding 5% of datum depth. Borehole yields were comparable for both types.

In Kenya, drillers lobbied Government for six years to relax the drilling specifications and thus drilling and rig costs but did not succeed. Part of the rational for this are plans to upgrade these sources to motorised pumps with small piped distribution systems at a later stage. However, given the enormous challenge of meeting the MDGs, the paucity of finance and difficulty in maintaining existing rural water supplies, such thinking may be too advanced for many countries. Higher levels of abstraction also raise questions with respect to water resources.

Drilling beyond the optimum yield depth is common, with examples documented in Ethiopia (Carter et al., 2006), Kenya (Doyen, 2003) and Nigeria (Adekile and Olabode, 2008b). Doyen (2003) estimates that cost savings of 25% could be made in Kenya if drilling was not beyond the optimum yield depth. In the basement complex, a geophysical survey can provide a good indication of depth requirements; for sedimentary formation existing drilling records can be used to determine realistic drilling depths. There is need for close on-site supervision, with the supervisor having the confidence and authority to decide when depth is sufficient. It is envisaged that the increased cost of better supervision would ultimately be offset by reduced drilling costs and improved construction quality.

14.4.4   *Drilling equipment—smaller and less costly rigs*

It is preferable that smaller, less costly equipment be used to match fit for intended purpose borehole designs. Manual drilling should be brought into the mainstream of water supply programmes, with appropriate quality control.

Borehole costs are affected by the type of equipment used, with cheaper and lighter equipment resulting in lower mobilisation costs. Box 2 shows the drilling component for equipment costing US$ 170,000. However, the total borehole cost also includes mobilisation, installation and pump test. Ball (2004) compares drilling with equipment costing US$ 470,000 and US$ 95,000 and estimates that the price per borehole (including overheads) for the larger rig is $ 8,837, while boreholes with the smaller rig cost US$ 2,652 (a factor of 3.3).

In many countries (e.g. Kenya, Ethiopia, Mozambique, Niger), the rigs in use are oversized for the purpose of drilling rural handpump boreholes (Doyen, 2003; Carter et al., 2006; WE consult, 2006; Danert 2007). In Mozambique, only NGOs use light rigs while private enterprises use large conventional rigs (WE Consult, 2006) whereas in Nigeria, half of the rigs encountered on a study by Adekile and Olabode (2008b) were classified as light to medium and 30% were locally manufactured.

There is a tendency to overestimate required well depth and over-drill, or specify large rigs which have a bearing on the equipment that drilling enterprises decide to buy (Carter et al., 2006). If a contractor can only invest in one rig, he may purchase the largest possible rig, to provide flexibility. Discussions with Government stakeholders and drillers in Niger (Danert, 2005) and Ethiopia revealed a lack of awareness of new light conventional rigs on the international market. Stakeholders may be aware of equipment exists but unsure its capability and wary of claims made by manufacturers. Improved access to reliable information on drilling equipment is essential.

Baumann et al. (2005) state that most drilling operators in Tanzania use old equipment, with the result that breakdowns are frequent and the performance is slow. Most of the drilling equipment in Senegal, Burkina Faso, Mali and Mauritania is old (some over 30 years) and lacks adequate maintenance (ANTEA, 2007). 68% of drilling rigs in Ethiopia are older than 15 years (Carter et al., 2006). Maintaining ancient equipment is costly and time consuming and the wide variety of rigs in use means that spares need to be sourced from all over the world. However, lack of initial capital can seriously limit ones options with respect to drilling equipment purchase. In Nigeria, a drill rig manufacturing industry is growing, with conventional rigs available at a much lower cost than for imported equipment.

Manual (or hand) drilling techniques can provide a viable alternative in particular environments (soft formation and shallow groundwater). A preliminary analysis of the potential for hand drilled wells in terms of geology and hydrogeology estimates that 12% of the total population of sub-Saharan Africa (SSA), or 18% of the rural population of SSA, could be served with hand-drilled wells (Danert, 2007). Adekile and Olabode (2008b) found that the cost of a manually drilled hole in Nigeria was about one third of a conventionally drilled hole. While Practica (2005) claim that in Chad, they cost a tenth of machine-drilled wells.

Manual drilling techniques are used in Niger, Benin, Burkina Faso, Nigeria, Chad, Ethiopia, Mozambique, Malawi, Madagascar, South Africa, Senegal and Tanzania. In Nigeria, there are an estimated 30,000 hand drilled wells in existence (for domestic and irrigation water supply). Apparently in some parts of Chad and Nigeria, conventional drillers win contracts and sub-contract the work to hand drillers.

A concern raised repeatedly with respect to hand drilled wells it that of construction quality, as well as water quality. It is essential that these concerns are taken into concern with appropriate quality assurance mechanisms, as well as water quality testing and remedial action.

14.4.5 *Procurement process*

Systematic, transparent and timely processes of advertising, pre-qualification, tendering, evaluation and award need to be established and followed.

Tendering procedures for private sector drilling in many countries are still weak and procedures can take a long time. This is not good for business and unnecessarily increases costs which in turn raises drilling price or compromises construction quality.

- Adekile (2007) found that in Nigeria, contracts are often awarded to non-professionals who then sub-contract to the drilling contractor, lowering the profit margin and sometimes compromising technical standards. In Nigeria, numerous drillers complain of not being able to tender for Government as they do not stand a change (Adekile and Olabode, 2008b).
- In other countries (e.g. Malawi and Uganda), there are companies which will not tender for work with certain District Governments (Danert, 2008a).
- Baumann et al. (2005) found that there was no pre-qualification of bidders in Tanzania and that tender evaluations did not find out inconsistencies in the capabilities of different bidders.

Table 14.4.   Summary of contract packaging arrangements in different countries.

Country	Contract packaging
Kenya	Doyen (2003) estimates that costs could rise by as much as 25% if drilling campaigns are not in economic lots of 50 wells or more.
Nigeria	Many contracts are packaged as one or two boreholes (Adekile, 2007). Up to 2008, UNICEF contracted in lots of 5, but paid for separate mobilisation on each bill of quantities rather than one mobilisation fee and payment for movement between sites (Adekile and Olabode, 2008b).
Tanzania	There are cases where a contractor had to enter five or six contracts to drill nine or ten wells (Baumann et al., 2005).
Uganda	Each of 80 Districts annually contracts out its own boreholes: numbers of wells drilled are small (ranging from 1 to 20 in 2007; average 9.5 in 2008 (MWE, 2007b; MWE, 2008).

- In Ethiopia, considerable procurement is "unplanned", which means that is rather sudden, and driven by the availability of funds. In such cases the sequence of steps followed for open and limited tenders are not adhered to.

14.4.6   *Contract packaging*

Transport is a major cost component for borehole drilling, which can be reduced by clustering wells to limit expenditure (Box 2). Unfortunately, small contract packages are common in many countries (Table 14.4). Not only do these raise costs (and prices), but they do not allow for long term planning and investment by private enterprises.

Community mobilisation efforts and response to the demand driven approach by end users should be reconciled with clustering of wells to achieve economies of scale. This is not always easy.

14.4.7   *Program and contract management*

It is essential that drilling programmes (whether national or more local) have sufficient skills to design and manage the programmes or bring in expertise. Payment for works must be timely.

As more countries move over to national programmes, or adhere to sector wide approaches, there is a danger that expertise with respect to programme management as well as drilling contact is insufficient. Where governments are changing role from implementer to that of service provider, or as more responsibilities are given to District level, skills may be lacking. In Tanzania, for example, model documents for tendering, evaluation and contracts were lacking and there were no contract management guidelines (Baumann, 2005). Although poorly documented, anecdotal evidence suggests that this is an area which is particularly weak in many countries. To make matters worse, understaffed ministries and local Government offices are not uncommon.

Payment systems for water well drilling vary considerably. In Nigeria, Malawi and Mozambique it is common for drilling contractors to be paid for a geophysical survey and only to be paid for successful wells. In Uganda, payment is theoretically against a bill of quantities, but this is not always followed (Danert, 2008a). It has been strongly argued that such a mechanism increases prices, as drillers take into account of risk. In Mozambique, payment delays of three months are common in Government projects but there are examples where delays have been for several years (WE Consult, 2006).

14.4.8   *Siting practices*

Appropriate siting practices should be utilized.

Improvements in knowledge of hydrogeology (see section 14.4.11 and 14.4.12) and enhanced experience in site survey can increase drilling success rates, and reduce the disparity between anticipated and actual drilling depths. Professional siting involves desk and field reconnaissance, but does not always require the use of geophysics (MacDonald et al., 2005). In many countries in the region, drillers themselves undertake the siting, and are subsequently only paid for a successful well.

In many countries there is a tendency to specify geophysics on drilling sites, even where it is not necessary. Adekile and Olabode (2008b) point out that on some of the consolidated sediments in Nigeria, a review of existing borehole data would be more applicable in determining depths than geophysics. In Tanzania, when siting, consultants are required to undertake a geophysical survey using at least two methods, including a VES resistively survey, which is not always necessary (Baumann, 2005). Doyen (2003) reports on a Kenyan drilling programme where blind drilling and use of geophysical techniques achieved 51% and 89% success respectively.

However when trying to locate water in fractured bedrock, geophysical techniques may significantly improve success rates. In the challenging hydrogeological conditions of Mauritania, there are between two and three reconnaissance wells drilled per successful well (ANTEA, 2007).

14.4.9   *Supervision*

High quality and timely construction supervision needs to be emphasised.

Doyen (2003) states: "over-drilling is roughly inversely proportional to the degree of supervision of drilling operations". The quality of drilling supervision (including knowledge of the local physical and hydrogeological environment) and on-site authority are important. Degree courses in geology and hydrogeology do not provide graduates with a solid foundation in drilling supervision. It is not uncommon for drillers to complain about being supervised by inexperienced hydrogeologists, straight out of university. Some drillers use their monopoly on knowledge and exploit this. When supervisors are not able to take a prompt decision, drillers will incur waiting time, which can significantly raise the cost of drilling. Unfortunately, supervision capacity is extremely limited in much of sub-Saharan Africa and is a key reason for borehole failure e.g.:

- In Nigeria "the capacity for proper supervision, in terms of experienced personnel and equipment is limited at State level" Adekile and Olabode (2008b). Kaduna State Ministry of Water Resources realised that they did not have sufficient competence to supervise their drilling programmes and invested in training (Adekile, 2007). The Nigerian Federal Government and external support agencies engage consultants to carry out drilling supervision (Adekile and Olabode, 2008b).
- In Malawi, there are only a handful of hydrogeologists in the country. Supervision of test pumping is often the only professional supervision that takes place. Communities are expected to undertake a certain amount of drilling supervision (for which they are given no more than two days training) (Baumann and Danert, 2008).
- In Uganda supervision is either undertaken by private consultants or by District Government depending on who is financing the work.
- In Ethiopia, supervision is undertaken directly by the Water Bureaux or through hired consultants with variation regarding the level of supervision and strictness. Contractors cite lack of timely decision-making by supervisors as a frequent problem (Carter, 2006).

14.4.10   *Pumping test*

Test pumping requirements should be matched to borehole purpose while taking into account the importance of data to improve the understanding of hydrogeology and water resources.

Doyen (2003) estimates that 7% savings would be possible in Kenya if a 3-hour, rather than a 24-hour discharge and 12 hour recovery was used to test pump rural handpump wells. The high standards test pumping requirements are intended to obtain as much hydrogeological information about the aquifer in the vicinity of the borehole as possible. Doyen (2003) states that although per meter drilling costs in Kenya fell by 35% between 1988 and 1996, the increased standards for well development, pump testing and well design increased costs by as much as 36% with the result that there were no net savings. Tanzania specifies a 24-hour pumping test (Baumann, 2005). In Nigeria, pumping tests have been matched to borehole purpose for several years, both the Federal Ministry of Water Resource and State project usually specify pumping tests of 2 to 6 hours for handpumps and 8 to 24 hours for motorised schemes.

14.4.11   *Groundwater resources monitoring and evaluation*

Rigorous evaluation of groundwater resources should be undertaken and information made available.

MacDonald and Davies (2000) point out that: sustainability of groundwater supplies; overexploitation in sedimentary basins; variations in natural water quality and contamination of groundwater demand more attention. There is an urgent need for improved groundwater resources monitoring:

* Groundwater levels appear to have fallen in some parts of Nigeria and it has been suggested that intensive drilling in the urban areas of Lagos and Kano State could lead to water level decline (Adekile, 2007; Adekile and Olabode, 2008b).
* Arsenic has been reported in some parts of Nigeria but it is not tested for in water supply projects.

14.4.12   *Hydrogeological data*

Hydrogeological data collection and storage should be undertaken.

MacDonald and Davies (2000) provide an overview of the four main hydrogeological environments in SSA (crystalline basement—40% of land area; volcanic rocks—6%; consolidated sedimentary rocks—32% unconsolidated sediments—22%) and the different methods for finding and abstracting groundwater from each. Different hydrogeology requires different levels of technical capacity for development, and much is still not known about groundwater in Africa (MacDonald and Davies, 2000).

Hydrogeological data is extremely important and insufficient attention to the storage, analysis and utilisation of drilling data is a lost opportunity. Unfortunately coordinated research and data collection on groundwater in SSA has become increasingly difficult. Mistakes are repeated, while information from thousands of boreholes is not collected. In Tanzania, for example only 60% to 70% of boreholes drilled by the Parastatal are recorded in the central database and records from industry and mining are not included at all (Baumann et al., 2005).

However, knowledge of hydrogeology in Nigeria has improved considerably over the years and data has been collected with a view to publishing hydrogeological maps (Adekile and Olabode, 2008b) and hydrogeological mapping is underway in Ethiopia and Uganda.

Simple techniques for the collection and analysis of high value data from drilling programmes exist, but are inadequately used. This is a missed opportunity for significantly enhancing the knowledge base of groundwater Africa, and enabling issues for specific research to be identified and targeted. MacDonald and Davies (2000) advocate for the dissemination of simple techniques on groundwater resource assessment to stakeholders involved in rural water supply.

14.4.13 *Regulation and professionalism of the private sector*

A strong public sector is needed to oversee and regulate the private sector. The private sector needs better access credit and should professionalise.

The public sector in many sub-Saharan countries is still struggling to fulfil its emerging regulatory role. Regulation on number of employees and equipment is demanding in some countries, e.g. Ethiopia (Carter et al., 2006) and lacking in others, e.g. Nigeria (Adekile, 2007). Although drilling permits are issued in Tanzania, they are not based on consistent professional assessments of the companies, and quality is not monitored in a regular basis (Baumann, 2005). If 35,000 wells are to be drilled annually in sub-Saharan Africa, and each rig drills 100 wells per year, the continent needs some 3,500 drilling rigs. Even if as many as 20% of these are owned by NGOs and Government, this still leaves a requirement of some 2,800 privately owned rigs. However, the private sector has nowhere near this capacity. Some countries such as Nigeria and Uganda have considerable national expertise while others are still heavily reliant on foreign companies. Costs of expatriate staff (from Europe, Australasia, Japan and North America) are more expensive than local staff, i.e. four to eight times as much in Burkina Faso, Senegal, Mali and Mauritania (ANTEA, 2007).

Productivity rates are often low due to the use of old equipment, challenges of obtaining spares and lack of maintenance skills as well as lack of steady work. Obtaining regular work is essential to enable capital-intensive drilling enterprises to remain in business, and be cost-effective. However, contractors generally have to tender for work every year, and for many different projects or local authorities. Only one documented case of a drilling concession, running over several years has been found in the literature (Robinson, 2006). Low productivity of the private sector flies in the face of arguments against use of Government equipment due to low productivity.

Setting up in business can be extremely difficult which makes it very difficult for enterprises to enter the sector. There are cases in Mozambique where it has taken three years for a company to establish itself (WE Consult, 2006).

There are many examples of people with the skills, but not the finances to invest. Conventional drilling is a very capital-intensive undertaking. There are challenges with the banking sector across the continent. Interest rates on loans are high, e.g. 20–40% in Mozambique (WE Consult 2006); 18% in Tanzania (Baumann, 2005). Repayment periods can be short, e.g. 3 years in Tanzania (Baumann, 2005). In Nigeria, people generally use their own savings and those of relatives as start-up capital. There are major difficulties of showing sufficient collateral to obtain credit throughout the region. Commercial banks in Tanzania require a security of 125% and the assurance of continual Government work (Baumann, 2005). Existing, and potential drillers are often cash-strapped (Baumann, 2005). Delays in payment for work completed (see section 14.4.7) exacerbate this problem.

Importation of equipment and spares can be very difficult if contractors do not have foreign connections (Carter et al., 2006; Robinson, 2006; Adekile, 2007).

The capability and availability of skilled personnel (professionals and technicians) is an issue for both the public and private sector. Many drillers, supervisors and technical staff were originally working for Government and trained within projects. Given the shift in emphasis to decentralised service delivery by the private sector, there are serious questions regarding adequate opportunities for training and skills development. Ethiopia is a case in point, where an estimated 4,000 technicians are needed to enable the MDG water target to be met (Carter, 2006). However, there is only one training school where 200 are trained per year. Contractors in Nigeria and Ethiopia face problems in retaining personnel due to skills shortages (Adekile, 2007; Carter et al., 2006).

Networking, collaboration and lobbying are recognised as important mechanisms to professionalize organisations and bring about policy shifts. Drillers Associations in Mozambique and Nigeria have recently been established, initially with donor support. In Mozambique, the association successfully lobbied for more realistic contract terms and conditions. The Uganda Drillers Association had collapsed by 2003, although drillers have recently collaborated to demystify tax procedures. The Project Management Unit in South Sudan provides an interesting example of drilling enterprises which are collaborating with each other. Documentation and analysis of the success of networking and collaboration of drillers is lacking, but evidence from other sectors indicates that it could be instrumental in bringing about positive change.

14.5   CONCLUSIONS AND RECOMMENDATIONS

Simple comparisons of borehole costs between countries and programmes can be misleading. In order to better understand cost variations, a standard accounting framework is needed, as well a methodology for modelling key variables. The costing of boreholes needs to be demystified to sector stakeholders so that they can better understand how they are calculated. This could improve tender evaluation. A simple but robust tool for sensitivity analysis regarding depth, rig amortization, distance and drilling time could prove very useful.

The conceptual framework set out in this chapter provides insights into the issues that affect borehole costs and prices, as well as construction quality. There are no single, simple magic bullets. Each particular country and specific project has its own strengths and weaknesses with respect to cost-effective borehole provision. The chapter shows a number of initiatives which are already taking place (e.g. drillers associations in Nigeria and Mozambique). In addition, steps are being taken to develop national codes of practice for cost-effective boreholes (e.g. Nigeria).

In order to better move towards improving the cost-effectiveness of borehole drilling in specific context, it is recommended that as a first step, stakeholders use the conceptual framework to analyse borehole costing, appreciate the core factors, and undertake a preliminary analyse of the key elements at national level and for specific programmes. This should enable aspects that can be dealt with relatively quickly and easily, and those which need longer term efforts to be identified. In some cases, the scope for improvement is closely confined within the narrow confines of a particular project, while in others, national consensus or change of legislation may be required. In the case of very large countries, or those where there is considerable decentralisation, prioritisation and action is likely to be required at a sub-national as well as at national levels.

It should be well appreciated that underlying all of the elements set out in this chapter are inherent structural strengths and weaknesses. In general, there is need for concerted and long term investment in human resources, institution-building and better monitoring and information systems as well as strengthening the regulatory framework. Improved transparency in terms of reporting and publishing inputs and programme outputs is also critical to enable better scrutiny of programmes. However, without sufficient financial resources decision makers will be faced with very hard decisions such as whether to focus on groundwater resources monitoring, improve supervision capacity or to develop robust operation and maintenance systems.

It would be prudent to utilise the thirteen elements as a basis for benchmarking the drilling sector in a particular country or for a particular programme. Such benchmarking could be undertaken under the umbrella of a generic and national code of practice for cost-effective boreholes. However, there is need for political and technical buy-in at international as well as national level to enable such an initiative to have a significant impact.

ACKNOWLEDGEMENTS

The authors extend thanks to WSP-AF, SDC and UNICEF which have supported the Cost-Effective Boreholes flagship of the Rural Water Supply Network (RWSN), thus enabling this chapter to be prepared.

REFERENCES

Adekile, D. (2007). The Drilling Environment and Establishing a Drillers Association in Nigeria. Summary report RWSN/WSP. Nairobi, Kenya.

Adekile, D. and Olabode, O. (2008a). Study of Public and Private Borehole Drilling in Nigeria—Executive Summary. Consultancy Report for UNICEF Nigeria Wash Section. Abuja, Nigeria.

Adekile, D. and Olabode, O. (2008b). Report on comparison of Cost-Effective borehole Drilling in the Project Sates and other Programmes. Consultancy Report for UNICEF Nigeria Wash Section. Abuja, Nigeria.

Adekile, D. and Olabode, O. (2009). Study of Public and Private Borehole Drilling in Nigeria Consultancy Report for UNICEF Nigeria Wash Section.

ANTEA. (2007). Etude sur l'optimisation du coût des forages en Afrique de l'Ouest Rapport de Synthèse. Juin 2007. Banque Mondiale Programme Pout L'Eau et L'Assainissement—Afrique (PEA-AF).

Ball, Peter. (2004). Solutions for Reducing Borehole Costs in Africa. Field Note. WSP/RWSN/SKAT.

Baumann, E., Ball, P. and Beyene, A. (2005). Rationalization of Drilling Operations in Tanzania. Review of the Borehole Drilling Sector in Tanzania. Consultancy report of World Bank.

Baumann, E. and Danert, K. (2008). Operation and Maintenance of Rural Water Supplies in Malawi—Study Findings. Final Draft Consultancy Report for UNICEF Malawi. Lilongwe, Malawi.

Carter, R.C., Desta, H., Etsegenet, B., Eyob, B., Eyob, D., Yetnayet, Ne, Belete, M. and Danert, K. (2006). Drilling for Water in Ethiopia: a Country Case Study by the Cost-Effective Boreholes Flagship of the Rural Water Supply Network. Federal Democratic Republic of Ethiopia/WSP/RWSN.

Doyen, J. (2003). A comparative Study on Water Well Drilling Costs in Kenya. Unpublished Report. Research commissioned by UNDP- Water and Sanitation Programme of the World Bank.

Danert, K. (2005). Cost-effective Boreholes Scoping Study. Visit Report for RWSN/WSP.

Danert, K. (2007). Niger – RWSN Focus Country, Report 1. Consultancy Report for RWSN/WSP

Danert, K. (2008a). Personal Communication with Private Drilling Contractors in Uganda (September 2008).

Danert, K. (2008b). Personal Communication with Private Drilling Contractors in Malawi (November 2008).

MacDonald, A., Davies, J., Calow, R. and Chilton, J. (2005). Developing Groundwater. A guide for Rural Water Supply, ITDG Publishing.

Mthunzi, M. (2004). Monitoring of Partially Cased Boreholes. Research Report of Concern Universal.

MWE. (2007a). Uganda Water and Sanitation Sector District Implementation Manual. Ministry of Water and Environment, Government of Uganda.

MWE. (2007b). Uganda Water and Sanitation Sector Performance Report 2007. Ministry of Water and Environment, Kampala, Uganda.

MWE. (2008). Uganda Water and Sanitation Sector Performance Report 2008. Ministry of Water and Environment, Kampala, Uganda.

Practica. (2005). Report Phase 1. Assessment of the feasibility of manual drilling in Chad. Practica Foundation. Consultancy Report for UNICEF.

Rowles, R. (1995). Drilling for Water—A Practical Manual, Avebury, Ashgate Publishing Company, Aldershot, UK 1995.

Robinson, A. (2006). Who is going to drill the African Boreholes? Field Note. WSP/RWSN

Wurzel, P. (2001). Drilling Boreholes for Handpumps. SKAT Working Papers on Water Supply and Sanitation, No. 2. SKAT, St Gallen, Switzerland.

WE Consult. (2006). Assessment of the National Drilling Sector Capacity for Rural Water Supply in Mozambique. Executive summary. Consultancy Report Prepared for WSP and DNA.

15

Water supply and sanitation in the Democratic Republic of the Congo

Josué Bahati Chishugi
Department of Earth Sciences, University of the Western Cape, South Africa
Département de Géologie et Minéralogie, Université Officielle de Bukavu, R.D. Congo

Yongxin Xu
Department of Earth Sciences, University of the Western Cape, South Africa

ABSTRACT: The 7th goal of the Millennium Development Goals (MDGs) concerning the environmental sustainability envisages halving, by 2015, the proportion of people without sustainable access to safe drinking water and basic sanitation. Unlike many African countries where fresh water is unevenly distributed both spatially and temporally, the Congo River accounts for nearly 30 percent of Africa's surface water reserves with the second worldwide largest equatorial forest and high groundwater potentialities assessed based on its climate and geology. Although having these resources and having consumed the three quarters of the allocated time for the 2015 MDGs target, DRC is still severely behind schedule in satisfying the drinking water supply and sanitation targets. This situation is unquestionably unrelated to the lack of water resources. The failure of the management policy of the Drinking Water Supply and Sanitation (DWSS) services observed since 90s, the political and military unrest since 1994, leading to stoppage of investments, degradation of facilities and weakening of the managerial capacities of the sub-sector entities, seems to be the major cause.

15.1 INTRODUCTION

The legendary slogan, Water is Life, has been approved world wide since the sustainability of human life and the ecosystem in humid- and arid regions are associated with regions where water resources are easily accessible. These natural resources are hidden (groundwater) or exposed (atmospheric and surface waters) but their accessibility depends mostly on the climatic conditions, geological setting and economic strength of each country. Unlike many African countries where fresh water is unevenly distributed both spatially and temporally, the Congo River accounts for nearly 30 percent of Africa's surface water reserves (UNEP, 2008) with the second largest equatorial forest in the world. From its climate and geological settings, groundwater resources are also expected to significant.

Predictions revealed that by 2010 more than 17 African countries, including the DRC, will experience water scarcity which will severely affect more than 400 million of the African population. This scarcity will constrain food production,—environmental protection and economic development.

In September 2000, 189 UN Member States adopted the Millennium Development Goals (MDGs), where the 7th goal concerning the environmental sustainability envisages to halve, by 2015, the proportion of current people without sustainable access to safe drinking water and basic sanitation. The baseline for most of the MDG targets, including that on water and sanitation, has been set in 1990. As reported by the WHO/UNICEF (2008), DRC is not on track to meet the MGD drinking water and sanitation targets. Assessing the state of water supply and sanitation at present is of extreme importance owing to the fact that we seriously behind schedule and we are seventy five percent through the allotted time; hence it would be prudent to evaluate the achievements done thus far and decide what still has to be done within the remaining time. This chapter intends to give a general view on the situation of water supply and sanitation (WATSAN) in the DRC.

15.2   OVERVIEW OF THE STUDY AREA

15.2.1   *Location and climate*

Situated in Central Africa region, the DRC is the third largest country in Africa (2,345,000 km²). It borders the Central African Republic and Sudan on the North; Uganda, Rwanda, and Burundi on the East; Zambia and Angola on the South; the Republic of the Congo on the West; and it is separated from Tanzania by Lake Tanganyika on the East (Figure 15.1).

Due to its location inside the equatorial belt, different parts of the Congo basin receive substantial rainfall throughout the year; with a decreasing trend of rainfall with latitude. The northern and central portions of the basin have two major rainfall seasons which begin in March and October each year (Kazadi, 1996). In the south, the two rainfall seasons gradually merge into a single season beginning in December and lasting for six months each year. In the far southern part of the basin—at latitude 12° S, in the Katanga region—the climate becomes definitely Sudanic in character, with marked dry and wet seasons of approximately equal length and with mean annual rainfall of about 1245 mm. The rainfall peaks are associated

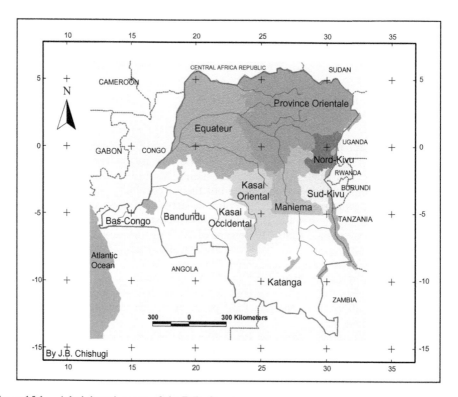

Figure 15.1.   Administrative map of the D.R. Congo.

with the passage of the Inter-Tropical Convergence Zone (ITCZ), which is a large zone of low pressure caused by excessive heating from an overhead sun.

15.2.2   *Geology and tectonics*

The Geology of the RDC is characterized by two large structural units separated by discordance and/or a significant gap: the superficial formations (Phanerozoic), un-metamorphosed, generally fossiliferous, and of age ranging between the Upper Carboniferous and the Holocene; and the Basement terrains (Precambrian shield) that are highly metamorphosed, shielded and contouring continuously the basin.

15.2.3   *Soils and land use*

Generally, there are two types of soils in the study area: those of the equatorial areas and those of the drier savannah (grassland) regions. The equatorial soils occur in the warm, humid lowlands of the central basin, which receive abundant rainfall throughout the year and are covered mainly with thick forests. This soil is almost fixed in place because of the lack of erosive forces in the forests. In the shore areas, however, swamp vegetation has built up a remarkably thick soil that is constantly nourished by humus, the organic material resulting from the decomposition of plant or animal matter. Although in the savannah regions the soils are constantly endangered by erosion, the river valleys contain rich and fertile alluvial soils. The highlands of eastern Congo in the Great Lakes region are partly covered with volcanic lava that has been transformed into exceptionally rich soil.

Land use distribution comprises 3.33% (75,511 km²) of Cultivated Land; 2.86% (64,853 km²) of arable land; 0.47% (10,658 km²) of Permanent Crops and 96.67% (37,853 km²) of Non cultivated area estimated in 2005); the irrigated land was estimated to 110 km² in 2003.

15.2.4   *Hydrography*

The hydrography of the DRC is characterised by that of the Congo basin. This basin is drained by the Congo River which flows cross the whole central Africa region and eventually discharges itself at the Atlantic Ocean. The Congo River has many tributaries in both the northern and southern hemisphere, consequently flowing throughout the year. Its mighty flow has for years remained between the high level of 65,411.92 m³/s (in 1908) and the low level of 21,407.54 m³/s (in 1905). During the unusual flood of 1962, however, it is by far the highest for a century, with the flow probably exceeding 73,623.8 m³/s (Encyclopaedia Britannica, 2007). At Kinshasa, the river's regime is characterized by a peak high flow at the end of the year and a second high flow in May, as well as by a major low level during July and a secondary low level during March and April. In reality, the downstream regime of the Congo represents climatic influence extending over 20° of latitude on both sides of the equator, a distance of some 2253 km. Due to its immense area, the weather pattern in one particular region would have little effect on the river's overall performance.

15.3   SOCIO-DEMOGRAPHY

The population of the DRC has been increasing since the first administrative census whereas the last administrative census took place in 1984. Data from 2000 to 2006 (Figure 15.2, FAOSTAT, 2009) up to 2008 show an increasing trend of population. In 2007 and 2008, the population was estimated at 62.6 million and 66.5 million inhabitants (www: World Population Datasheet, 2009), respectively. Projections for 2010, 2015 and 2025 are 68.88 million, 78.0 millon and 109.7 million inhabitants, respectively. About 70% of the population live in the rural areas.

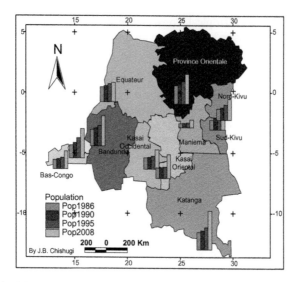

Figure 15.2.   Provincial population growth distribution in the D.R. Congo, from 1986 to 2008.

During the last decade, the political and military instability that prevailed in the DRC affected the country's socio-economic indicators, which remains a cause for concern, very often falling below international standards. More than 80% of the population lives on less than one dollar a day, with the life expectancy being about 51 years. Approximately 45% of adults are illiterate, and only 22% of the population have access to drinking water and 9% to adequate sanitation services. According to the human poverty indicator, the probability for children born between 1995 and 2000 to die before the age of 40 is estimated at 34%. About 30% of the population lives in chronic malnutrition. In general, the population is vulnerable to diseases, notably water-borne diseases which account for 80% of all the diseases, as well as HIV/AIDS, with a prevalence rate sometimes reaching 20%. Access to health and education facilities is also very limited. Gender parity is not respected (ADB, 2007).

The population distribution over the country is depicted in Table 15.1 and Figure 15.2. The population given in Table 15.1 (1986 to 1995) has been calculated by the National Institute of Statistics. Those of the year 2008 where obtained from the World Food Program (WFP, 2008). The 2008 data's should be looked at in term of percentages rather than population number due to inconsistency comparing to other reports. The country's most populous province is the Katanga with a population of 4.2 million (in 1986), 4.6 million (1990), 5.4 million (1995) and 9.66 million (2008) followed by the Oriental province and then by Kinshasa. Kinshasa is the smallest of the eleven provinces but it has the highest population density.

15.4   WATER RESOURCES

The DRC is rich in surface and groundwater resources. The country controls the largest river of Africa (Congo River) and is nearly a river basin by itself. It is assumed that overlap between surface water basin and groundwater basin is nearly 100%. Most of the groundwater is drained by rivers in the form of the base flow of water courses. Some groundwater escapes and flows out into the sea. The total flow of the Congo River at its mouth is 1283 km^3/yr (FAO/AQUASTAT, 1995). The figure for Internal Renewable Water

Table 15.1.   Population distribution by province in 2008.

| Provinces | Population | | | | Population 2015 |
	1986	1990	1995	2008	
Kinshasa	3,0	3,7	4,9	7,27	–
Bas-Congo	2,1	2,3	2,6	4,24	–
Bandundu	3,9	4,3	5,0	7,44	–
Equateur	3,6	3,9	4,4	4,75	–
Orientale	4,4	4,9	5,5	7,39	–
Nord-Kivu	2,6	3,0	3,65	5,10	–
Sud-Kivu	2,2	2,7	3,4	4,42	–
Maniema	0,8	0,9	1,0	1,79	–
Katanga	4,2	4,6	5,4	9,66	–
Kasaï Oriental	2,6	2,8	3,1	4,76	–
Kasaï Occide ntal	2,3	2,5	2,6	5,81	–
R.D. Congo	31,5	35,6	41,6	64.8	78.0

Source: 1. WFP et al. (2008) and 2. Institut Royale des Sciences Naturelles de Belgique et Centre d'Echange d'informations de la R.D. Congo (2004).

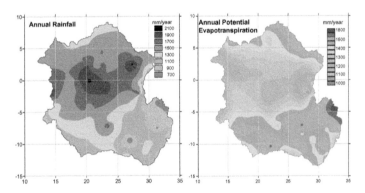

Figure 15.3.   Atmospheric Water resources of the Congo basin from the monthly time series FAO database (1961–1990): Rainfall (left), Evapotraspiration (right) (Bahati Chishugi, 2008, Bahati and Alemaw, 2009).

Resources (IRWR) of the country has been estimated at about 900 km³/year (FAO/AQU-ASTAT, 2005).

15.4.1   *Surface water resources*

With respect to surface water resources, they come from the vast system of rivers and lakes which account for 52% of overall reserves of the continent and cover approximately 86,080 km², representing 3.5% of the country's surface area. Located at the very centre of Africa and straddling the Equator. DRC is among the wettest countries on the African continent, with high average rainfall of 1,200 mm distributed throughout the year (Figure 15.3).

The Congo River Basin covers 3.8 million km², 3/4 of which are in the DRC. Its mean flow of 42,000 m³ per second, with minimum- and maximum flows of 23,000 and 80,000 m³ respectively makes it the first African river and world second in terms of discharge rate. The Congo River is 4,700 km long and has around twenty tributaries draining both the northern and southern hemispheres (Figure 15.4). The chemical characteristics of raw surface waters

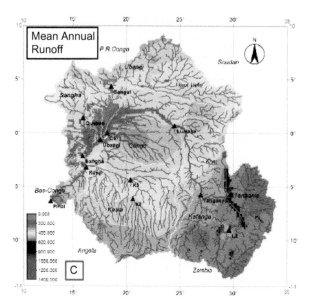

Figure 15.4.   Grid total surface runoff (mm/year) in the Congo basin from the monthly time series (1961–1990). (Bahati Chishugi, 2008; Bahati and Alemaw, 2009).

in the DRC, except in cases of localized pollution, make them completely fit for consumption after treatment.

15.4.2   *Groundwater resources*

The hydrogeology of the DRC is still not well understood and hence the data, information and knowledge of the geometry and hydraulic properties of aquifers remain scarce, constraining the sustainable use and management of the resources. Groundwater data and information management system is at a poor state, leading to reduction in yields and groundwater pollution from on-site sanitation, industry, and agricultural practices.

15.4.2.1   *Hydrogeology of major aquifers in the country*
Groundwater resources of DRC occur in six geological formations (Nile IWRM-Net, 2007 and Hakiza, 2002):

a. The Equateur and Bandundu formations, which represent very important aquifers in North West of DRC;
b. Central Cuvette Sand with average thickness of 120 m covering larger proportion than the previous in North West of DRC. It extends to the central part of DRC sharing the boundaries with heterogeneous and anisotropic materials. The hydrodynamic characteristics related to this type of geological formation include normal yield range from 25–450 m³/h, static waterlevel ranges from 2 mbgl to artesian, dynamic level from 5.7 to 41.6 mbgl with drawdown of 6.2 to 27.5 m. The submersible depth of GMP is about 18 m and the normal yield using GMP ranges from 18 to 250 m³/h;
c. Low potential Sand Aquifer of thickness 80 m occurs in Southwest of DRC and extends to Southeast in Kinshasa, Kasai Occidental and Lubumbashi. The hydrodynamic characteristics related to this type of geological formation described as Normal yield range from 15 to 40 m³/h, static level range from 10 to 131.9 m, dynamic level from 23.3 to 147 m with

drawdown of 0.43 to 15 m. The submersed depth of pumps varies from 30 to 171 m and the normal yield using GMP ranges from 15 to 40 m³/h;

d. Heterogeneous and anisotropic materials with rapid infiltration and presence of salt water in their deepest part are located in central part of DRC essentially in Kasai Occidental and Kasai Oriental, extending to Bandundu, but is also found in Kinshasa. The hydrodynamic characteristics related to this type of geological formation range from 42.5 to 55 m³/h Normal yields, the static level approximately 97.6 m and the dynamic level 113.68 m with the drawdown of 16.8 m. The immersed depth of GMP varies between 30 to 120 m and the Normal yield using GMP ranges from 30 to 50 m³/h. The aquifer transmissibility is 2.26×10^{-3} m²/s and the coefficient of storage S $\approx 1 \times 10^{-5}$. The quality analysis of sampled water from this geological formation shows a pH of 6 (in situ) and 6.4 (in laboratory). The CO_2 (equilibrium) is 0.17 mg/ℓ, CO_2 (free) is 123.12 mg/ℓ with a pH of equilibrium equal to 8.75;

e. Very heterogeneous and anisotropic materials with compact cretaceous rocks containing very important water resources essentially found in Katanga. This type of geological formations is characterized by a normal yield of 80 to 180 m³/h, a static level of 14.7 to 21 m and a Normal yield using GMP of 40 m³/h;

f. Fractures conditioning the development of aquifer zones are the major geological formation found in DRC. It extends from Northeast in Orientale province to Southeast in Katanga including Kivu. It is also located in Equateur, Kasai oriental and Bas Congo. The Normal yield of this geological formation range from 30 m³/h in Bas Congo (Moanda) to 60 m³/h in the Orientale Province. The static level is about 2 m and the artesian level is about 45 m with a drawdown of 43 m and a normal yield using GMP ranging from 10 to 45 m³/h.

15.4.2.2 *Groundwater data availability*

Most of data monitored in DRC are still kept in hard copy format, thus the data are not readily available to the public. The Groundwater resources database in DRC is almost non-existent; few existing information on hydrogeological maps, Geographical and climatic maps, reports and publications are stored in hard copy which render the access to them difficult (Nile IWRM-Net, 2007). Table 15.2 summarise the state of data availability in DRC.

Table 15.2.   Groundwater data availability (Nile IWRM-Net, 2007).

Data type	Data availability
Groundwater maps (piezo-metric, deepness, thickness…)	No recent inventory (scattered)
Geological maps (lithologic and stratigraphic nature of aquifers)	Not recent survey available
Hydrogeological & Hydrological reports (memories, thesis…)	Reports available at educational institutions and at REGIDESO
Analysis of the major groundwater systems (recent surveys)	No recent analysis has been carried out. Existing analysis is done during the 50s or earlier
Climatic data, rainfall and ET maps	Yes
Groundwater chemistry data	Not recent survey available
Groundwater level, EC, pH data etc…	Not recent survey available
Groundwater recharge rates	Not recent survey available. Lack of an adequate groundwater database renders difficult the evaluation of groundwater potential, dynamics including recharge rate, yield and vulnerability
River runoff data; rainfall and climatic data	Partially available

15.5   WATER SUPPLY

Due to lack of data on population projection for 2015 for each province, it is not possible to correctly estimate water demand for each province. However, a general projection for the country gave a figure of 68.88 million by 2015.

In rural areas, the Drinking Water Supply and Sanitation (DWSS) Master Plan 1991–2010 was based on objectives aimed at providing drinking water coverage rate between 80% and 100% so as to ensure access to hygienic sanitary facilities by year 2000 and 2010 respectively. These objectives are currently being reviewed under the MDGs (ADB, 2007).

15.5.1   *Water supply sources and access to drinking water*

In rural areas, drinking water supply in the DRC was, until independence in 1960, provided by the "Fonds Belge Indigène" (FBI). After independence, the activities of FBI were taken over by a Belgian NGO (AIDR) until 1964. From 1964 to 1977, some NGOs and churches constructed certain specific drinking water structures in rural areas. With its 17 stations established nationwide, the National Service for Rural Hydraulic (SNHR) has since its creation in 1983 up to 2004, protected 5,084 springs, rehabilitated 543 wells, equipped 659 boreholes with pumps, and simplified 65 water supply systems serving a total population of 2,451,200 inhabitants, representing about 8% of the rural population. The water points constructed with the assistance of the other actors cover 9% of requirements implying that. Only 17% of the rural population have access to drinking water (WFP et al., 2008).

As illustrated in Figure 15.5, the sources of water in rural area are mostly from "unprotected wells" which is used by a mean of 48.3% of rural population in the DRC. This figure hides many disparities given that the proportion of the households using unprotected well is very high in certain provinces. These provinces include Bandundu (60.3%), Eastern (62.4%), Equator (66.3%), and finally Kasaï Oriental (69.8%). The second source of drink water is the surface water (to 25% of the households), which is followed by the protected well (24%). In rural areas, only 4.8% of the population has access to the water from the tap. With regard to drillings, they feed only 2% of the households in rural medium in RDC (WFP et al., 2008). Figure 15.6 illustrates different means of Rural Water Supply in DRC. A: from Borehole in Nduruma, B: from a captured spring in Kadutu, exposed to anthropogenic pollution; C: communal water pump (Indian model) in Lulimba (Southern Kivu).

Urban areas in the DRC are mostly under the responsibility of **REGIDESO** which is currently facing serious technical and financial difficulties. The status of water supply services in the urban areas is particularly worrisome. In the 1970s and 1980s, **REGIDESO** was one of the most effective public enterprises in DRC and achieved remarkable growth, financed

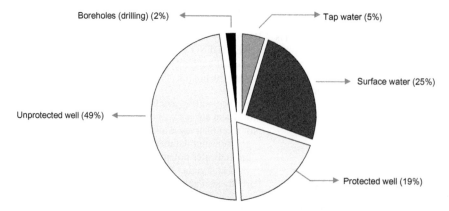

Figure 15.5.   Principal sources of water in rural area of the D.R. Congo (WFP, 2008).

Figure 15.6. Pictures showing some means of Rural Water supply in D.R. Congo. A: Left hand, a Borehole in Nduruma; B: Upper right hand, from a captured spring in Kadutu (South Kivu), in volcanic aquifer exposed to anthropogenic pollution due to household located in the recharge and protection area of the spring; C: A communal water pump (Indian model) in Lulimba (South Kivu).

largely with external grants and loans. Since 1990, its operational performance has declined considerably. Most of the REGIDESO systems serving secondary cities are not operating today, due to the combined effects of war, lack of investment and maintenance, and the suspension of aid. Despite the strong urbanization growth in the country in the past 17 years between 1989 and 2006, the number of connections decreased by 8 percent; and the volume of water sold decreased by 10 percent. Over the same period, the number of connections per km decreased from 36 to 18 and REGIDESO's labour productivity has clearly deteriorated from 14 to 22 staff per 1,000 connections.

In the urban areas currently about 8 million city dwellers have access to drinking water and this represents 37% of the total urban population in the DRC and 15% of the total population of the country. For instance, the town of Mbuji Mayi (Kasai Oriental), with 3 million inhabitants, does not have a drinking water service and represents nearly 30% of the unserved people. In Bukavu (South Kivu), with a population that currently exceeds one million inhabitants as a result of the influx of refugees, the situation is also problematic. In Tshikapa (Kasai Occidental), the diamond producing town, 1% of the population is served (ADB, 2007).

Concerning improved water supply, an increasing trend was observed from 1990 up to 2002 and starts to slow down until 2006. This could be explained by the military and political unrest observed in during this period since 1996. The government concentrated its budget on peace and other issues than the development of water sector affecting the REGIDESO administration and consequently the urban water supply sector. Whilst the REGIDESO reduces its activities in the urban area which is under its responsibility, important progress have been observed in the rural water supply sector. Indeed, the International Organizations accentuated their efforts in the distribution of water supply in order to attend to the rural population who were the most affected during the armed conflicts and the post- conflicts period.

15.6   SANITATION

15.6.1   *Current situation in the DRC*

Physical infrastructures and regional administrative structures for urban sanitation are still underdeveloped. Concerning wastewater, the few collective networks in big towns have not been maintained on a regular basis, and most of them are clogged, blocked and out of function. All the treatment plants are out of service or have disappeared. Sanitation for individual households is left to private initiative, which very largely dominates this sector. It however uses techniques that are not controlled, and most often very rudimentary, owing to users' lack of know-how (ADB, 2007).

The drainage of septic tanks is a market shared between the private sector and the National Sanitation Program (PNA) in Kinshasa, the only town where PNA is operational. Operators discharge waste products in disregard of any rules, generally into nearby rivers flowing within the country. Concerning household refuse collection, only Kinshasa, Lubumbashi and, to a lesser extent, Kisangani and Bukavu, have a minimum collective service. However, there are ongoing initiatives supported by NGOs to develop household refuse collection systems whereby the populations of the country can be self-reliant. Concerning storm water, the drainage networks are degraded all over the country. Although severe flood problems are relatively rare, those of erosion due to poor drainage are frequent and dramatic (ADB, 2007).

The fundamental problem of urban sanitation is mainly institutional one, which involves the lack of sound management of urban space in terms of its use and generation of resources required for its operation (territorial organization, and overall and operational urban policy). The same applies to inadequate organization of disease vector control and health education.

Regarding rural sanitation issue, it is difficult to evaluate the current status of facilities due to lack of information. It however seems that all the physical investments are mostly privately funded.

At the institutional level, this sub-sector is not covered by a specific body, and is shared primarily between Ministry of the Environment, Conservation of Nature and Forestry (MECNE), through PNA and the Ministry of Public Health (MSP). The key public missions involve activities relating to monitoring, guidance, education, training, sensitization, and financing assistance. The sub-sector does not seem to feature among the priorities of technical and financial partners (TFP). Only the health zones obtain assistance from aid programmes. There is also a gulf between sanitation service requirements and supply (ADB, 2007). Available data shows that access to appropriate excreta disposal system is limited. Overall access to sanitation services is estimated at 9%. Sanitation is not yet considered a priority by the population.

In rural areas of the DRC 70.3% of population uses "traditional latrine or simple pit latrine"; 17% uses "VIP pit-latrine"; open defecation method is applied by 12%, and only 0.1 uses the pour-flush latrines (Figure 15.7) (WFP et al., 2008).

15.6.2   *Water and health issues*

By linking the source of drink water to severe malnutrition cases in DRC, one would notes that 46% of the acutely malnourished children have traditional wells as principal source of drinking water and 24.3%, surface water. In general, more than two thirds of these children consume unclean water, which, as identified by NGOs, is the source of diarrheic diseases.

A report from the Humanitarian Aid Sector of 16 December, 2008, shows how the waterborne diseases can have a devastating affect on the DRC's population. Since early October 2008, high morbidity and mortality rates associated with a cholera epidemic outbreak have been registered in the Maniema, Katanga, North Kivu and South Kivu provinces. Ministry of Health (MoH) statistics show that over 25,503 cholera cases had been registered, including

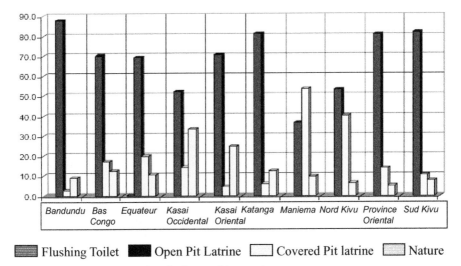

Flushing Toilet   ■ Open Pit Latrine   ☐ Covered Pit latrine   Nature

Figure 15.7. Principal means of sanitation used by province.
Source: WFP et al., 2008.

515 deaths in the localities mentioned below. The following are statistics from the Congo Ministry of Health:

a. In Katanga province, over 10,214 cases and 229 deaths have been registered;
b. In some villages of the Maniema province, 189 cases and 11 deaths have been registered just the past weeks;
c. In North Kivu, the ICRC has identified nearly 8,826 cases and over 229 deaths in the most affected health zones of Binza, Bwambizo, Goma, Karisimbi, Kirotche, Masisi, Rutchuru and Walikale, with registered out of a total population of 1,272,981 inhabitants;
d. By the same period, well over 5,000 cases have been registered in South Kivu, with the most affected health zones including Minova, Nundu, Baraka/Fizi (which is an endemic zone), Kalehe, Ruzizi, Katana, Kabare, Kadutu and Bagira.

Despite the numerous organisations operating in WATSAN sector building up a strong cluster (UNICEF, FHI, IRC, ASSATE, ADI KIVU, BDD, ICG, ASF/PSI, ACF, TEAR-FUND, SNHR, CAB, APEO, UNICEF, BRAPTE, ACTED, CORDAID, CTB, CROIX ROUGE DE LA RDC, APIDE, CICR, IMC, IPS/B9. etc) in southern and northern Kivu, as example, the sanitation situation remains alarming mostly due to the unawareness of the population in protection of water sources methods, the mismanagement of the land alloca-tion services. Figure 15.8 shows how water sources are surrounded by waste dumping and sewages sites in a volcanic environment where water easily through fractures and basaltic joins. This leads to recurring endemic waterborne diseases in the town.

15.6.3 *Financial requirements*

In order to tackle the MDGs goals, the financial requirements of the DWSS sector in the DRC for the 2005–2015 decade were estimated at US$ 1.71 billion, that is about US$ 214 million per year, with US$ 171 million for drinking water and US$ 43 million for sanitation. This substantial financing should allow for an average access rate of 49% for DWS and 45% for sanitation. It is still to be evaluated to what extent the Congolese economy, the institutional system of the sector, and the contributions and pledges of donors can finance such a pro-gramme. However, it should be noted that the link between development of the sector and the central planning and budgeting process is fragile (ADB, 2007).

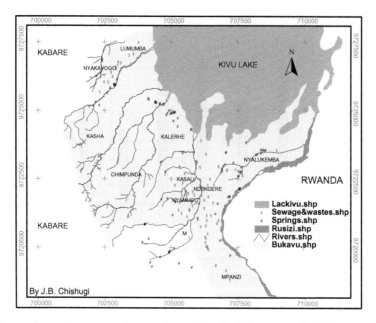

Figure 15.8.   Springs and sewage and waste disposal sites in Bukavu town (Sud Kivu province).

15.7   INSTITUTIONS AND LEGISLATION

15.7.1   *Laws and regulations*

The main instruments governing the water and sanitation sector in the DRC are:

a. Ordinance of 1 July 1914 on the contamination of springs, lakes and rivers, which provides for the demarcation of protected areas for the collection of drinking water;
b. the Decree of 6 May 1952 on the award of concessions and administration of water and determining the rights of way of groundwaters, lakes and waterways, as well as their use;
c. Ordinance No. 52-443 of 21 December 1952 on measures to protect springs, water tables, lakes and waterways, to prevent water pollution and wastage, and to controlling the exercise of user rights and conceded rights of occupancy;
d. Ordinance No. 071-079 of 26 March 1971 defining State action as regards storm and waste water;
e. Ordinance No. 74/345 of 28 June 1974 on hygiene measures in built-up areas, as supplemented by Inter-Ministerial Decree No. 120/89 of 6 September 1989 on public health protection measures in cities and urban, commercial, industrial, agricultural, and mining centres, as well as rural built-up areas;
f. Ministerial Decree No. 0014/DPT-MINER/86 of 2 September 1986 prohibiting the use of natural water other than that supplied by REGIDESO;
g. Ordinance law No 91-348 of 27 December 1991 fixing the rate and modalities of taxes and charges for the ministry of Energy;
h. Ministerial act No E/SG/O/0133 C2/93 0f 7 Mars 1993 fixing conditions for obtaining the groundwater and surface water permit, conditions for carrying out a drilling of aquifer and exploiting groundwater in DRC; and
i. Decree No. SC/073 of April 2005 on sanitation and public hygiene measures in the city of Kinshasa.

Most of these instruments which date back to the colonial period have become obsolete and need to be revised. Adapting legal framework to the provisions of the new Constitution, which stipulates that water is managed exclusively by the Provinces (Article 204, §20), and adopting new laws and regulations for water production and distribution within the current context of tracking the MDGs are requested. Hence, the support of development partners, the Government has embarked on strengthening the legal and institutional framework and implementing reforms in order to ensure a balanced development of the urban drinking water supply (DWS) sub-sector (Nile IWRM-Net, 2007). In 2007, a new Water Code was under preparation and should have been tabled before the National Assembly for adoption before end 2007.

15.7.2 *Institutions*

The Institutional Framework of the sector is marked by the involvement of seven ministries and several organizations in its management, resulting in overlapping and conflicts of juris-diction. It is the governmental responsibility of the Ministry of the Environment, Nature Conservation and Forestry (MECNE) to manage the water resources as provided by Ordi-nance No. 75-231 of 22 July 1975 which lays down its duties (ADF, 2007). Other institu-tions/authorities are:

a. Ministry of Environment, Nature Conservation, Water and Forest—MECNE (Directo-rate of Water Resources);
b. Ministry of Energy MINE (The national company of water supply and sanitation—REGIDESO; National Commission of Energy—NCE);
c. Ministry of Rural Development—MDR (The national Service of Rural Hydraulics—SHNR);
d. Ministry of Planning—MINIPLAN (Congo National Action Committee for Water Sup-ply and Sanitation—CNAEA).
e. Ministry of Heath—MSE (The rural health zone—ZSR).
f. Ministry of public works and Infrastructure—MTPI (The service of drainage: Office de Voirie et Drainage—OVD).

15.8 GOVERNMENT POLICIES AND STRATEGIES TO IMPROVE WATSAN SECTOR

In 2006, after receiving enough funds from the International Community and the World Bank to support the WATSAN sector, the DRC authorities have defined their priorities for the next 5 years, in which the water and sanitation sector occupies a prominent place. This priority was confirmed in the Growth and Poverty Reduction Strategy Paper (GPRSP) of the coun-try for 2006–2008, and the drinking water and sanitation sector is among the five projects of the Congolese Head of State. The sector goal, as contained in the GPRSP approved by the Government in July 2006, is to increase: (a) the drinking water service rate from 22% in 2005 to 26.9% in 2008 and 49% in 2015, and (b) the sanitation service rate from 9% in 2005 to 15% in 2008 and 45% by 2015.

Consequently, the existing infrastructures will be rehabilitated to facilitate access by the greatest number of users, increase the capacities of water production units, improve the management of water points by promoting community and private sector participation, strengthen the existing sanitation programmes and extend them nationwide. The actions will consist of: (a) reforms in the water and sanitation sector, (b) an inventory of urban and rural water requirements, (c) preparation of the Water and Sanitation Code, which should include, in particular, aspects concerning the protection and integrated management of water resources, definition of the roles of private operators of the sector, and clarification of their areas of action: large and average cities, small semi-urban centres and rural areas, and (d) creation of a water and sanitation development fund (ADB, 2007).

To that end, in the medium term, under the urban and semi-urban water sub-sector programs for 2006–2015, efforts will be made to: (a) implement the REGIDESO Ten-Year Plan and the Kinshasa Water Supply Master Plan; (b) open up the water and sanitation sector to civil society initiatives, private operators, NGOs and associations, and the beneficiaries themselves; and (c) implement the Kinshasa city sanitation Action Plan. As such, the impacts of the strategy should be noticed only as from 2009. For the semi-urban and rural areas, there are plans to put in place autonomous community management systems under the decentralization and rural development policy, as well as local public works and engineering enterprises. These policies will be accompanied by the search for and promotion of appropriate technologies, while popularizing good practices in community management of the autonomous facilities. A training programme will be developed in order to strengthen the operational capacities of the technicians of the sector, as well as health workers of the zone, including sanitary engineers, field workers, and drinking water and sanitation tradesmen (ADB, 2007).

15.9   CONSTRAINTS TO THE WATSAN DEVELOPMENT IN DRC

The development of the water sector is faced with institutional, technical and financial constraints, including (a) a multitude of stakeholders in the WATSAN sector, often with unclear roles, overlapping of functions, and total absence of coordination; (b) the state of degradation of infrastructures or their under-utilization; (c) inadequate funding; and (d) water sector institutions weakened, with an unmotivated staff whose skills have been eroded.

Consequently, the urban drinking water service rate is only 35%, as against 70% in 1990, while the service rate in rural areas stands at 17%, as against 24% in 1991. REGIDESO is facing serious financial difficulties due to insufficient production and very low collection rates. The sanitation policy is non-existent. The current resources of the stands to National Service for Rural Hydraulic (SNHR) and National Water and Sanitation Committee (CNAEA) are extremely limited, making it impossible for them to play their roles (ADB, 2007).

One should notice that the sanitary degradation of the environment is not due to lack of legal instruments applicable to the protection of natural resources, since efforts are being made to improve all the laws. It is rather due to lack of the will, on the part of the public authorities to enforce existing regulations, to take recommended remedial measures and to check the illegal and destructive exploitation of renewable and non-renewable natural resources (ADB, 2007).

15.10   WATSAN DONORS

Key donors for the DWSS sector are: the European Union (EU), the French Development Agency (AFD), the World Bank (BM), the Belgian Cooperation, the Kreditanstalt für Wiederaufbau (KFW) and the Department for International Development (DFID). Under the PMURR, the German Co-operation, Deutsche Gesellschaft für Technische Zusammenarbeit (GTZ), UNICEF, etc. Most of the activities of these key donors are located in the Kinshasa area and the semi-urban areas of some hinterland towns where they prefer to work with REGIDESO. In provinces and rural areas, UNICEF is the key donor in WATSAN sector.

In order to achieve the MDGs on WATSAN, the financial requirements are estimated to US$ 171 million for drinking water and US$ 43 million for sanitation. This substantial financing should allow for an average access rate of 49% for DWS and 45% for sanitation.

15.11 CONCLUSION

As reported by the UN/UNICEF Joint Monitoring Programme for Water Supply and Sanitation during the AMCOW summit of 30 June to 2 July 2008, DRC still on SEVERELY OFF-TRACK and OFF-TRACK levels concerning Water supply and Sanitation, respectively. Its WATSAN state remains detrimental with no hope of achieving the 2015s objectives of the MDGs. Multiple reasons could explain this situation.

Since 1994 up to the present the country has been unstable due to repetitive wars and conflicts and with a direct impact mostly on the Eastern Provinces of the country. The 1994 genocide in Rwanda has constrained millions of Rwandan population to move to the eastern part of DRC and later on toward the rest of the other provinces. The violent raids of different rebel troops since 1996 up to the present prompted mass rural displacement into overcrowding urban areas. This overpopulation has considerably exacerbated the further deterioration of the infrastructures and ecosystem, which began its steady decline after independence of the country in 1960 due to a mismanagement of water resources and infrastructures.

Access to quality DWSS services has dropped considerably since the '90s following plundering and political and military unrest, leading to stoppage of investments, degradation of facilities and weakening of the managerial capacities of the sub-sector entities. The key management-related constraints are those faced by these entities (REGIDESO, SNHR and PNA). Hence, the dire need for rehabilitation of existing infrastructures and replacement of those which were destroyed. The means of communication are degraded, making it difficult to supply inputs, materials and spare parts in DRC.

A key recommendation to the DRC authorities is to finance a population census as well as a hydrocensus, in order to estimate the water need of the country. A complete and detailed National Water Master Plan (NWMP) will assist to catch-up the millennium goals concerning water and sanitation.

REFERENCES

ADB (Africa Development Bank) (2007). Democratic Republic of the Congo: Semi-urban Drinking Water Supply and Sanitation Project. An appraisal report, Water and Sanitation Department, OWAS 2007.

AQUASTAT and Margat, J. (2005). Computation of renewable water resources by country, in km³/year, average. www.fao.org/nr/water/aquastat (accessed on Nov. 2007).

Bahati Chishugi, J. and Alemaw, B.F. 2009 The Hydrology of the Congo River Basin: A GIS-Based Hydrological Water Balance Model. Starrett, S. (eds): Proceedings of World Environmental and Water Resources Congress 2009: Great Rivers, May 17–21, 2009, Kansas City, Missouri; *ASCE/EWRI*, 978-0-7844-1036-3, 2009, pp. x1–16 [doi 10.1061/41036(342)593]

Bahati Chishugi, J. 2008 Hydrological Modeling of the Congo River Basin: A Water Balance Approach. MSc. Dissertation, University of Botswana, Botswana.

FAO/AQUASTAT. 1995. Water resources of African countries: a review.

FAOSTAT (2005). Population-Estimates 2006. FAO Statistics Division 2009 (http://faostat.fao.org, accessed on the 20th Jan. 2009).

Hakiza, G. (2002). Potentialités aquifères de la Plaine de la Rusizi (au Nord du lac Kivu), Thèse. Doct., Université de Liège.

Institut Royale des Sciences Naturelles de Belgique et Centre d'Echange d'informations de la R.D. Congo (2004). Etat de la diversité biologique en République Démocratique du Congo, Annexes techniques: Volet 15: Données démographiques et socioculturelles, Kinshasa.

Kazadi, S. and Kaoru, F. (1996). Interannual and long-term climate variability over the Zaire River Basin during the last 30 years, Journal of Geophysical Research, 101: 351–360.

Nile IWRM-Net (2007). Capacity Building Actions in Groundwater Management Issues as an Aspect of IWRM for the Nile Region. Short term Consultancy Report.

United Nations Environment Programme (Content Partner); Marty Matlock (Topic Editor). 2008. "Opportunities from freshwater in Africa." In: Encyclopedia of Earth. Eds. Cutler J. Cleveland (Washington, D.C.: Environmental Information Coalition, National Council for Science and the Environment)

UNEP (2002). African Environment Outlook, in 4th World Water Forum: African Regional Document, Water Resources Development in Africa, Mexico, 2006.

WHO/UNICEF (2008). A Snapshot of Drinking Water and Sanitation in Africa: A regional perspective based on new data from the WHO/UNICEF Joint Monitoring Programme for Water Supply and Sanitation. African Union Summit on Water Supply and Sanitation, 30 June 2008.

World Population Data Sheet (2009). Web: www.Scribd.com.

WFP et al. (2008). Republique Democratique du Congo: Analyse globale de la sécurité alimentaire et de la vulnérabilité (CFSVA). Données: Juillet 2007 et Février 2008.

16

Rural water supply and sanitation in Malawi: Groundwater context

Thokozani O.B. Kanyerere
Department of Earth Sciences, University of the Western Cape, South Africa
Department of Geography and Earth Sciences, University of Malawi, Malawi

Macpherson G.M. Nkhata & Timothy Mkandawire
Department of Water Development, Ministry of Irrigation and Water Development, Malawi

ABSTRACT: This chapter reviews current situation of rural water supply and sanitation (RWSS) in Malawi from groundwater context in terms of sources, coverage and functionality. The chapter presents trends in RWSS sector and suggests possible way forward. A holistic assessment determines variations that exist between groundwater and surface sources in terms of functionality rates. This chapter also evaluates progress about water monitoring activities and sanitation services. Since the target for the Malawi Government to a achieve the Millennium Development Goals as a UN Member state by 2015 is to reduce by half the proportion of population without access to sustainable safe drinking water and reduce by half the proportion of people without access to improved sanitation, there is need to assess both initiatives that are working and those that are weak in RWSS sector. The review focused on groundwater points and surface water stand pipes. Records on water resources and RWSS were reviewed and few interviews with key officers in water sector ware conducted. The 2007 water point mapping survey revealed that, out of 48,055 documented water points, 73% were groundwater points. Out of 27 districts, 23 districts (85%) depended more on groundwater as main source of water supply than surface water. This shows that groundwater serves majority of people in rural areas of Malawi, hence requires intensive scientific assessment of the resource. In terms of functionality, 72% of groundwater points were operational compared to 49% of surface water points. This indicates that groundwater is more reliable source than surface water. For both sources, the 2007 survey estimated that access to safe rural water supplies was 75%. The non-functional rate of 31% of rural water points is considered unacceptable by the Malawi Government if investment in the drilling more boreholes is to raise further coverage. This chapter suggests that department of water development should enforce a sustainable operation and maintenance system about broken-down boreholes. The review found that groundwater monitoring assessment is largely conducted by drilling contractors. Surface water monitoring activities are conducted on an ad hoc basis. This chapter recommends addressing the observed situation about water monitoring. The review revealed the existence of weak operation and maintenance aspects of rural water supplies due to lack of coordination, comprehensive framework, monitoring and evaluation among others. The review found that sanitation services for rural areas remain unacceptable and unsatisfactorily low (6%). This chapter recommends addressing the observed negative trends proactively.

16.1   INTRODUCTION

UNICEF (2008) observed that the main improved rural water supplies in Malawi are boreholes and shallow wells installed with hand pumps and piped gravity fed schemes (GFS) while unimproved sources comprise unprotected springs, streams, rivers and shallow open wells. (MoIWD, 2006) estimated the coverage of improved water supplies in rural Malawi at 71% which is higher than countries such as Tanzania 46%, Zambia 41% and Mozambique 26% within the Southern African Development Community (SADC) region. However, in 2007, it was estimated that 31% of improved water points were not functioning. This negative observation reduced effective coverage to 55% among the rural population in Malawi.

The rural water development context in Malawi is governed by the National Water Policy (2005), National Sanitation Policy (2008), National Irrigation Policy (2000), National Environmental Policy (2004), Water Resources Act (1969), Water Works Act (1995), Decentralization Policy (1998) and Local Government Act (1998). The current efforts of the Malawi Government are in two-folds; firstly, to develop the Water Supply and Sanitation Services and Water Boards Compensation Act and secondly, to develop a regulatory framework. The absence of these two is not desirable. The overall water policy goal is sustainable management and utilization of water resources in order to provide water services of acceptable quality and sufficient quantities that satisfy the requirements of every Malawian and enhance the country's natural ecosystem. The overall objective of the rural water supplies sub-sector is to achieve sustainable provision of community owned and managed water supply services that are equitably accessible to and used by individuals and entrepreneurs in rural communities for socio-economic development at an affordable cost, Ministry of Irrigation and Water Development (MoIWD, 2008). Specific objectives are to increase access of potable water to 80% by 2011 and reduce non-functional water points from 31% to 25% by 2011.

Provision of improved rural water supplies in Malawi started in 1968 with the Malawi Piped Scheme Program. Construction of a small gravity piped scheme was followed by a larger scheme. By 1980, 32 schemes had been constructed which were designed for a population of 640,000 and by 1988, a total of 63 schemes had been completed or were under construction that were designed for a population of 1.5 million (Baumann and Danert, 2008). However, from 1989 onwards, the program retracted because USAID support was reduced. The 1995–2003 National Water Development Project invested in water supply infrastructure and community development and also acted as an instrument for implementing sector reforms and institutional restructuring. The financing trends in the rural water supply project have been donor driven/supported and hence not been stable. In agreement with this observation, the MoIWD (2008) states that financing of the sector has not been well coordinated, with implementation taking place in discrete and fragmented projects with different objectives and conditionalities from funding agencies thus threatening sustainability of the system.

Currently, several documents exist that sets out investment requirement for new rural water supply infrastructure. For example, it is estimated that about 0.21 million rural people per year will need to be provided with services at an annual cost of USD8.28 million (USD40 per capita). MoIWD (2008) estimates a total investment requirement of USD193 million for the investment planning for rural water and sanitation to reach 98% coverage in 2025. It is further estimated that there are some 25,000 boreholes fitted with hand pumps as well as about 12,500 GFS taps in Malawi. This estimate equates to a capital investment of 24.65 billion Malawi Kwacha (USD170 million) in rural water supplies already. Analysis of various investment scenarios shows that they all lack realistic figures on investment that needed to operate and maintain such supplies. Nevertheless, it is important to note that investing in the management of the Malawi's rural water supply assets is a priority of the state, development partners and other service providers as well as the water users themselves.

In rural areas, the older practice of using the bush as toilets is declining because of increased hygiene education, increased health related interventions and desire of the rural population to be progressive. Health related interventions have both sanitation and water aspects. However, the fact remains that diarrhea disease is the fifth leading cause of death even though more than 60% of the population have latrines and 65% have safe water. The Malawi Population and Housing Census of 1997 found that 64% of households had some form of toilet (94% urban and 61% rural). The report showed that 61% of households had pit latrines (74% urban, 59% rural). Most rural latrines are unimproved and in urban areas 36% of households use shared pit latrines. Qualitative preliminary results of the Malawi Population and Housing Census of 2008 shows an improvement of such a situation (NSO, 1998, 2008).

16.2   STUDY AREA CHARACTERIZATION

This chapter presents the review on RWSS for the entire country. Malawi is part of the Great Rift Valley with 75% of the land surface being between 750 m and 1,350 m above sea level. Highland elevations are over 2,400 m and the lowest point on the southern border is 37 m above sea level.

Most of the country receives 760–1,150 mm of rainfall per annum. Almost 90% of the rains occur from December to March, with almost no rains from May to October in many parts of Malawi, (FAO 2008). The country experiences good rainfall from November to April. The mean annual rainfall is 1037 mm with 63.1%, 17.1% and 19.8% of the country receiving annual rainfall ranges of 650–1000 mm, 1001–1200 mm and greater than 1200 mm respectively. The mean monthly temperature ranges from 10° to 16°C in the highlands, 16° to 26°C in the plateau areas, 20° to 29°C along the lakeshore, 21° to 30°C in the Lower Shire Valley. The mean annual pan evaporation ranges from 1500–2000 mm in plateau areas and is highest 2000–2300 mm along lakeshore and Shire Valley. Extreme variations over the years in terms of rainfall and temperature exist which result into floods due to heavy rains and droughts due to low rainfall (Malawi Government, 1998b, 2008).

In terms of topography, Malawi has a wide range of relief which strongly influence climate, hydrology, movement of groundwater and population distribution. As a result, the country is divided into water resources units largely based on its topography. The four major topographic descriptions include: a) Plateau areas: These are peneplain, gently undulating surfaces with broad valleys and large level areas on the interfluves. These areas are on both sides of the rift valley and are generally 900–1,300 metres above mean sea level (m.a.m.s.l). The plateau areas are largely covered by a thick weathered material which forms an extensive and important aquifer; b) Mountainous areas: These are upland areas with several mountain areas and small uplands rising abruptly from the plateau where the underlying strata are more resistant to erosion. These have an elevation of 2,000–3,000 m.a.m.s.l; c) Rift valley escarpment areas are areas that fall steeply from plateau areas and slopes are commonly dissected. There is a considerably faulting in association with the development of the Malawi Rift Valley System which is the southern end of one limb of the East Africa Rift Valley System. In these areas, there is little potential for groundwater development except perhaps very locally. The aquifers are poor and discontinuous because the weathering products are often stripped away by erosion; d) Rift valley plain areas on the other hand are the alluvial plain areas of the rift valley floor which are generally sloping with low relief. The elevation is below 600 m.a.m.s.l. These areas have considerable potential for groundwater wherever suitable sedimentary sequences are found (Malawi Government-UNDP, 1986; Malawi Government, 2003a and 2003b).

In term of the geology suffice to say that knowledge of rock types facilitates understanding of local hydrogeological context of groundwater in different catchments. About 80% of Malawi is underlain by crystalline metamorphic and igneous rocks of Precambrian to Lower Palaeozoic age, commonly known as Basement Complex which is tectonically stable shield area such as gneisses. Rocks of sedimentary origin are also common. Along the Lakeshore plain and large parts of the Shire Valley, the Basement Complex/bedrock is covered by unconsolidated quaternary alluvium such as clays, silts, sands and gravels. These are derived from weathering of the escarpment area. Proper knowledge on these lithological and structural variations of rock types is important because they control the occurrences, movement, quality, availability of groundwater resources among other factors (Carter and Bennett, 1973; Malawi Government, 1986, 2003, 2008).

Malawi has relatively abundant natural water resources, with an estimated average of 3,000 m³ of surface water resources per capita, renewed annually. However, the spatial and seasonal distribution of the resource remains uneven. Few parts of the country have surface water available throughout the year. Most areas experience seasonal fluctuations, with water scarcity and pronounced shortages during the dry months of the year (Malawi Government, 1998). Malawi is heavily dependent on run-of-the-river schemes for drinking, irrigation among other uses. There are about 700 small to medium dams in Malawi with reservoir capacities ranging from a few cubic meter to about 5 million cubic meters. The total storage of these dams is estimated at about 100 million cubic meter or 0.1 km³. Most small dams were built in the 1950s by the then British colonial government. By 1990s, private individuals and companies were undertaking almost all dam projects. Since 2000, government intensified small community dam construction. They have been constructed under the auspices of the Ministry of Water Development and Agriculture, the Malawi Social Action Fund and other Non-Governmental Organizations. These dams supply water for domestic purposes such as drinking and small scale agricultural activities (Keating, 1995).

NSO (2008) reveals that Malawi has an estimated population of 13,066,320 of which 80% live in rural areas where they generally depend on water supplies for their livelihood options. The current approach of Malawi Government is to develop sector wide approach on planning, implementing, monitoring and evaluating water and sanitation issues. This is in line with the decentralization process which increases autonomy to district assemblies on the above issues. The Ministry of Irrigation and Water Development (MoIWD) had developed a second National Water Development Program which aims at building capacity of key stakeholders; improving access to potable water and improved sanitation facilities; creating effective management structures for piped water supply systems; improving support systems for regulation and monitoring; and strengthening coordination in the water and sanitation sector (Baumann and Danert, 2008).

16.3   METHODS

The objective of this chapter is three-fold: firstly, the chapter describes trends that exist between groundwater and surface water points as source for rural water supply in terms of coverage and functionality; secondly, the chapter evaluates current activities in the sanitation sector in terms of water monitoring and service provision; thirdly, the chapter highlights current initiatives that are working and not working within the rural water supply and sanitation sub-sector.

The selected variables for discussion in this chapter include: water sources, water coverage (water access); water point functionality, water source reliability, aquifer systems for water quantity, water quality monitoring, sanitation services and initiatives within RWSS. The analysis of these selected variables was envisaged to demonstrate current patterns in the RWSS in a crude holistic view to initiate further analytical approaches in assessing each variable with a more robust methodology.

The data presented and discussed in this chapter were obtained through record reviews and interviews. Different documents such as study reports, project reports and policy were obtained and critically synthesized. In addition to record review, interviews were conducted with different selected few key stakeholders at national, regional and district water offices on rural water supply and sanitation related aspects. The criteria for selecting key informants depended more on working within RWSS sub-section rather than being a mere officer in the water office. Familiarity with RWSS activities in the water office was the main criterion for choosing key informant respondents.

Many different methods have used to show distribution of rural water supplies with mapping methods being the most efficient technique to visualize the spatial coverage of rural water supplies. Particularly, exploratory data analysis is often used because the resulting tables and graphs can be easily understood. Statistical descriptive analysis using raw data obtained from the department of water development was used to analyze quantitative data presented in tables. Such analysis together with interpretative methods of document analysis and discourse analysis that were used to discuss results from qualitative data revealed the current situation in the rural water supplies and sanitation sub-sector in Malawi.

16.4   RESULTS AND DISCUSSIONS

16.4.1   *Water sources and water point coverage*

Improved water-points refer to piped supplies (taps), boreholes and protected shallow wells. The 2005 National Water Policy (MoIWD, 2005) does not consider unprotected shallow wells, springs, rivers or lakes as a means of providing potable water. Potable water means water which is free from disease-causing organisms, harmful chemical substances and radio-active matter, tastes good and is aesthetically appealing and free from objectionable colour or ordour. In line with this situation, groundwater is the largest source of rural water supply with 34,970 (73%) water points compared to surface water which had 13,085 (27%) water points by the end 2007 evaluation (Malawi Government, 2007). It can be said that groundwater covered larger parts of rural areas compared to surface water points. Coverage is defined as total number of water points both functional and non-functional multiplied by the respective standard served population figure and divided by the total rural population. The Ministry of Irrigation and Water Development (MoIWD, 2005) had defined access to safe water as people with a minimum quantity of 36 litres per capita per day within a maximum distance of 500 meters. Access is measured by assuming that a borehole serves a population of 250 people while a communal standpipe serves 120 people. The MoIWD does not consider shallow wells as safe water points. Since data on 36 litres per capita per day, maximum distance of 500 metres, 250 people per borehole and 120 people per stand pipe were not available for analysis, little is known on operational status of such variables. Quantitative analysis on these parameters is desirable.

16.4.2   *Functionality status of rural water points*

Groundwater sources have the highest functionality rate of 72% compared to 49% of surface water. UNICEF (2008) defines functionality as the number of working water points in a specified area such as Traditional Authority (TA) or District or national (minimum flow of 0.2 ℓ/s from a water point). Details for each district are presented in Table 16.2. Raw and unprocessed data sets from the Department of Water Development were accessed, processed and analyzed for the purpose of this chapter (Malawi Government, 2007). Results from such data are shown in Tables 16.1 and 16.2. Reasons for functionality versus non-functionality are complex for both groundwater and surface water sources as presented in the initiative

Table 16.1.   Coverage versus functionality rate for water sources.

Type of sources	Functionality (%)	Non-functionality %	Total coverage
Total surface water sources (stand pipes)	49 (6,464)	51 (6,621)	13,085
Gravity fed stand pipe (surface water)	48 (5,836)	52 (6,339)	12,175
Motorized stand pipe (surface water)	69 (628)	31 (282)	910
Total groundwater sources (hand pumps)	72 (25,189)	28 (9,781)	34,970
Mechanical drilled boreholes	75.5 (20,304)	24.5 (6,594)	26,898
Hand drilled boreholes	57 (680)	43 (504)	1,184
Shallow wells with hand pumps	61 (4,205)	39 (2,683)	6,888

Source: Malawi Government, 2007.

Table 16.2.   District level comparative analysis of rural water supply in Malawi.

		Comparative analysis of rural water supply between surface water and groundwater in Malawi					
		Total of stand pipes			Total of hand pumps		
	District names	Functional	Non-functional	Total	Functional	Non-functional	Total
1	Balaka	504	293	797	762	127	889
2	Blantyre	23	3	26	1,186	189	1,375
3	Chikwawa	70	271	341	609	426	1,035
4	Chiradzulu	33	141	174	798	165	963
5	Chitipa	253	282	535	686	325	1,011
6	Dedza	386	187	573	1,213	692	1,905
7	Dowa	9	5	14	964	458	1,422
8	Karonga	152	324	476	975	344	1,319
9	Kasungu	4	42	46	1,421	581	2,002
10	Likoma	5	3	8	21	9	30
11	Lilongwe	105	39	144	1,834	1,316	3,150
12	Machinga	307	629	936	902	265	1,167
13	Mangochi	74	51	125	1,844	476	2,334
14	Mchinji	96	54	150	690	191	881
15	Mulanje	623	1,240	1,863	786	234	1,020
16	Mwanza/Neno	41	21	62	542	210	752
17	Mzimba	690	266	956	2,596	1,079	3,675
18	Nkhata Bay	174	99	273	590	315	905
19	Nkhotakota	79	142	221	866	314	1,180
20	Nsanje	28	15	43	571	274	845
21	Ntcheu	387	651	1,038	1,007	452	1,459
22	Ntchisi	14	8	22	501	264	765
23	Phalombe	339	731	1,070	407	160	567
24	Rumphi	879	199	1,078	476	167	643
25	Salima	89	54	143	933	221	1,154
26	Thyolo	169	–	169	893	208	1,101
27	Zomba	931	871	1,802	1,116	305	1,421
	Grand total	6,464	6,621	13,085	25,189	9,781	34,970

Source: Malawi Government, 2007.

section of this chapter but needs further analysis cause-effect scenarios because the present overview is crude in nature.

Table 16.2 presents a detailed comparative analysis of rural water supply between groundwater and surface water scenario on one hand and among the 27 districts in Malawi on the other hand. Results show that countrywide, groundwater is widely supplied and is more reliable than surface water in terms of functionality. At district level, only 15% (Mulanje, Phalombe, Rumphi and Zomba have more surface water points) compared to 85% that depend more on groundwater as their main source of water supply putting more demand on groundwater abstraction for various uses. These findings strengthen the reasoning to have working groundwater monitoring system in order to know the quality and quantity of borehole water being used throughout the country and also to sustain the operation of such boreholes in the country.

The Water Point Mapping for Rural Water Supplies was carried out by a number of sector players between 2004 and 2008 and such work provides inventories of sources, functionality and provides the basis for estimating coverage. However, (Baumann and Danert, 2008) observe that update mechanisms for this data are still weak and Non-Governmental Organizations (NGOs) rarely report their new constructions to the District Assemblies or District Water Offices, even though they may monitor facilities, which they constructed. There seems to be no diagnostic data providing information on why sources are non-functional or abandoned. The Village Health Book represents another data source, but does not contain sufficient information for district planning. Lack of coordination between Government and NGOs offices at one hand and within Government Office such as District Water Office and Regional Office and National Office on the other hand makes the situation worst on water information. Baumann and Danert (2008) observed that NGOs work in isolation and do not provide information in terms of their plans or outputs. There are apparently numerous cases of the District Water Office learning about new water facilities by chance, or when the supply breaks down and the community seeks assistance from the Government for repair (Baumann and Danert, 2008).

16.4.3   *Groundwater availability and its associated quality pattern*

Some of the basement aquifer systems in Malawi have also been affected by intense lifting associated with the Great East African Rift system. Kaluwa (1998) reports that the existence of major geological lineaments and faults in most crystalline aquifers in Zomba area especially the North East trending faults are associated with the East African Rift Valley which marks the Zomba Plateau's west border. These major faults and lineaments identified on the plateau, together with the weathered portions of the crystalline basement rocks constitute the aquifer system of the plateau. Kaluwa (1998) further reports that these geological features, coupled with high precipitation in areas like Zomba Plateau and Mulanje Mountain make them good recharge areas for many aquifers in the Southern Region of Malawi. These conditions explain the availability of groundwater for rural water supply and sanitation uses.

Groundwater quantities available in the Basement Aquifer Systems (marked blue in Figure 16.1) are usually very limited. However, even these limited resources can be reduced in exploitation value by the presence of undesirable natural hydrogeochemicals or by introduced contaminants. However, most of the basement aquifers to a larger extent do not suffer from natural hydrogeochemical problems in that the rock materials are relatively inert. In Malawi, just as many developing countries, groundwater contamination is not normally pronounced as is the case in many developed countries. This possibly explains the non-existence of routine monitoring of groundwater quality by officials.

Hydrogeochemical composition of groundwater in the basement aquifers relate to a complex sequence of processes, initially dependent upon atmospheric input but dominated by soil water and biological reactions in the unsaturated zones and saturated zone. Hydrogeochemistry is further complicated with occurrence of contaminants. The basement aquifer are characterized by water which is dominated by alkaline earth in the cation group and by the carbonates in the anion group. Total dissolved solids content values are generally less than 1000 mg/ℓ and typically around 350 mg/ℓ. Many areas in Malawi continue to be extensively drilled to exploit groundwater for water supply regardless of routine assessment of such resources. However, borehole drilling contributes to availability of data about groundwater resources in terms of physical, chemical and microbiological status of the water at the time of borehole drilling and installation.

Generally, the trend is that sodium-magnesium bicarbonate groundwater predominates throughout the country. There is very little variation in the chemical composition whether the groundwater is from colluvium, weathered gneiss or fracture gneiss. This implies that the aquifers are in continuity with each other. Cationic exchange and the mineralogy of the aquifer are the main factors determining the water composition even though chemicals from human activities have potential to change such expected pattern especially with current increased agricultural activities.

At national level, groundwater quality is generally acceptable for domestic use despite widespread occurrence of high iron concentration (Malawi Government-UNPD, 1986). However, there are some problem areas where groundwater quality is not suitable for human consumption although most of these local areas have still not been delineated. Nevertheless, extensive

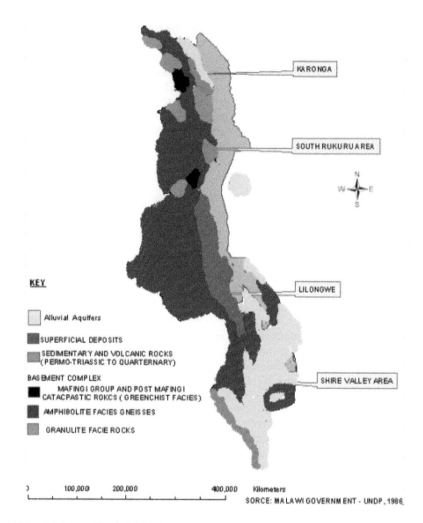

Figure 16.1.   Major aquifers in Malawi.

studies have been done on concentration of fluoride, iron and sulphate in groundwater. Concentration of fluoride, is still restricted to Karonga Lakeshore area, Lilongwe, Nkhotakota, Mwanza and Chiradzulu. Similarly, analyses of sulphate and iron concentrations have mostly been confined to boreholes in Dowa district. Further research is therefore needed for other parts of the country. The existing data reveal high concentration (0.6–0.8 mg/ℓ) of fluoride in groundwater above what is recommended by Ministry of Health (Malawi Government, 1998b). Such high concentration levels are not fit for human consumption. Such unacceptable fluoride levels have been reported in Karonga, Lilongwe, Mwanza, Nkhotakota and Chiradzulu (Malawi Government, 1998).

Alluvial aquifers (marked yellow in Figure 16.1) are generally high yielding with recorded yields in excess of 11 litres per second (Stanley International 1983). These aquifers are commonly sedimentary rocks and layered volcanic rocks (Malawi Government-UNDP, 1986). These are found along the Lake Malawi Shore, in the western side of the Shire River Valley and the Lake Chilwa basin on the outer slopes of Zomba Plateau. The alluvial aquifers are fluvial and lacustrine in nature. They are highly variable in character both in vertical sequence and lateral extent. Groundwater in the alluvial aquifers is more mineralized than in the basement aquifers. Its routine monitoring would have desirable regardless of its current inexistence.

16.4.4 *Water quality monitoring*

It is the responsibility of the Water Quality Division in the Department of Water Development to be monitoring the quality of water both surface and groundwater resources and ensure conformity to the specified standards. The Division evaluates applications for discharge effluent permits and monitors effluent discharge to ensure compliance with established standards. Water quality monitoring started in the early 1970 s with emphasis on analysis of sediment in rivers. This was undertaken at the Lilongwe Central Water Laboratory. Over time, the laboratory started to carry out analysis of other parameters and other water sources including groundwater. To ensure countrywide coverage, in 1993, two regional laboratories (Mzuzu in Northern Region and Blantyre in Southern Region) were also established. However, up to today, the Northern Region laboratory remains non-functional. Financial resources, equipment and technical expertise are lacking. The same applies to the Blantyre laboratory (Malawi Government 1998a, 2001).

Since 1980 s, the Department of Water Development has been conducting analysis of different parameters from stream, river, lake, shallow well, spring and borehole water. Monitoring of surface water sources involves major rivers and lakes in the country. Sampling points are mainly located in the River Gauging stations. These have been established by for hydrological monitoring. In addition, effluent samples are collected from 42 wastewater treatment plants that discharge into water bodies. About 5 of the total surface water are of regional and international interest for Southern African Development Community (SADC) Water Sector Projects. No monitoring boreholes are established yet even though proposals are underway. However, Table 16.3 shows summarized parameters that are tested for physical, chemical and microbiological status of both surface and groundwater for human consumption in relation to established guidelines by MoIWD about drinking in Malawi.

Despite the fact that there has been an increase in the use of agrochemicals for agricultural activities country wide, surprisingly the Department does not conduct pesticide and herbicides analysis. This is despite the fact that the department has acquired equipment for a Gas Chromatography (GC). Discussions with the Department revealed that lack of other accessories for the equipment, skills to maintain it and inadequate training on its use make the equipment not being used. Currently, the monitoring of surface water sources and

Table 16.3. Summarized parameters tested in drinking water sources.

Groundwater sources	Surface water sources
PH value	PH value
Electrical conductivity	Electrical conductivity
Temperature	Temperature
Suspended solids	Suspended solids
Turbidity	Turbidity
Dissolved oxygen	Dissolved oxygen
Biochemical oxygen demand (BOD_5)	Biochemical oxygen demand (BOD_5)
Chemical oxygen demand (COD_{cr})	Chemical oxygen demand (COD_{cr})
Nitrate	Nitrate
Total phosphorus	Total phosphorus
Fluoride	Fluoride
Faecal coloform (count/100 ml)	Faecal coloform (count/100 ml)
Faecal streptococci (count/100 ml)	Faecal streptococci (count/100 ml)

Source: World Bank, 1992.

wastewater treatment works are actually conducted when the need arises or on request by the client.

Water quality assessment of groundwater is normally carried out soon after pump installation of a newly constructed water source or after rehabilitation. It is mainly carried out by drilling constructors as it is part of their contract agreement. However, it is not always carried out. This is a particular problem when borehole drilling is driven by political influence or religious objectives or other NGOs motivations where poor or lack of coordination in the water sector exist.

16.4.5   *Current sanitation service provision*

The 1998 Population and Housing Census and Malawi Growth and Development Strategy Document for 2005 estimated that 74% and 83% of the population have access to some form of sanitation, although only 5% are served through a sewer connection, septic tank or improved latrine. This situation is not even and in some rural villages the percentage drops to 40% while in some urban areas it can rise to about 95% (Malawi Government, 2001). By definition a safe toilet facility should be that with a flush to a sewerage system, a septic tank or a latrine with sanitation platform (san-plat) located less than 50 meters from user's dwelling. The existing difference in sanitation coverage between levels and services for rural and urban areas need to be narrowed. In urban areas, 89% of the population has access to some form of sanitation system, 22% are served by sewer connection, septic tank or an improved latrine, and 36% households share facilities (Malawi Government, 1995).

In rural areas, 70% of people have access to some form of excreta disposal but only 4% are served by an improved latrine. The standard of most latrines constructed in rural areas is generally poor with average latrine life of 3.9 years (Malawi Government, 1998). These toilets are major sources of breeding of flies which transmit various diseases. If unimproved latrines are considered unacceptable and require upgrading through the application of a san-plat or other improved technology, then only 6% of rural population are considered to have a satisfactory level of sanitation service (UNICEF, 2005).

Out of the 13,066,320 million people that live in Malawi, about 11 million are in need of an improved system for excreta disposal. Over 7 million of these are rural residents (NSO, 1998, 2008). Strategies to increase the supply of potable water and to alleviate sanitation problems must, therefore, focus on rural areas and the squatter or peri-urban areas if coverage is to be significantly improved (UNICEF, 2005).

Different reports about RWSS show that sanitation and hygiene promotion has lagged behind because emphasis has been placed on water supply. Before 1996, sanitation and hygiene lacked clear policy direction on which ministry had the overall responsibility for sanitation services. Currently, responsibility seems to have become shared by the Ministries of Health, Water Development, Natural Resources and Environmental Affairs and the Department of Local Government. There is no clear leader and method of co-ordination. Improved sanitation coverage is currently below 10% and this leads directly to a high incidence of diarrhoea related cases, especially during the rainy season. Department of Water Development took a lead to initiate the process that saw the birth of a sanitation policy (Malawi Government, 1998, 2001).

16.4.6   *Current initiatives within the rural water supply and sanitation sub-sector*

Several reports including that of Baumann and Danert (2008) observe that efforts to improving access to safe water in the rural areas of Malawi are in the positive direction as evidenced by higher coverage and functionality rate figures when compared with some neighbouring countries of Tanzania, Mozambique and Zambia. In addition to high coverage, standardization policy on hand pumps is effectively operational and the country has good quality pumps mainly of Afridev and Malda type and spare parts for such pumps are available within the country in local specialized shops. Functionality rate of the supplied water facilities is quite high even though further improvement is needed to significantly benefit from the current investments being made in the rural water supply sector. Communities in rural areas are aware about the ownership concept of their water facilities even though they are struggling in revenue collection towards maintenance work of their water pumps. Generally, the Malawi Government, NGOs and donor organizations had been investing financial resources significantly in the rural water supplies. Water point mapping has been initiated even though it lacks coordinated approach for updating the mapped water supplies (UNICEF, 2005 and MoIWD, 2005, 2008).

Regardless of the positive strides that the country experience, the rate of non-functioning water points slows down the progress on water point coverage and access. The operations and maintenance of water points are inadequate and inconstant which contributes to the increased number of non-functional water points because sometimes communities do not know whom to turn to on when their water points break down or dry up. Lack of a common approach to accessing safe water supplies to rural communities is another problem. The current situation is that each NGO or Government project or religious grouping works on their own without collaborating with each other. The result has been that some rural areas are oversupplied with water facilities while others are under served. The present coordination is that District Water Offices are bypassed with data, information and activities about water supplies in their districts. With their few staff numbers and lack of other resources such vehicles and stationary, the situation is worst. District water offices are not aware of number of organizations working on water-related projects, number of water supplies, working and not non-working water points. General water quality and water quantity assessment activities are weak in the water sector. Integrated water resources management approach remains a theoretical concept with other district water offices not even being aware of the concept itself (Nkhata, Mponda, Personal Communication, 2008).

16.5 CONCLUSIONS

Results of this study have shown that regardless of substantial investment in rural water supplies by both the Malawi Government and NGOs, figures on investment scenarios lack realistic comprehension where financing system is not well coordinated with implementation taking place in discrete and fragmented project with different objectives and conditionalities. Even though this chapter found that one of the major reasons for functionality and non-functionality status was due to management style, our analysis reveal more complex reasons beyond the scope of the current chapter.

Consistent with previous reports (MoIWD, 2008; Malawi Government 1995; Kleemeier, 2000 and UNICEF, 2008) our study found that comprehensive nationwide water resource assessments are not being carried out in terms of quality and quantity. Coupled with lack of updated regulatory framework, it is difficult to properly develop, abstract, utilize and manage such waters. Our review on sanitation services revealed than only 6% of rural population are considered to have a satisfactory level of sanitation service and this sub-sector lags behind rural water supply sub-sector. On information availability about water resources, our findings agree with Baumann and Danert (2008) that update mechanisms coupled with lack of coordination within the water sector cloud the realist knowledge on rural water supply and sanitation situation.

Lastly, our findings should be interpreted within the context of the study's limitations. In addition to common limitations associated with responses from key informants' interviews, raw data obtained from the department of water development within the MoIWD might be influenced by the data cleaning process by Government Officers. However, our analyses show convincing patterns in terms major water sources, extent of water coverage, proportions of functionality and non-functionality rates between groundwater and surface water in all the districts in the country (Tables 16.1 and 16.2). It shows wich physical, chemical and microbiological parameters are tested when water is being supplied to people from groundwater and surface water sources.

Our description on sanitation services clearly demonstrates the importance to scale up the investment in this sub-sector. The analysis deliberately disentangles urban from rural situation, thus providing a guide to designing a more appropriate approach to reducing the present gap between sanitation and rural water sub-sectors. Future research is needed to gain a better understanding of water resource management and utilization using a multilevel analytical approach to provide important insights and to identify multidimensional aspects within communities. These results led us to conclude that we are dealing with rural water supply and sanitation trajectories in time. Therefore, static approaches in dealing with such issues will lead to erroneous interpretations of the current and future efforts or initiatives in the RWSS dynamics.

ACKNOWLEDGEMENT

Authors would like to thank NUFU Project on Water Sciences for the financial support of this work. Authors are grateful to Departments of Water Resources and Water Supply and Sanitation in the Ministry of Irrigation and Water Development in Malawi for providing raw data and reviewing some of the research findings.

REFERENCES

Baumann, E. and K. Danert. (2008). Operation and Maintenace of Rural Water Supplies in Malawi; SKAT Study Findings, UNICEF, Lilongwe, Malawi.

Carter, R.T. and Bennett, J.D. (1973). The Geology and Mineral Resources of Malawi. Bulletin, Geology, Survey, Malawi. Malawi Government Printer, Zomba, Malawi pp. 262.

FAO (2008). http://www.fao.org/ag/AGP/AGPC/doc/Counprof/Malawi.htm#2.1, Food and Agriculture Organisation, Italy. Viewed on 15th December 2008.

Kaluwa, P.W.R. (1998). Water Resources Management and Policy in Malawi, Paper presented at a workshop on Water Resources Development and Vector Borne Diseases, Ryalls Hotel—Blantyre 9th to 13th November, 1998.

Keating, M. (1995). Agenda for Change: "A Plain Language Version of Agenda 21 and the Lake Malawi and Shire River system, "Lake Malawi level control stage 2: Feasibility Study, Inception Report." Norconsult, March.

Malawi Government (2008). The Second National Communication of the Republic of Malawi to The United Nations Framework Convention on Climate Change; Climate Change Project Office; Department of Environmental Affairs, Ministry of Lands and Natural Resources, Lilongwe, Malawi.

Malawi Government (2007). Raw Unprocessed Data and Unpublished Data, Ministry of Water Department, Lilongwe, Malawi.

Malawi Government (2003a). Integrated Water Resource Development Plan for Thindwa, Harriet, (1999), Red locust population monitoring and control in Malawi, 1988–1998." Ibsect Sci. Applic. 19:351–353.

Malawi Government (2003b). Research and Systematic Observation in Malawi; Environmental Affairs Department, Ministry of Natural Resources and Environmental Affairs, Lilongwe, Malawi.

Malawi Government (2001). Joint Review of Malawi Water and Sanitation Sector, Ministry of Water Development. Lilongwe , Malawi.

Malawi Government (1998a). Water and Sanitation Sector Program up to the Year 2020.

Malawi Government (1998b). State of the Environment, Report for Malawi, Environmental Affairs Department, Lilongwe, Malawi.

Malawi Government-UNDP (1986). National Water Resources Master Plan, Report and Appendices; Department of Water Resources, Ministry of Works and Supplies, Lilongwe, Malawi.

Malawi Government (1995). Rural Water supply and Sanitation in Malawi Sustainability through Community Based Management, Government of Malawi.

Ministry of Irrigation and Water Development (MoIWD) (2005). The 2005 National Water Policy, Ministry of Irrigation and Water Development August, 2005, Malawi Government.

MoIWD (2006). Implementation Guide Manual for Establishment of Area Mechanics, Revised Draft, November 29, 2006, Ministry of Irrigation Water Development, Malawi Government

National Statistical Office (NSO) (1998). 1998 Population and Housing Census; Preliminary Report, Government Printer, Zomba, Malawi.

National Statistical Office (NSO) (2008). 2008 Population and Housing Census; Preliminary Report, Government Printer, Zomba, Malawi.

Personal Communication (2008). Key Informant Interviews, Lilongwe (Marcpherson Nkhata); Mzuzu (Kondwani Mponda) and Nkhata-Bay Water Offices (Kingsely Mdhuli), Malawi.

Stanley International (1983). Working Paper on Hydrogeology in Lilongwe Region: Lilongwe Water Supply and Sanitation Master Plan, Lilongwe, Malawi.

UNICEF (2005). Water, Environment and Sanitation (WES) Malawi Country Profile, UNICEF Malawi.

UNICEF (2008). Operation and Maintenance of Rural Water Supplies in Malawi; Proposed Operation and Maintenance and Action Plan, SKAT Report, Lilongwe, Malawi.

World Bank (1992). Economic reports on environmental policy, Malawi, Volumes I and II, Lilongwe, Malawi.

17

Groundwater and sanitation in an arid region

M.A. Wienecke
Habitat Research & Development Centre, Windhoek, Namibia

ABSTRACT: Water in an arid region should be considered as a valuable resource, due to its scarcity. Consequently it should neither be polluted nor wasted. However, certain practices and technological applications contribute to groundwater pollution. Examples are the overflow oxidations dams after heavy rainfall and the consequent pollution of the groundwater downstream; fertilisers used in irrigation schemes may flow into a river; or the dumping of effluent on the solid waste site from where the liquids permeates into a river. To avoid groundwater pollution, several approaches could provide solutions, utilising dry or wet sanitation systems.

A dry sanitation system developed in Namibia is the Otji-Toilet, which is a desiccation unit utilising solar energy. It is a low-cost solution, which is suitable in urban as well as rural areas, as the units make use of locally available materials and skills. The promotion of wet systems by means of decentralised waste water treatment systems has been accomplished in Lesotho. The most important feature is the recovery of resources, such as biogas, fertiliser and water, for gardening and food production. No moving equipment or electricity is required. Various modules are offered to treat effluent to any required water quality standard.

17.1 INTRODUCTION

Water is one the most essential components of what is termed life on Earth. Leonardo da Vinci (quoted by Miller 1996:117) summarised the importance of water: "Water is the driver of nature", because the other nutrient cycles would not exist in the present form and life on earth, consisting of water containing cells and tissue, would not be possible. There are two important aspects: humans cannot produce water, and potable water is a limited resource, in particular in an arid region. However, this resource is often treated as if there are no limits. Indigenous peoples in arid areas have adapted to these conditions. However, the process of modernisation has and had negative consequences, such as pollution, over-abstraction or misuse, e.g. in the case of waterborne sanitation.

Water scarcity and population growth are crucial issues in Africa. Other water resource issues are (Houghton-Carr 2006:51): extreme variability of water resources, as droughts are common, shared resources in the form of international basins. In arid or semi-arid regions, groundwater is often the only source of water locally available. In Namibia about 45% is sourced from groundwater sources, whereas in Lesotho, due to the mountainous character of the country, the inhabitants in the past never made "extensive use of surface water found in rivers and dams for household purposes, nor were they engaged in the digging of shallow wells, due to the fact that supply was met through the use of springs" (Molapo 2005:43). This changed with the introduction of boreholes, which made it possible for settlements to expand into more arid parts of the country and to populate areas, which otherwise would not have been occupied (Chakela quoted by Molapo 2005:43).

Contamination of groundwater can occur as a consequence of private, municipal, agricultural or industrial activities, "particularly the discharge of waste water effluents" (Christelis et al. 2001:39). Another possibility is the overflow of oxidation ponds during the rainy season, resulting in contaminating potable groundwater. Sanitation is therefore an important aspect in protecting, conserving and limiting the risks of polluting groundwater sources, in particular where population growth rates are high. The BGR (2008) summarises the linkages between groundwater and pollution:

> "*Groundwater is worldwide the major source of drinking water supply, especially in arid regions. The protection of groundwater resources from pollution thus is a key element of sustainable human development. Natural groundwater, unaffected by human activities, is free of pathogenic germs. Once such germs have infiltrated into the groundwater, e.g. through leaking sewerage systems, it takes about 50 days until 99% have vanished. In cases where drinking water wells are located in direct neighbourhood to a pollution source (e.g. cesspits without any further treatment), travel times of the groundwater will be much shorter. This way, water users face increased health risks*".

Two case studies from Namibia and Lesotho will illustrate the relationship between ground-water/aquifers, sanitation and water consumption. Aranos is located above the Stampriet aquifer, which forms part of the Kalahari/Karoo Basin, whereas the Kingdom of Lesotho and the Maloti mountain range are located above the Karoo Sedimentary Aquifer, which is shared with South Africa. The underlying question is: how vulnerable are these resources?

17.2   VULNERABILITY OF GROUNDWATER SOURCES

Sanitation options, such as pit latrines, have to be considered as potential hazards, due to their possible contamination of groundwater. Sudgen (2006) warns that,

> "*The contamination of boreholes and shallow wells from on-site latrines is an issue that is generally poorly understood and irrationally assessed by organisations imple-menting water supply and sanitation programmes. This should not be the case as the health risks are often lower than popularly anticipated*".

The two most widespread groundwater contaminants attributed to latrines are nitrates and microbiological contamination (Cave & Kolsky 1999:18). Cairncross and Feachem (Cave & Kolsky 1999:19) describe how pit latrines in arid Botswana, with a population density of 63 people/hectare, can lead to nitrate concentrations far in excess of the WHO guidelines of an upper limit of 10 mg-l as nitrogen, or 50 mg-l as NO_3 (Cave & Kolsky 1999:18). The prob-lem arose because of the limited infiltration available to dilute the human nitrogen loading.

The reason for this is that pathogens are moved in the medium in which they live. They cannot travel further or faster than the medium in which they are suspended (Cave & Kolsky 1999:12). This is an important aspect in understanding water point contamination (Sugden 2006). In addition there are two factors that are closely related to the contamination of water sources: size of the pathogen and die-off rate. The size of Helminth eggs and Protozoa are relatively large and are efficiently removed through the physical filtration process in the soil (Lewis, Foster et al. cited by Sugden 2006; Cave & Kolsky 1999:9). However, the smaller bac-teria and viruses may travel more unrestricted through the subsoil. Faecal micro-organisms have a limited life span in the environment. Their die-off rate varies considerably, for example E. coli bacteria have an estimated half life in temperate groundwater of 10 to 12 days, with survival of high numbers up to 32 days. Some salmonella species may be active for up to 42 days. The number of organisms surviving in an aquifer declines at an exponential rate, which depends upon a range of chemical, physical and biological processes, indigenous groundwater micro-organisms and water chemistry (Toze et al. quoted by EPA 2005:17). The key factor in the removal and elimination of bacteria and viruses from groundwater is the maximisation of the effluent residence time between the source of contamination and the point of water abstraction (Cave & Kolsky 1999:2).

Groundwater hydrology has a major effect on the risks of aquifer pollution (Cave & Kolsky 1999:2). Lewis et al. (quoted by Cave & Kolsky 1999:17) state that, the linear travel of bacterial pollution is dependent on the velocity of the groundwater flow and the viability of the organisms. Sugden (2006) has identified six factors, which have an effect on pathogen transmission from a latrine to a nearby water point: the amount of liquid in the pit, the nature of the unsaturated zone below the pit, the distance between the base of the pit and water table, the nature of the saturated zone (aquifer), the horizontal distance between latrine and water point, and the direction and velocity of the groundwater flow. In this regard the author points out that, the "greater the hydraulic gradient towards the water point, the higher the risk of water point contamination". Furthermore, the longer the time needed to travel from the pit to a water point, the higher the natural die-off rate. In order to assess the risk of groundwater pollution from on-site sanitation, the British Geological Survey indicated that if the time is more than 50 days, it represents a low risk, whereas a time of less than 25 days, it would represent a high risk (Sugden 2006). The type of aquifer influences the permeability, for example, in silt a permeability of 0.01–0.1 m/d can be expected, in clean sand 10–100 m/d, and fractured rock is difficult to generalise, but can be thousands of metres per day. Sugden (2006) suggests a number of methods to reduce the risk of contamination:

- Increase horizontal separation distances between latrine and water point
- Move water point higher than latrines
- Change to a drier form of latrine
- Increase vertical separation between bottom of pit and water table by using shallower pits or vaults latrines
- If a borehole is being used, site the screens lower in the water table
- Treat water supplies or encourage use of home water treatment

These recommendations are also important guidelines for the implementation of the Millennium Development Goals.

17.3   MILLENNIUM DEVELOPMENT GOALS (MDGs)

Goal 7 of the MDGs aims at reducing by half the proportion of people without sustainable access to safe drinking water and basic sanitation. However, the original declaration states:

> *"To halve, by the year 2015, the proportion of the world's people whose income is less than one dollar a day and the proportion of people who suffer from hunger and, by the same date, to halve the proportion of people who are unable to reach or to afford safe drinking water"* (United Nations 2000:5).
> *"To stop the unsustainable exploitation of water resources by developing water management strategies at the regional, national and local levels, which promote both equitable access and adequate supplies"* (United Nations 2000:6).

Sanitation **was not** included. Although the various goals should form an integrated framework in order to provide a comprehensive basis for implementation, this is usually not the case (as seen in the declaration above). MDG "country programmes in Africa remain constrained by insufficient as well as unpredictable financing and do not spell out the full set of policies and supporting public expenditure needed to achieve sustained economic growth and the MDGs" (MDG Africa Steering Group 2008:26). To promote alternative sanitation in an arid country, in the case of Namibia, the most arid country south of the Sahara, requires appropriate technologies for the poor AND the affluent strata, implemented by municipalities and communities. An enabling policy environment is also essential, as well as acceptance of the alternatives offered, as perceptions may hamper the implementation of appropriate technologies. Practical examples will illustrate the following:

1. affordable sanitation projects,
2. avoiding pollution of water resources, including groundwater, and
3. locally available resources to produce food.

17.3.1   *Namibia*

Based on the 2001 Census, sanitation systems in urban and rural areas differ widely through-out Namibia. The bush, as the most common dry sanitation method (54.2%), is basically a squatting method, which has often been described as unhygienic, unsuitable, and unaccept-able. The government claims that progress has been made with regard to the proportion of rural population's access to basic sanitation. With respect to the urban population, access to basic sanitation is worsening (Republic of Namibia 2004:34).

Namibia has published a progress report on the Millennium Development Goals (MDGs) in 2004. Among the challenges and opportunities, concerning Goal 7, the following is men-tioned (Republic of Namibia 2004c:34):

> *"Providing access to safe water and basic sanitation is severely complicated by the size of the country, its arid climate and the dispersion of the population. A particular challenge is to provide for the additional water needs of people and families affected by HIV/AIDS".*

Under the heading *"Supportive Environment"*, the document elaborates on State of the Environment reports, the Namibia Water Resource Management Review, the 2004 Water Act, adequate shelter for all and sustainable human settlement development (Republic of Namibia 2004c:35). Sanitation is not mentioned. The summaries of achievements in connec-tion with Goal 7 are (Republic of Namibia 2004:34) discussed in Table 17.1.

17.3.2   *Lesotho*

In 1981 only 21% of the population in Lesotho (84% in urban areas and 15% in rural areas) had any sort of sanitation (World Bank 2002:2). The Bank points out that the urban statis-tics were misleading, as only 22% of the population had a sanitation system that effectively isolated human excreta from the environment. Many urban households had highly unsat-isfactory bucket latrines, which had to be emptied manually once or twice a week. In 2002, 53% of the population had adequate sanitation (World Bank 2002:1), after the government implemented water supply and sanitation improvement programmes. The MCC (2006:6) reinforces this statement by referring to the absence of waterborne diseases in either the urban or rural areas, which indicates that sanitation is relatively good.

In Lesotho about 74% of households have access to treated piped water. Another 6% has access to what is called "reasonably safe water", which refers to water from boreholes (UNDP Lesotho 2007:20). This is regarded as safe, due to the depth of the boreholes and if protected from contamination by residents and surface run-off. Rural water supply systems are typically hand pumps or small-piped systems, utilising water from springs and boreholes (UNDP Lesotho 2007:21).

Table 17.1.   Summary of indicators (Republic of Namibia, 2004:34).

Indicator	1992	2003	2006 target	Progress toward target
Proportion of rural population with access to safe drinking water	45	80	80	Good
Proportion of urban population with access to safe drinking water	99	98	+95	Good
Proportion of rural population with access to basic sanitation	15	21	50	Slow
Proportion of urban population with access to basic sanitation	89	82	–	Worsening

Table 17.2.   Prospects for achieving the MDGs (UNDP, Lesotho, 2007:61).

Goal	Target	Will the goal/ target be met?	State of supportive environment
Ensure environmental sustainability	Integrate the principles of sustainable development into country policies and programmes and reverse the loss of environmental resources	Potentially	Weak but improving
	Halve, by 2015, the proportion of people without sustainable access to safe drinking water	Potentially	Fair
	Halve, by 2015, the proportion of people without sustainable access to sanitation	Potentially	Weak but improving

However, the MCC (2006:26–27) states:

"The majority of villages in Lesotho are served by gravity water systems based on spring sources. In many parts of the lowlands the yields of the springs have also been affected negatively by reduced recharge of ground water resources. The decline in spring yields coupled with population growth means these systems no longer meet the required standard of 30 liters per capita per day ...".

"Wetlands are among the most important eco systems in the country as they act as groundwater recharge points, control floods and erosion and most importantly contribute to the maintenance of the required water quality and quantity in streams and springs" (UNDP Lesotho 2007:21). The wetlands in the highlands of Lesotho purify water, store it, regulate stream flow, recharge groundwater and retain nutrients (MCC 2006:53). However, UNDP Lesotho (2006:12) points out that the degradation of wetlands and water sources, rapid urban sprawl, accompanied by unplanned human settlements, air and water pollution, as well as ineffective solid waste management, are among the serious environmental problems the country faces. Therefore the Millennium Challenge Account invests in the protection of wetlands to support water purification, erosion control, flood attenuation, groundwater recharge, ecosystem support, sediment trapping and microclimate maintenance (MCC 2006:52). The summary of MDG goals[1], relating to water and sanitation, according to UNDP Lesotho (2007:61) is given in Table 17.2.

As part of the Millennium Challenge Account and to achieve the MDGs, sanitation facilities and water provision project anticipate to improve the life of 75,000 people in rural areas of Lesotho (MCC 2006:27). The Poverty Reduction Strategy made provision for the construction of water-borne sewerage systems. Government intended to assist poor households in Maseru "to up-grade their dry sanitation systems (latrines) to ensure that these provide adequate protection against disease and do not present a risk to ground water" (Kingdom of Lesotho no date:xv). The reason for this is that in Maseru, spillage from septic tanks and the high concentration of latrines may contaminate groundwater resources (MCC 2006:48). These water sources are utilised by many poor households depend through the use of hand pumps.

17.4   SANITATION PRACTICES

The provision of sanitation can be achieved by employing either a wet or a dry system. Two examples will illustrate both options. The first case study is found in Namibia, where a

---

[1] The official MD+5 Status Report 2007, to be published by the United Nations Country Team, is "In Progress" (UNDP 2007). UNDG (2009) shows that so far no report has been submitted (accessed 27 June 2009).

locally made dry sanitation option is promoted and the second comes from Lesotho where decentralised wet systems are propagated.

17.4.1   *Dry sanitation: The Otji-Toilet*

In Namibia over 100,000 boreholes have been drilled. About 50,000 are regarded as production boreholes. About 45% of the water supplies to urban areas and farms are groundwater, and 45% of the water used in agriculture is also groundwater (Heyns & Struckmeier 2001:12). Only around 1% of Namibia's total area is suitable for seasonal or permanent crop production (Schneider et al. 2001:26). Therefore water availability, especially in the long term, is a major concern. Heyns and Struckmeier (2001:11) have summarised the sources in Table 17.3.

The artesian aquifers at Aranos are shared with Botswana and South Africa, but are predominantly used in Namibia where they are also recharged (Bockmühl 2001:90). The annual mean precipitation in this region is about 200 millimetres, whereas evaporation is around 3,300 millimetres. The aquifer has a relatively small recharge area and is of importance to the many artesian wells, which are important to agricultural activities in this semi-arid region (Hemming et al., 1998:598). Water samples from the Stampriet-Auob aquifer range in age from modern to approximately 45,000 years. The Stampriet aquifer is a primary aquifer covering an area of 65,000 square kilometres (Lindgren 1999). The aquifer is artesian and confined between impervious rocks, resulting in high pressure. In some parts, saline water overlies the freshwater and poses a contamination threat to the freshwater.

The Stampriet aquifer is enclosed by a surface limestone plateau that rises 80 metres above the Fish River plain (Bockmühl 2001:88). Large storage dams constructed in the upstream river system cut off major floods that would otherwise feed the Stampriet Artesian Basin (Bockmühl 2001:90). The aquifer "seeps through the overlying Kalahari sands until it is trapped by a layer of Karoo sandstone" (Mendelsohn, et al., 2002:65). This aquifer is described as porous and productive. However, groundwater flowing into the south-eastern section becomes unpalatable water, due to the high TDS levels. This area is also known as the saltblock, as the TDS is above 5,000 mg/ℓ.

Groundwater in the Stampriet aquifer does not require pumping, due to the high pressure head, which forms artesian wells (Christelis et al., 2001:34). However, modelling results indicate that current "development" projects overutilising this resource and a reduction of 30% in abstraction is required (Bockmühl 2001:91). Porous aquifers are vulnerable to pollution. Bockmühl (2001:91) points out that, within weeks after heavy rainfall events, water levels in boreholes sunk into the confining layer of the aquifer some 50 kilometres from the recharge area, begins rising. Hemming et al. studied *Sr* isotopes and possible sources thereof to the aquifer, which include rainwater, dry precipitation and reactions with the aquifer rock. The authors conclude (1998:599) that climate (rainfall, wind direction) must have a large influence on the prevalence of the isotope. This seems to indicate that pollutants on the surface could also enter the aquifer, for example, as a result of inadequate sanitation in informal settlements.

Table 17.3.   Water availability in Namibia.

Source	Volume (Mm³/a)	Remarks
Groundwater	300	Estimated long-term safe yield
Ephemeral surface water	200	Full development at 95% assurance of supply
Perennial surface water	150	Presently installed abstraction capacity
Unconventional	10	Reclamation, reuse, recycling
Available resources	660	

The Aranos Village Council was faced with a situation of providing services to about 800 families who were resettled from an informal area to a formalised township. The provision of water and sanitation were a particular concern, due to the low incomes. During 2005 the Habitat Research and Development Centre (HRDC) in Windhoek was approached by the Council to obtain advice on sanitation systems suitable for low-income households. Council then budgeted for a limited number of toilets and requested the HRDC in 2006 to visit the village in order to discuss various sanitation options with the community. Various dry systems were demonstrated and explained. During the following discussions, several community members indicated that they could build a certain type of toilet, i.e. the Otji-Toilet. The necessary local capacity and skills were available therefore no contractor had to be appointed. This ensured that the funds would be utilised in the local economy. In addition the local brick factory was able to supply most of the building material.

The initiative in Aranos received support from the Clay House Project (CHP) in Otjiwarongo and the Global Environment Fund (GEF) to build two demonstration toilets. One requirement was that a beneficiary supplies and transports 460 cement bricks to the site where the toilets are to be built and to provide basic accommodation for three workers on site during the construction period of one week. Consequently, a team from Otjiwarongo trained locals in Aranos and constructed two units. In the following year, another 18 toilets were completed by the local authority, followed by an additional 40 units in 2008, using local labour.

One problem encountered by the Aranos Village Council was the hard rock (calcrete) found on the sites. Therefore a jackhammer had to be used to excavate the pit. However, if this equipment is not available, the toilet structure could be constructed as an elevated unit, i.e. on top of the rock. This requires stairs, which could make access difficult for older residents. Access to the containers would be much easier, as the containers are level with the surface.

The Otji-Toilet support structure consists of a pit lined with bricks and an earthen floor. A concrete slab cover is placed on the top of the pit to provide for the construction of the toilet building. The Otji-Toilet is considered as a low risk sanitation option, as excessive liquids merely trickle onto the slab above the floor of the pit, where it evaporates due to the build-up of heat and the flow of air through the space. A painted black access cover and a ventilation pipe enhance the latter. The pit is accessible from the back, which should be oriented towards the north in order to use solar energy in the drying out of the content in the pit. This prevents water pollution in regions with a low water table. Inside the pit two containers are placed, one to collect all matter, the other either as a stand-by or, if filled, to compost the contents. Schönning (2003:401) provides advice on inactivating pathogens:

> "*Storage is probably most beneficial in dry-warm climates where the low moisture content will result in desiccation of the material and aid in pathogen inactivation. If the faecal material looks dry, and also is dry right through, the risk of viable pathogens being present has decreased significantly. Regrowth of bacterial pathogens may however occur after application of moisture (e.g. by contact with moist soil)*".

The Otji-Toilet was conceived as a contribution to the provision of safe sanitation in an arid environment, using local building materials and labour. In addition, the national reliance on groundwater is reduced, as no water is required for the operations. It is an on-site system with low operation and maintenance costs, does not require water or municipal sewer connection, no chemicals and electricity are required. Some capital is needed to pay for the construction. Construction skills are also essential, therefore practical training is offered by the CHP. In addition the maintenance of units is crucial to ensure that no contamination of the natural environment takes place. This involves educating beneficiaries and the officials of the local authority. The Otjiwarongo municipal council decided to make the CHP responsible for the maintenance of the dry sanitation units. A team employed by the CHP is regularly emptying the containers (see CHP, no date). The municipality is charging a monthly amount for this service from the beneficiaries. Part of the money is paid to the CHP, whereas the remainder is used to cover the administrative costs.

Figure 17.1.   Otji-Toilet.

The Otji-Toilet, though inexpensive in comparison to a flush toilet, still requires an initial amount of money to be constructed, which can be an issue for low-income residents of informal settlements (UNDP 2008). Substantial savings are made on water usage (water abstraction and conservation), reduction of environmental pollution, maintenance and municipal service costs. However, modifications to the design would be necessary in localities with a high groundwater table, e.g. the pit has to be made watertight.

The Otji-Toilet has shown that forward-looking authorities, in cooperation with partners, can provide affordable sanitation services. Aranos and Otjiwarongo are evidence that, once the participants have understood the technology and its operations, suitable options can be promoted. In both cases, the local authorities continue to support the construction of these sanitation alternatives. Furthermore, the Otji-Toilet concept is also utilised in Latin America.

17.4.2   *Wet sanitation: Decentralised Wastewater Treatment System (DEWATS)*

In the 1930s the bucket latrine system was introduced by the British protectorate authorities to prevent water resources from becoming contaminated by the pit latrines, which were being built in urban areas. The concern related to the fact that water was drawn from shallow wells and the dangers of water pollution were real (Blackett 1994:3). This changed in the 1970s, when water pollution was no longer a major concern as most urban dwellers were using piped water supplies or deep boreholes.

The transboundary area between South Africa and the lowland western parts of Lesotho consists of different geological groups and comprise horizontal to sub-horizontal dipping sedimentary rocks (Cobbing et al. 2008:1211). According to the authors (Cobbing et al. 2008:1212), due to the low transmissivities and consequent low borehole yields of these Karoo Supergroup rocks, the transboundary impact of groundwater abstraction is likely to be insignificant. The depths of the formations vary between 50 and 250 metres.

Lesotho has in principle abundant water resources, enough to export water to South Africa. However, the distribution of water in the country is disproportionate, due to the rainfall pattern and physical factors (UNDP Lesotho 2007:21). Despite the natural abundance of water, seasonal shortages occur, "because of the inadequate development of the water distribution network, institutional and management constraints (UNDP Lesotho 2007:64). Between 1996 and 2002 improvements in water provision are attributed to the work done in rural areas. Access to safe drinking water improved from an estimated 62% of the population in 1996 to 74% in 2002/03. In the same year 8% of the urban population had no access, compared with 34 percent of the rural population (UNDP Lesotho 2007:xi). "Improvements in both public sanitation and drinking water supply are closely linked because the lack of sanitation precludes the proper treatment of human waste, which is, in turn, one of the main sources of unsafe water" (UNDP Lesotho 2007:19). The increasing pollution of fresh water sources by untreated household sewage, industrial effluent, agricultural run-offs and inappropriate land-use patterns are becoming a threat to water sources. Another threat to water sources is the problem of serious water pollution, which is a consequence from the activities of large-scale industries, such as textiles, which are in need of water as an input. Other sources of water pollution include: agricultural chemicals (pesticides, herbicides), pit latrines, uncontrolled urban drainage, and landfill sites (UNDP Lesotho 2007:64).

During the introduction of the National Rural Sanitation Programme in 1987, "Technology choice was a simple issue" (World Bank 2002:3). Households typically built their own latrines using various materials. The ventilated improved pit (VIP) latrine, invented in Zimbabwe in the 1970s, became the accepted and superior form of on-site sanitation. It overcame the two major disadvantages of traditionally designed pit latrines: smells and fly infestation, as a screened vent pipe in the design prevented this nuisance. Besides the high costs for a water-borne system, the non-availability of electricity, and the freezing temperatures in winter would pose another technical problem for an outdoor water toilet, and this may partly explain why people are not eager to wash in an outdoor bathroom (Blackett 1994:7). Around 38,000 VIPs were constructed in the urban centres (Maseru and thirteen towns), and in the rural areas approximately 36,000 new VIP latrines have been built, 19,000 ordinary pit latrines have been upgraded to VIP latrines, in addition to 30,000 ordinary pit latrines (World Bank 2002:4).

The VIP latrine concept was readily adopted as standard and existing latrine construction techniques were incorporated into its design, and a national standard was adopted for both urban and rural programmes (World Bank 2002:8). However the bank points out that, the biggest remaining technical and financial problem is pit emptying. Among the difficulties experienced are the emptying of double-pit VIPs by hand, after three to five years maturing, has been problematic in areas where a high groundwater table is found.

Despite the fact that Lesotho exports water to South Africa, there is a growing problem of water shortage in urban areas (Lall 2005:1013), which affects industries and residents. Traditionally, the Basotho have regarded water as "a valuable and readily available resource that is a free gift from God as they often call it" (Chakela quoted by Molapo 2005:43). However, this situation of an essentially rural population with an abundant supply of water has changed remarkably over the years. Major factors include urbanisation, the rapid population growth and the migration from the mountain areas to the lowlands. Therefore government had to intervene to ensure the supply of water to both the rural and urban population. This resulted in investments in water resources development, a process that has made water a costly resource and not longer a free gift (Molapo 2005:43). The changes in the provision of water that occurred between 1986 and 1996 are illustrated by Molapo (2005:44) in Table 17.4.

Despite having a relative abundant rainfall, Lesotho is facing ecological problems with regard to water. The population growth of inhabitants and animals, and the free-for-all attitude in land-resource management, has accelerated the degradation of fragile ecosystems (Makhoalibe 1999:460). Animals invade bogs, fens, marshes and wetlands in search of green

Table 17.4.   Water provision in Lesotho.

Main source of drinking water	Total % (1986)	Total % (1996)
Piped water	32.3	51.0
Catchment tank	0.6	1.4
Boreholes	12.6	0.3
Springs	52.6	44.4
Rivers	1.9	0.8
Other	0	0.1

vegetation and the water sources in the wetlands, which have been eroded and became gullies. Serious erosion can be observed, for example between Maseru and Mafeteng, where deep gullies are visible, as a result of deforestation. The water retention capacity has been lost and rainfall is lost immediately as surface runoff. Any contamination of the runoff water, such as animal or human excreta, could eventually pollute local drinking sources. Another possible source of contamination is found in the case of dysfunctional technologies. As observed in 2006 in Roma, semi-purified water from oxidation ponds enters a nearby stream.

The government of the Kingdom, has as part of the construction of a cultural centre on the outskirts of Maseru, which is located next to a river, incorporated a decentralised treatment plant for the main centre and the accommodation facilities. The cultural centre can accommodate up to 2,000 visitors and therefore requires a reliable sanitation option, which is also affordable. The centre's sanitation infrastructure is utilising a decentralised wastewater treatment system (DEWATS), which established an approach between the high-cost first world standards (waterborne sanitation) and the no-cost open defecation. Among the benefits of a DEWATS are (BORDA, no date):

- Establishing of multi-stakeholder networks to combat water pollution
- Fulfilment of discharge standards and environmental laws
- Providing treatment for domestic and industrial effluent
- Low primary investment costs as no imports are needed (materials/inputs used for construction are locally available)
- Efficient treatment for daily effluent flows up to 1000 m³
- Modular design of all components (for example settler, anaerobic baffled reactor, anaerobic filter, planted gravel filter, and biogas digester)
- Tolerant towards inflow fluctuations
- Reliable and long-lasting construction design
- Expensive and sophisticated maintenance not required
- Low maintenance costs
- Resource efficiency
- Non dependence on energy supplies
- Resource recovery through effluent re-use and biogas generation

Decentralised systems utilise various modules that can be combined as applications to provide treatment for domestic and industrial sources. The technical options offered are based on a modular and partly standardized design. These include the following (Ulrich 2005);

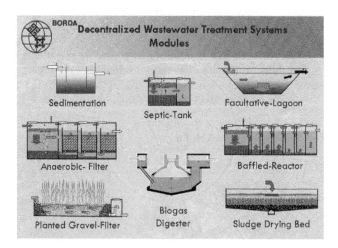

Figure 17.2.   DEWATS modules.

Figure 17.3.   DEWATS construction in Maseru.

The design and the costs of a DEWATS project are determined by the final usage of the water and the resources to be recovered. In countries with a lot of streams and rivers, such as Lesotho, the water quality had to be high to prevent water contamination, if the treated water is released back into a water course. In cases where the water is utilised, for example, in a garden project, the water quality can be lower as the nutrients in the effluent are beneficial to plants. The addition of a biogas digester enables institutions and households to produce fuel for cooking purposes.

In Lesotho smaller DEWATS projects are also carried out, for example, individual households or institutions, which experience problems such as: septic tanks with soak-aways that are either not allowed by authorities or do not work due to the ground texture; and regular emptying of the septic tanks is expensive and the commercial utility (WASA) can hardly meet the demand (Schmidt 2008). Therefore a system has been developed which consists of a biogas digester to produce gas for cooking purposes. The digester is connected to an anaerobic baffled reactor and a planted gravel filter. The water can be utilised for gardening and food production by the households. However, the placing of a treatment plant on a plot is often challenging in addition to finding a convenient solution for re-using of the treated water (Schmidt 2008).

Not all sites are suitable for the treated water to reach the garden by gravity flow. Therefore the water has to be taken out by bucket or to be pumped regularly. Due to the mountainous character of Lesotho, the high organic loads and low biological activity in winter are challenges in the promotion of DEWATS approach. In cooperation with the local partner TED (Technologies for Economic Development), but also the public and private sector, thousands of septic tanks can be converted into Biogas/DEWATS plants and thus to change significantly the sanitation condition in urban areas of Lesotho (Schmidt 2008).

17.5   RECOMMENDATIONS

The possible contamination of groundwater sources in countries with limited potential or seasonal variations, as described by Sugden above, should be a major concern in any utilisation of water. Taken the limited human resources available in Africa, hi-technology "solutions" do not necessarily work. This is one possible cause of water contamination, first in river systems, and secondly groundwater sources, when water permeates into the ground. Furthermore, as Rothert (2000:2) states, very few water conservation and demand management measures have been implemented in southern Africa to date. According to Lange and Hassan (2006:x), between 1995 and 2001, **environmental accounts** were introduced in Botswana, Namibia and South Africa. In each country, a partnership between agencies that compile accounts, such as a national statistical office, and agencies that would use the accounts, was established. This information needs to be disseminated in order to educate decision-makers and provide a basis for the future planning of water provision. Groundwater sources are a particularly sensitive source, as the volume stored underground depends on the annual recharge rate, which in the case of Namibia, is only one percent of precipitation.

Feitelson and Chenoweth (2002:264) sum up the issues surrounding water availability:

> *"The key word in discussing scarcity is availability. In other words, water scarcity is in essence a scarcity of readily available fresh water. Availability problems are an outcome of the spatial variability and temporal fluctuations in fresh water supply and demand, as fresh water is not uniformly distributed in time and space and does not conform to the patterns of human demand".*

The importance of **disseminating information** also relates to education emphasising suitable alternative sanitation systems and the availability of water. Besides the theoretical aspects of alternatives, it is important that practical experiences reinforce the theory. An understanding of the benefits of alternative systems is essential, in particular when unworkable hi-technologies are regarded as the standard without considering the implications of a lack of qualified personnel to operate these systems and as a consequence, the pollution of water bodies. Increasing pressure on abstracting from available water sources necessitates an awareness that inappropriate sanitation may contribute to the mixing of pollutants with water. For example, water in shallow wells could become contaminated, "particularly where these are affected by local pollution sources, such as badly sited latrines or agricultural activities" (Briggs 2003:8).

Experience has shown that **demonstrations units** can play an important role, as part of awareness raising and education. This is demonstrated by the experiences in Aranos and the CSIR office in Stellenbosch: "As toilets and toilet behaviour are sensitive issues, the three-year pilot project will also look at people's perceptions in an opinion survey of CSIR staff using the no-mix toilets for the first time and again later, after people have become more used to the idea" (CSIR 2009).

Political will is always a challenge if alternatives are introduced or important issues have to be attended to. The two case studies have shown that an understanding of locally available options, by decision-makers as well as beneficiaries, supports the usage of suitable sanitation systems. These alternative options are not only exclusively for the poor, they are suitable for all strata of society. Winblad and Kilama (1978) have described numerous alternative sanitation options from Europe, Asia, America and Africa.

Another important issue, especially in Africa, is the will to address water insecurity. Beekman et al. (2003:19) caution that, "Southern Africa is among the world's most drought prone regions". An awareness regarding the possible impacts of climate change is very limited among decision-makers and society as a whole. Variations in precipitation could reduce the ostensible reliability of groundwater resources, when surface water is becoming increasingly scarce or polluted. As Beekman et al. (2003:20) point out:

> *"When the rains stop, and surface water sources dry up, groundwater can become the only water source available. As a result boreholes and wells that were previously utilised within a sustainable level are typically over-used at a time of diminished recharge. As a result water levels drop. The extent of aquifer depletion in such a situation is controlled by the aquifer's permeability".*

Consensus usually maintains that the majority of victims of degrading water supply conditions will be the less affluent members of society. They often have inadequate access to sanitary facilities and clean water sources. To improve conditions, policies may incorporate subsidies in supply programmes, to ensure the provision of services. Samanta and van Wijk (1998:80) warn that, "Programmes providing latrines with high subsidies have inadequate funds to meet the large requirements for sanitary facilities by vast numbers of beneficiaries and are unable to keep pace with population growth".

The **management of sanitation facilities** is one of the most important requirements in the protection of water sources. The case studies illustrate that appropriate systems are available, which can be managed locally and that do not demand highly skilled operators. In addition, water sources are not contaminated when utilising suitable sanitation systems. In addition water resource management is essential, especially if the impacts of climate change are taken into consideration, which are not limited to a country alone. This requires international networking, collaboration and coordination to study the vulnerability of the regions relating to the protection of groundwater sources. Integrating water supply and appropriate sanitation delivery also has to involve beneficiaries. As described above, Lesotho faces increased degradation of wetlands and water sources, inter alia, due to untreated household effluent. The problems of emptying pits have been pointed out, which is not a serious issue in the case of the Otji-toilet. Managing the dry content of these toilets is easier and safer than those with wet content.

17.6   CONCLUSION

Arid regions, such as Southern Africa, have to be cautious in their water consumption and the effects utilisation of a scarce resource could have on the potential contamination of essential sources, as in the case of industries, agriculture and domestic sanitation. According to Falkenmark (1990:182), "pollution of the atmosphere, which will change the chemical composition of precipitation and subsequently the chemical composition of the root-zone water, the groundwater, and the runoff, which will have impacts on both aquatic and coastal ecosystems". Increasing water stress and the pollution of surface water sources could result in an increasing abstraction of groundwater. This could become a major concern should climate changes affect the availability of water in the region, coupled with the still rapidly increasing population numbers. These factors could undermine efforts of water conservation.

Richter et al. (2006:298) claim that "ecological science is not yet being adequately integrated into water decision-making in most parts of the world—most water management decisions or plans continue to be made on the basis of engineering considerations alone, with little or no scientific input concerning the water needs of freshwater ecosystems".

The three case studies discussed, have shown that the countries in southern Africa are vulnerable. A large percentage of the area is arid or semi-arid, but experiences a considerable population growth rate. Urban areas are growing at a rate, which cannot be sustained. Providing every resident with potable water is becoming increasingly difficult, as the affordability and the availability of water is limited. To provide all inhabitants with safe water, under these circumstances, could be an illusion.

Namibia is at present in the fortunate position of having a relative small population, but the increasing water stress is nevertheless a fact that has to be addressed. The availability and affordability is a concern throughout the region. Cooperation is essential, if the problems relating to groundwater have to be addressed, whether on an international scale or on a local scale. The cases from Namibia have illustrated what is being done to protect groundwater and how this is related to sanitation. In Lesotho households and the government are supporting technologies and approaches suitable to local conditions. However, the local capacity to maintain the existing infrastructure and to extend the infrastructure, to avoid threats to groundwater sources, remains a concern. Cleaning polluted water is a very expensive process.

The arid areas in southern Africa seem to have relatively few problems with respect to groundwater pollution, but this could be the result of not knowing what happens beneath the surface. Current attention is often focussing on surface water conditions. This could mean that the water supplies in countries face an uncertain future, if the groundwater sources are being polluted. Furthermore, the uncertainties relating to the impact of climate change will add to the vulnerability in Southern Africa. The UNDP (2007:25) emphasises that, in the context of climate change "emerging risks will fall disproportionally on countries already characterized by high levels of poverty and vulnerability". Groundwater resources will remain a serious ecological, economical, political and social concern, especially in arid regions.

REFERENCES

Ashton, P.J. (2002). *Avoiding Conflicts over Africa's Water Resources*. http://0-www.jstor.org.wagtail. uovs.ac.za/stable/pdfplus/4315243.pdf

Beekman, H.E. Saayman, I. and Hughes, S. (2003). *Vulnerability of Water Resources to Environmental Change in Southern Africa*. http://iodeweb1.vliz.be/odin/bitstream/1834/352/1/unep55.pdf

BGR. (2008). *Coupling Sustainable Sanitation and Groundwater Protection*. http://www.bgr.bund.de/ cln_092/EN/Themen/Wasser/Veranstaltungen/symp_sanitat-gwprotect/symp_san_gwprot.html

Blackett, I.C. (1994). *Low-Cost Urban Sanitation In Lesotho*. http://www-wds.worldbank.org/exter-nal/default/WDSContentServer/WDSP/IB/1994/03/01/000009265_3961006080020/Rendered/PDF/ multi_page.pdf

Briggs, D. (2003). *Environmental pollution and the global burden of disease*. http://0-bmb.oxfordjournals.org. wagtail.uovs.ac.za/cgi/reprint/68/1/1?maxtoshow=&HITS=10&hits=10&RESULTFOR\MAT=& fulltext=Environmental+pollution+and+the+global+burden+of+disease&searchid=1&FIRSTINDE X=0&resourcetype=HWCIT

Bockmühl, F. (2001). Stampriet artesian basin, in Christelis, G. & Struckmeier, W. *Groundwater in Namibia an explanation to the Hydrogeological Map*. Windhoek: MAWRD & MME.

BORDA. (No date). *Decentralized Waste Water Treatment—DEWATS*. http://www.borda-net.org/ modules/cjaycontent/index.php?id=29

Cave, B. and Kolsky, P. (1999). *Groundwater, latrines and health*. Task No: 163. http://www.lboro.ac.uk/ well/resources/well-studies/full-reports-pdf/task0163.pdf

Christelis, G. and Struckmeier, W. (2001). *Groundwater in Namibia an explanation to the Hydrogeological Map*. Windhoek: MAWRD & MME.

Christelis, G., Heyns, P. and Struckmeier, W. (2001). Essentials of groundwater, in Christelis, G. and Struckmeier, W. *Groundwater in Namibia an explanation to the Hydrogeological Map*. Windhoek: MAWRD & MME.

Clay House Project. No date, ca. (2006). *The Otji-Toilet*. (Pamphlet) http://home.arcor.de/clayhouse/ downloads/otji-toilet-leaflet-s.pdf

CSIR. (2009). *CSIR employees use no-mix toilets in pilot study*. http://ntww1.csir.co.za/plsql/ptl0002/ PTL0002_PGE157_MEDIA_REL?MEDIA_RELEASE_NO=7522708

Cobbing, J.E., Hobbs, P. J., Meyer, R. and Davies, J. (2008). *A critical overview of transboundary aquifers shared by South Africa*. http://0-www.springerlink.com.wagtail.uovs.ac.za/content/00w5272x7704356q/ fulltext.pdf

Desanker, P. and Magadza, C. (2001). *Africa*. http://www.grida.no/climate/ipcc_tar/wg2/pdf/wg2TAR chap10.pdf

du Pisani, P.L. (2006). *Direct reclamation of potable water at Windhoek's Goreangab reclamation plant*. http://www.desline.com/articoli/6994.pdf

EPA. (2005). *Strategic Advice on Managed Aquifer Recharge using Treated Wastewater on the Swan Coastal Plain*. http://www.epa.wa.gov.au/docs/2125_B1199.pdf

Falkenmark, M. (1990). *Global Water Issues Confronting Humanity*. http://0-www.jstor.org.wagtail. uovs.ac.za/stable/pdfplus/423575.pdf

Feitelson, E. and Chenoweth, J. (2002). *Water poverty: towards a meaningful indicator*. http://0-www. sciencedirect.com.wagtail.uovs.ac.za/science?_ob=MImg&_imagekey=B6VHR-46 NXR91-1-5&_ cdi=6073&_user=736898&_orig=search&_coverDate=12%2F31%2F2002&_sk=999959996&view= c&wchp=dGLzVzz-zSkzS&md5=064546f6a9a42702f2e8ca0f345c1432&ie=/sdarticle.pdf

Foster, S. (2008). *Urban Water-Supply Security In The Developing World groundwater use trends and the sanitation nexus*. http://www.geozentrum-hannover.de/cln_092/nn_324952/EN/Themen/Wasser/Veran staltungen/symp_sanitat-gwprotect/present_foster_pdf,templateId = raw,property=publicationFile. pdf/present_foster_pdf.pdf

GTZ. (2004). *Ecosan—closing the loop*. Proceedings of the 2nd international symposium, 7th–11th April 2003, Lübeck, Germany. GTZ: Eschborn.

Hemming, N.G., Stute, M. and Talma, A.S. (1998). *Secular variation in the 87Sr/86Sr composition of groundwater, Stampriet artesian aquifer, Namibia*. http://www.minersoc.org/pages/Archive-MM/ Volume_62 A/62a-1-598.pdf

Heyns, P. (2008). *Water Supply and Sanitation Sector Policy Review. Situation assessment.* Presented at the National Stakeholder Consultative Workshop, 17 July 2008, Windhoek.

Heyns, P. and Struckmeier, W. (2001). Introduction, in Christelis, G. and Struckmeier, W. *Groundwater in Namibia an explanation to the Hydrogeological Map.* Windhoek: MAWRD & MME.

Houghton-Carr, H.A. (2006). Southern Africa FRIEND, in Servat, E. and Demuth, S. (eds.). 2006. *FRIEND—a global perspective 2002–2006.* http://unesdoc.unesco.org/images/0014/001498/149889e.pdf

Kirchner, J. and van Wyk, A. (2001). An overview of the Windhoek city water supply, in Christelis, G. and Struckmeier, W. *Groundwater in Namibia an explanation to the Hydrogeological Map.* Windhoek: MAWRD & MME.

Kingdom of Lesotho. No date (ca. 2003). *Poverty Reduction Strategy 2004/2005–2006/2007.* http://www.lesotho.gov.ls/documents/PRSP_Final.pdf

Lall, S. (2005). FDI, AGOA and Manufactured Exports by a Landlocked, Least Developed African Economy: Lesotho. *The Journal of Development Studies,* Vol. 41, No. 6, 998–1022.

Lange, G.-M. and Hassan, R. (2006). *The Economics of Water Management in Southern Africa.* http://www.columbia.edu/~gl2134/docs/5%20%20%20INTRO00%20 LANGE%20PRELIMS.pdf

Lindgren, A. (1999). *The Value of Water A Study of the Stampriet Aquifer in Namibia.* http://www.econ.umu.se/MFS/annali.pdf

Makhoalibe, S. (1999). *Management of Water Resources in the Maloti/Drakensberg Mountains of Lesotho.* http://0-www.jstor.org.wagtail.uovs.ac.za/stable/pdfplus/4314930.pdf

MDG Africa Steering Group. (2008). Achieving *the Millennium Development Goals in Africa.* http://www.mdgafrica.org/pdf/MDG%20 Africa%20Steering%20Group%20Recommendations%20-%20 English%20-%20HighRes.pdf

Mapani, B.S. (2005). *Groundwater and urbanisation, risks and mitigation: The case for the city of Windhoek, Namibia.* http://0-www.sciencedirect.com.wagtail.uovs.ac.za/science?_ob=MImg&_imagekey=B6X1 W-4H877MD-5-7&_cdi=7253&_user=736898&_orig=search&_coverDate=12%2F31%2F2005&_ sk=999699988&view=c&wchp=dGLzVtz-zSkWb&md5=33bdf316cefab53eb1b0e 192995315ec&ie=/sdarticle.pdf

Menge, J. (2006). *Treatment Of Wastewater For Re-Use In The Drinking Water System Of Windhoek.* Paper presented at WISA.

Millennium Challenge Corporation. (2006). *Lesotho Country Proposal to the Millennium Challenge Corporation (MCC).* http://www.mca.org.ls/documents/Lesotho_Program_MCA_Final.pdf

Millennium Ecosystem Assessment, (2005). *Ecosystems and Human Well-being: Desertification Synthesis.* http://www.millenniumassessment.org/documents/document.355.aspx.pdf

Miller, G.T. (1996). *Living in the environment.* 9th edition. Belmont: Wadsworth Publishing Company.

Molapo, L. (2005). *Urban Water Provision In Maseru (Lesotho): A Geographical Analysis.* http://etd.uovs.ac.za/ETD-db//theses/available/etd-11102005-081555/unrestricted/MOLAPOL.pdf

Niamir-Fuller, M. (2000). The resilience of pastoral herding in Sahelian Africa, In Berkes, F. and Folke, C. *Linking Social and Ecological Systems. Management Practices and Social Mechanisms for Building Resilience.* Cambridge: Cambridge University Press.

Republic of Namibia. (2004). *Namibia 2004 Millennium Development Goals.* Windhoek: Office of the President.

Rhodes University. (2008). *Southern Africa FRIEND.* http://www.ru.ac.za/static/institutes/iwr//friend/?request=institutes/iwr/friend/

Richter, B.D., Warner, A.T., Meyer, J.L. and Lutz, K. (2006). *A Collaborative And Adaptive Process For Developing Environmental Flow Recommendations.* http://www.nature.org/initiatives/freshwater/files/rra892_22_3_297_318.pdf

Rothert, S. (2000). *Water conservation and demand management potential in southern Africa: an untapped river.* http://74.125.113.132/search?q=cache:Zj4JdIgB2zsJ:www.environmental-center.com/magazine/inderscience/ijw/art8.pdf+%22 Water+Management+in+Southern+Africa%22&cd=9&hl=en&ct=clnk&gl=na

Samanta, B.B. and van Wijk, C.A. (1998). *Criteria for successful sanitation programmes in low income countries.* http://0-heapol.oxfordjournals.org.wagtail.uovs.ac.za/cgi/reprint/13/1/78?maxtoshow=& HITS=10&hits=10&RESULTFORMAT=&fulltext=Criteria+for+successful+sanitation+programm es+in+low+&searchid=1&FIRSTINDEX=0&resourcetype=HWCIT

Schmidt, A. (2008). *Biogas—DEWATS Challenge for On-site sanitation in Lesotho.* http://www.borda-sadc.org/modules/news/article.php?storyid=12

Schneider, G.I.C., Schneider, M.B. and du Pisani, A.L. (2001). Natural and socio-economic environment, in Christelis, G. and Struckmeier, W. *Groundwater in Namibia an explanation to the Hydrogeological Map.* Windhoek: MAWRD & MME.

Schönning, C. (2003). Recommendations for the reuse of urine and faeces in order to minimise the risk for disease transmission, in GTZ. 2004. *ecosan—closing the loop*. Proceedings of the 2nd international symposium, 7th–11th April 2003, Lübeck, Germany. GTZ: Eschborn.

Servat, E. and Demuth, S. (eds.). (2006). *FRIEND—a global perspective 2002–2006*. http://unesdoc. unesco.org/images/0014/001498/149889e.pdf

Sugden, S. (2006). *The Microbiological Contamination of Water Supplies*. http://www.lboro.ac.uk/well/ resources/fact-sheets/fact-sheets-htm/Contamination.htm

Ulrich, A. (2005). *Decentralized Wastewater Treatment Systems for Settlements & Small/Medium Enterprises—Experiences of a demand driven multi-stakeholder project implementation framework*. Presentation at a Stakeholders Round Table Meeting on Community Based Sanitation (CBS) and Decentralised Wastewater Treatment (DEWATS) for Urban Settlements and Small and Medium Industries, Windhoek, October 2005.

UNDP (2008). *United Nations Development Programme News*. http://www.undp.un.na/env/otjitoilet.htm

UNDP (2007). *Human Development Report*. http://hdr.undp.org/en/media/HDR_20072008_EN_ Complete.pdf

UNDP. (No date). *Goal 7: Ensure environmental sustainability*. http://www.undp.org/mdg/goal7.shtml

UNEP. (No date). *Vulnerability Assessments*. http://www.unep.org/dewa/assessments/EcoSystems/water/ Vulnerability/reports/drafts/South_Africa.pdf

UNDP Lesotho. (2007). *National Human Development Report 2006*. Lesotho. http://www.undp.org.ls/ documents/NHDR%202005%20 small.pdf

United Nations (2000). *United Nations Millennium Declaration*. http://www.un.org/millennium/decl aration/ares552e.pdf

WHO (2001). *WHO Expert Consultation On Health Risks In Aquifer Recharge Using Reclaimed Water*. http://www.euro.who.int/document/wsn/WSNgroundwaterrpt.pdf

Winblad and Kilama (1978). *Sanitation without Water*. Stockholm: SIDA.

WMO (2008). *WMO Greenhouse Gas Bulletin*. http://www.wmo.int/pages/prog/arep/gaw/ghg/documents/ ghg-bulletin-4-final-english.pdf

World Bank (2008). *Country Groups*. http://web.worldbank.org/WBSITE/EXTERNAL/DATASTATISTICS/ 0,,contentMDK:20421402~pagePK:64133150~piPK:64133175~theSitePK:239419,00.html

World Bank (2002). *The National Sanitation Programme in Lesotho: How Political Leadership Achieved Long-Term Results*. http://www.sulabhenvis.in/admin/upload/pdf_upload/af_bg_lesotho.pdf

18

Community-based groundwater quality monitoring:
A field example

Pamela E. Crane & Stephen E. Silliman
*Department of Civil Engineering and Geological Sciences, University of Notre Dame,
Notre Dame, IN, USA*

Moussa Boukari
*Département des Sciences de la Terre, Université d'Abomey-Calavi,
Cotonou, Bénin*

ABSTRACT: Discourse on the need to increase access to water in Africa increasingly includes focus on issues of characterizing and maintaining the water quality of improved sources. Monitoring of groundwater quality at a wellhead has the potential to provide critical information both at the time of implementation and over the useful lifetime of a groundwater well. However, there are a series of unique challenges to obtaining high-quality, high-temporal frequency data on water quality in rural regions of Africa. A unique approach to addressing the challenges associated with monitoring groundwater quality in rural regions of Africa was designed and implemented in Bénin, West Africa. Challenges of monitoring in rural Bénin were overcome by establishing water quality monitoring teams within the local communities who were trained to perform, on a weekly basis, basic water quality measurements using test strips and colorimetery. Over a period of three years the teams demonstrated that they could reliably measure water quality parameters on a weekly basis. This was further demonstrated through reliable reproduction of the concentrations of nitrate standards provided in unlabelled (e.g. blind) bottles. As such, this project has demonstrated how the challenges of monitoring water quality in rural Africa can be addressed through involving the local community as critical partners in the process of data collection. The principles behind this project have wide applicability to other rural water quality monitoring situations and offer a new strategy for pursuing water quality monitoring efforts in rural Africa.

18.1 INTRODUCTION

Discourse on the need to increase access to water in Africa increasingly includes focus on issues of characterizing and maintaining the water quality of improved sources. Monitoring of groundwater quality at a wellhead has the potential to provide critical information both at the time of implementation and over the useful lifetime of a groundwater well. However, there are a series of unique challenges to obtaining high-quality data on water quality in rural regions of Africa at sufficiently high frequency in time that often prevent this from occurring. This case study presents an approach to addressing the challenges associated with monitoring groundwater quality in rural regions of Africa that is based on involving local communities as project partners. This approach was designed, implemented, and assessed in Bénin, West Africa, but has potential applicability well beyond the conditions encountered in this region of rural Bénin.

18.2   BACKGROUND

Although the relationship between drinking water quality and mortality rates is widely accepted (e.g. Sepúlveda, et al., 2006, and Schoenen, 2002) and World Health Organization guidelines for drinking water quality have been established (WHO, 2006), a significant need for research and monitoring of water quality in Africa continues to exist. An article published by the United Nations Development Program highlights this through the statement, "Across much of the developing world, unclean water is an immeasurably greater threat to human security than violent conflict" (UNDP, 2006).

In tension with the need for clean water and international water quality standards created by the World Heath Organization are the operational challenges of performing water quality monitoring in rural regions of Africa. A few of these challenges are: inadequate laboratory facilities, significant distances between remote sites and laboratories, limited infrastructure, low education levels of local populations, and limited financial and personnel resources. Based on the disconnect between the need for, and difficulty of, performing water quality research in rural regions of developing nations, a program of study was initiated in Bénin, West Africa, to consider alternatives for water quality monitoring in a rural village. This program provides a foundation from which to expand groundwater research past the perceived limitations of data collection in rural Africa, and contains both theoretical and applied components.

The theoretical portion of this program is reported on elsewhere (Crane, 2007, and Crane and Silliman, 2009) and involves a numerical study of the relative worth of different forms of field measures of water quality (where measure is defined both by the instrument used and the expertise of the person using the instrument). Significant to the applied work discussed in the present chapter, the theoretical work demonstrated that for select field parameters such as mean concentration and total mass load, simple instruments such as test strips and colorimetry used by non-experts at high frequency can provide higher quality statistics (such as parameter mean and standard deviation) than more sophisticated instruments used by experts at relatively low frequency. This theoretical result provided incentive to investigate field strategies for monitoring water quality that allow data to be collected at high frequency in time (e.g. weekly), even if these strategies involve instruments of relatively low sophistication used by people with low levels of scientific expertise. These results justified pursuit of a unique collaboration with a local population in collecting scientifically meaningful water quality data that is detailed below. Comparison of our field applications with a recent case study on the use of test strips by private citizens in the United States (Nielsen et al., 2008) is also discussed below.

The applied portion of this research program directly addressed potential impacts on the value of water quality data sets given certain constraints on methods, sampling frequencies, and field personnel. These constraints are similar to those experienced in projects initiated in other developing nations (e.g. Hossain, 2006). A number of factors impacting groundwater monitoring in rural Bénin are similar to those experienced in many rural regions in Africa. As such, we believe that the conclusions drawn from both the theoretical and applied portion of this research effort are widely applicable. These factors include:

- the region used in the case study is a significant distance (approximately 5 hours of driving) from the nearest water-quality laboratory, thus collection of water samples for laboratory analysis on regular intervals from multiple villages in the study area was prohibitive from the view point of both logistics and resources,
- the water laboratory available in-country is not equipped to perform chemical analyses that require complex analytical methods (e.g. analysis for trace-metals, most organics, or isotopic analysis) thus requiring water samples to be shipped to laboratories in other countries,
- select analyses (e.g. nitrate) are time sensitive and therefore cannot be preserved for analysis in national or international laboratories without use of chemicals considered dangerous for handling by the local population,

- basic infrastructure (electricity, running water, computer access) is not available in the field except through equipment carried by project personnel, and
- limitations on financial and personnel resources in Bénin preclude extensive professional sampling efforts in the region of our case study.

This study began with research on groundwater quality in Bénin conducted by four organizations (Silliman et al., 2007): Direction Generale de l'Eau (a Bénin government agency with oversight for rural water), the Université d'Abomey-Calavi (a Bénin national university), the University of Notre Dame (a university in the United States), and Centre Afrika Obota (a Bénin NGO). Among the significant results from the interaction of these four organizations was identification of a region of Bénin, termed the Colline Department, with a significant number of hand-pump equipped boreholes (drilled wells) which have elevated concentrations of nitrate.

Elevated nitrate was of concern for a number of reasons. For example, the World Health Organization (WHO) set the drinking water standard for nitrate at 50 mg/ℓ NO$_3$ (equivalent to 11.3 mg/ℓ NO$_3$-N) (WHO, 2004). The WHO limit for nitrate is a result of health concerns, the primary of which is methaemoglobinaemia, or blue baby syndrome, which reduces the transportability of oxygen in the blood of infants (Knobeloch et al., 2000). Nitrate has also been associated with miscarriages, thyroid disease, central nervous system cancers, and has been linked to an increased risk of non-Hodgkin's lymphoma (Canter, 1997; L'Hirondel and L'Hirondel, 2002; and Nolan et al., 1997).

Normal levels of naturally derived nitrate in groundwater are less than 2 mg/ℓ (Mueller and Helsel 1996, and Masoner and Mashburn 2004). Beyond the health concerns specifically linked to nitrate are the sources from which the nitrate is derived. At levels above WHO standards, possible sources for elevated nitrate concentrations in groundwater are essentially all linked to human activity including: fertilizers from home or agriculture use, leakage from sewage lines or on-site sewage disposal, animal waste, and industrial sources (Wakida and Lerner, 2005). Significantly, each of these nitrate sources will commonly be associated with additional pollutants such as bacterial or organic contamination associated with human or animal waste, and chemical pollutants associated with fertilizers (and pesticides normally applied in the same locations as the fertilizers) or industrial waste (Wakida and Lerner, 2005; and Somasundaram, et al., 1993). Hence, the presence of nitrate represents a direct medical threat and also serves as an indicator of other possible contaminants. Given that nitrate is more easily monitored in field situations than fertilizers, pesticides or biological contaminants, nitrate has the ability to serve as an indicator of potentially more serious contamination in the groundwater and was therefore chosen as the focal point of the present case study.

18.3   METHODS

This case study is based on monitoring nitrate concentration in groundwater in the village of Adourékoman, which is located in the Colline Department of Bénin. This village was chosen due to the presence of three drilled wells equipped with hand pumps with three different levels of nitrate concentrations: the first (named well Ayewa) was below the WHO nitrate standard, the second (named Ayewa-Okouta) was consistently above, but within 5–10 mg/ℓ NO$_3$-N of, the standard, and the third (named Agbo) was consistently well above the standard. Significant for this case study, the village expressed interest in participating in this project.

In order to constrain the possible source(s) of elevated nitrate concentration in the Colline Department, it was determined that a combination of high-frequency and low-frequency sampling was necessary. The low frequency sampling was performed by project experts and was based on isotopic (N and O) analysis. The results of these isotopic analyses provided relatively convincing support both for the source of the nitrates being from human or animal waste and for the nitrates being impacted by denitrification in the subsurface. As the isotopic method was based on standard practice (collection and analysis), no further discussion of this method or the results is provided in this manuscript. The reader is referred to Crane (2006) for further details.

Although the low-frequency isotopic method supports a general source of the nitrate and a trend of denitrification, it was not designed to identify temporal variability in the nitrate concentration. Further, identification of temporal trends in the nitrate concentration was not possible with commonly accepted monitoring strategies because the research agencies involved in the study were unable to spend the resources (personnel, materials, or sample analysis) necessary to gather time-series data at the identified boreholes at frequencies sufficient to identify seasonal or higher-frequency variability. In order to identify alternative analytical methodologies with the potential to allow for high-frequency sampling of nitrate, we explored a spectrum of analytical methods involving a range of complexity, infrastructure requirements, level of expertise required to reliably use the method, detection limit, potential instrument bias, and precision. Methods analyzed ranged from test strips to cadmium reduction methods to nitrate electrodes to isotopic ratio methods.

Development of the approach to high-frequency sampling occurred in three stages so as to allow initial method design, implementation and assessment at a test location, and expansion to additional locations. The three stages are here termed: (i) development stage, (ii) initial implementation, and (iii) second level implementation. Although mention is later made of second level implementation (expansion to other locations), the focus of this study is on the development and initial implementation stages, which were completed in Adourékoman.

The development stage consisted of selecting the analytical methods most appropriate for use by lay individuals and establishing techniques for training these individuals to reliably use the selected analytical methods; the development stage was the focus of Crane (2006). As discussed in that thesis, the two analytical methods selected for the high-frequency sampling included a single-parameter nitrate colorimeter and nitrate/nitrite test strips. These two analytical methods were chosen based on the following desired characteristics:

- Field portable and does not require electricity (except standard batteries)
- Requires neither significant handling of hazardous chemicals nor precision measurement of chemical mass or volume
- Durable and either factory calibrated or has automated calibration not requiring the preparation/use of calibration standards
- Reagents can be safely delivered to country (or carried on commercial aircraft)
- Sufficient quality of concentration measure (in terms of bias/precision) in combination with appropriate detection limits and concentration bounds
- Availability of, or ability to, create training materials consistent with a population of low literacy and/or speaking a local dialect not conducive to translation of written materials
- Reasonable cost relative to the local population's or local government's ability to pay
- Prefer results provided on a continuous scale of concentration (versus results provided only in discrete ranges of concentration)

With the exception that test strips provide only concentration range (rather than a continuous concentration measure) both methods met all requirements, and were therefore considered acceptable for further study.

Following selection of these methods, significant effort was invested in understanding their strengths and limitations. The majority of this assessment effort was focused on the colorimeter due to complexities of the method. Initial evaluation of this instrument was conducted under laboratory conditions; the impact of interferences reported by the manufacturer was investigated and determined to be minimal under anticipated field conditions.

In application in the field in Bénin (by experts), it was noted on seemingly random occasions that the colorimeter indicated non-detect or extremely low concentrations despite the test strips (and prior measures using the colorimeter) indicating true concentrations above both instrument detection limit and WHO health standards. Tests to ascertain whether a chemical or physical interference was causing the underestimation of the concentration were performed in both the field and laboratory. These tests led to the identification of a previously unknown interference: particulate calcium carbonate apparently interferes with a necessary step in the cadmium reaction and, thereby, the necessary color development. This occurred regardless of the concentration of nitrate in the water or the concentration of calcium carbonate so long as particulate calcium carbonate was present. Specific results and additional discussion can be found in sections 2.3 and 5.1.1 of Crane (2006). While it was determined that acidification could be used to eliminate this interference, concerns regarding the use of acids by the local participants led to a decision to accept the occasional under estimation of nitrates with the colorimeter rather than require field acidification by local participants as the under estimation was identifiable from the parallel measures using the test strips.

Following initial testing and acceptance of the two analytical methods, the development effort was refocused on appropriate training materials for implementation in Adourékoman. Based on long-term interest in expanding this methodology to other regions of Bénin, and recognizing both that there are numerous languages spoken in different regions of Bénin and that significant portions of the population, particularly the adult population, have very low literacy levels, visual instructions (without required text) were created for each instrument. These instructions are based on a series of pictures taken of each test being performed and were improved throughout the project based on the suggestions of the local population. An example of these instructions is provided in Figure 18.1.

The final step of the development stage was the creation of a field data book. As with the visual instructions, design of the field data books used by the water monitoring teams required recognition of the desire to expand application to other regions of Bénin, as well as the literacy limitations of the local population. In order to address the language challenges, the analytical methods and the data book were color coded. Final design of the data sheets used swatches of color to identify the location on the data sheet to record results from a particular analytical method. Based on feedback from the local teams, a significant portion of space, approximately 2 cm in height, was provided on the data sheets to record the result for each test thus allowing for the larger handwriting of many of the participants. Further, the data sheets were printed on waterproof paper and boxes of pencils (required for use with the waterproof paper) were provided to the local monitoring teams. By working with the participants during the training period to design these sheets, all participants felt comfortable using them and were confident that the data sheets would be usable in other villages.

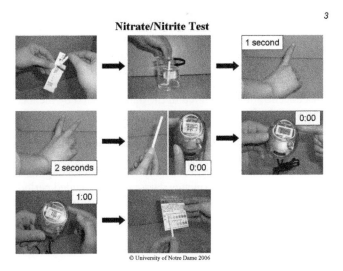

© University of Notre Dame 2006

Figure 18.1. Example of a page of the visual instructions created for the analytical methods. (See colour plate section).

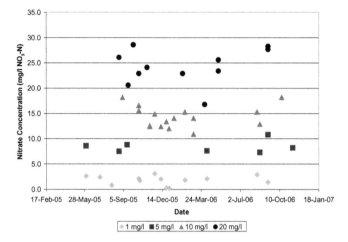

Figure 18.2.   Nitrate standards as analyzed in initial implementation of high-frequency sampling by groups in Adourékoman.

The second stage, initial implementation, specifically involved introducing and assessing the developed methods in Adourékoman. Trainees from the local population were formed into three teams, one team for each of the three wells in the village. Each of the teams was trained and then assigned to sample the well in their region of the village on a weekly basis. The three teams then performed weekly sampling with only periodic assessment of their methodology or oversight of their reliability. As of the writing of this manuscript, these three teams have been sampling (initially sampling was performed weekly and, more recently, biweekly based on early results) for a period of over 4 years.

In order to demonstrate that the trainees, who had no scientific background and little formal education, could accurately perform these methods (both immediately after being trained and throughout the period of sampling), each team was required to analyze blind nitrate standards (i.e., standards provided by the experts, but with concentrations unknown to the sampling teams) using the colorimeter on a monthly basis. Results from these standard measures (Figure 18.2 shows a typical portion of this data set) indicate that, while the measurements were subject to bias (at a level similar to that observed when the colorimeter was used by professionals), there is no obvious temporal trend in this bias. This is critical as it indicates that the local measurement teams, using the colorimeter, can provide consistent (if biased) measures of nitrate concentration over time, thus providing confidence in temporal trends in nitrate concentrations observed in the data collected by the local teams.

Based on the success of the first-level implementation, the third stage of the high-frequency monitoring, termed the second-level implementation, allowed expansion of the monitoring to an additional ten wells in four new villages during the summer of 2006. As with the initial implementation efforts, preliminary results from this expansion indicate local participants' ability to consistently perform the analytical methods through the regular analysis of blind standards and the collection of high-frequency data over an extended period of time. Results from the second-level implementation are not further discussed in this chapter.

18.4   RESULTS FROM THE INITIAL IMPLEMENTATION

Although three wells were included in the initial implementation, discussion of the utility of the high-frequency monitoring method is here based on results for only one of the wells

(Ayewa), the one with the lowest nitrate concentrations. This choice is made for two reasons. First, discussion of all three sets of results is beyond the scope and page limitations of this manuscript. Second, the results from the chosen well provide for the most interesting results in terms of both positive outcomes (for example, the identification of temporal trends) and negative outcomes (for example, the reason for, and impact of, gaps in the data). Hence, only data from well Ayewa are presented. Details of results for the other two wells (Ayewa-Okota and Agbo) are presented in Crane (2007).

The high-frequency colorimeter and test strip data for nitrate for well Ayewa are displayed in Figure 18.3. In order to understand the reported test strip data, it is important to note that the local sampling teams could report the test strip result in one of two fashions. For some samples, the team considered the particular result to be associated predominantly with one of the six concentrations listed on the color chart provided with the test strips (0.5, 2.0, 5.0, 10.0, 20.0, or 50.0 mg/ℓ NO_3-N). In this situation, the team reported a single value; that value was entered into our data sets. For other samples, the team felt that the result fell between two possible outcomes (e.g. between 5.0 and 10.0 mg/ℓ). In this case, the value entered into our data set was the average of the two possible outcomes (e.g. a 5.0–10.0 mg/ℓ entered on the field chart would be recorded as 7.5 mg/ℓ in our data base).

The results for the period July 2004, through July 2008, demonstrate that the data collected using both the colorimeter and test strips generally show similar magnitude in the nitrate concentration. Select points (particularly for the test strip data) which show very high concentrations are considered to likely be caused by transcription errors in the field records (in particular, the omission of a period in the recorded data would result in 5.0 mg/ℓ being transcribed as 50 mg/ℓ and 2.0 mg/ℓ as 20 mg/ℓ).

Figure 18.4 shows the colorimeter data plotted without the test strip results. The two gaps in the data relate to periods during which the well was not functional. This figure demonstrates an interesting long-term variation in the nitrate concentration in this well. Specifically, these data provide initial suggestion of some form of periodic variation that does not appear to coincide with seasonal variation in climate or related village routines (e.g. the rainy season versus the dry season, or periods of agricultural activity). Study of the source of this seasonality continues.

In addition to these technological methods, sociological methods involving surveys, focus groups, and observation were used to locate concentrated sources of human or animal activity (and wastes) throughout the village (Crane, 2006, 2007). This resulted in the identification

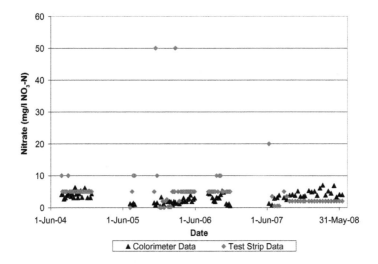

Figure 18.3.   Nitrate data from well Ayewa in Adourékoman as collected by the local team.

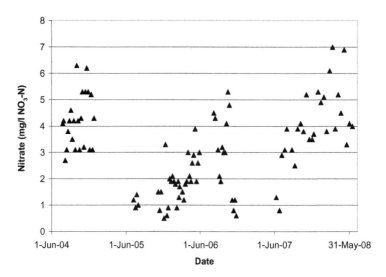

Figure 18.4.   Nitrate data from well Ayewa in Adourékoman as collected by the local sampling team using the colorimeter.

of four possible sources of nitrate contamination consistent with the isotope and high-frequency well data: (i) toilet practices (most people do not use latrines, rather they use certain geographical regions typically on the edge of the village; there are only four latrines in the village), (ii) animal shelters for pigs, cows and oxen, (iii) trash piles (also a toilet location for young children), and (iv) open hand dug wells. Further research (including use of direct push sampling methods) allowed for the elimination of one possible source—hand dug wells filled in with debris and soil by the local population. Hence, the results of the field measurement of nitrate have lead directly to identifying toilet practices, garbage disposal, and animal shelters as the most likely sources of nitrate in Adourékoman.

18.5   ANALYSIS OF THE INITIAL IMPLEMENTATION

While the data provided for a number of technical observations (see Crane, 2006, 2007), current discussion revolves around analysis of the strengths and weaknesses of working with the local population in Adourékoman to enable the collection of high-frequency nitrate data. Of interest at the beginning of this research were answers to the following questions:

1. Can a team of local residents be trained to properly perform field measures of nitrate concentration?
2. Will such a team reliably perform these measures on a weekly basis without substantial oversight of their efforts?
3. Will the team's technique remain adequate for times significantly beyond the original training period?
4. Will the resulting data be scientifically useful?

Initial implementation provides substantial insight into each of these questions. First, the training of the three monitoring teams in Adourékoman demonstrates that such training is possible, particularly when the training materials are based on the literacy and scientific knowledge of the local teams (in this case, through use of visual training materials). However, as discussed by Crane (2006), this training required both patience to develop a critical level of trust between the project experts and the monitoring teams, and the willingness of the experts to listen to, and incorporate, changes to the training materials and field methods as suggested by the local teams. Both trust and collaboration were critical in this project to develop ownership of the monitoring program within the three teams in Adourékoman. Project ownership was demonstrated by the local teams independently identifying and reporting problems within the field methods (in this case predominately related to failure of select reagents and test strips due to storage of materials in a tropical climate without benefit of air conditioning) as well as the teams' desire to continue the monitoring efforts after the official end of the research program (in summer 2007).

The observation that the teams were willing to perform sampling on a weekly basis for an extended period of time is demonstrated through the data provided in Figures 18.2–18.4. Two significant observations are made relative to the activities of the teams in the initial and second-level implementations. First, all but one of the 13 teams involved in the project (initial and second-level implementation) completed their measures at the appropriate frequency except during periods when individual wells were closed for repair. The one team that did not complete their weekly efforts involved a pair of local residents in which the female team member had a baby early in the monitoring period and was not able to support the effort; the other member of the team was not able maintain the monitoring. Second, the teams indicated interest in receiving compensation for the monitoring. However, despite this strong interest in compensation, each of the teams completed their assigned measurements without compensation throughout the agreed upon period of this study. Finally, the teams continued to demonstrate proper performance of the analytical methods as evidenced both by the continuing analysis of blind standards and through examination of their technique by observers during the summers of 2005, 2006 and 2007.

Given the demonstrated ability of the local teams to collect data, perhaps the most critical of the questions posed above is the final question, do the resulting data sets have scientific validity and utility? This question was addressed through analysis of the data derived from the initial and second-level implementations as well as through the theoretical study discussed briefly above with details in Crane and Silliman (2009). The assessment of the field data, as introduced above through analysis of the data for well Ayewa, provided substantial support for the validity and utility of the data. The consistent (although biased) results for the blind standards indicate that the resulting measures of nitrate in the well water samples were performed consistently throughout the study period. The consistency of the test strip and colorimeter data (specifically, the low rate of significant difference in between the two measures of nitrate) provides further support for the validity of the data.

In terms of utility of the data, the assessment of the well Ayewa data demonstrates that the high-frequency data provide insight into temporal variability in nitrate concentration that would not be possible based on sampling by professionals at a frequency of one or two samples per year. Hence, there is strong evidence that these data have significant utility, at least in terms of identifying temporal trends at time scales of weeks to months: it is difficult to imagine methods not involving participation by the local population allowing identification of trends at these time scales in rural Africa.

18.6 BRIEF COMPARISON WITH ANOTHER RECENT STUDY ON USE OF SIMPLE MEASURES

Following the completion of our study, an independent study on the use of nitrite/nitrate test strips was published in an epidemiology journal (Nielsen et al., 2008). Based on sampling for nitrite/nitrate in tap water in the state of Washington (USA), the study by Nielsen et al involved use of test strips to obtain single point in time measures of nitrate in the tap water by private residents randomly selected for the study. Beyond being asked to run duplicate test strip analyses on tap water, these residents were asked to collect, store, and ship tap water samples for analysis of nitrite and nitrate at a professional water quality laboratory.

The study by Nielsen et al. (2008) was performed with several major differences relative to the present study. These include: (i) the residents were not trained prior to use of the test strips (they were required to read instructions and then apply the method), (ii) the residents had only one opportunity to apply the test strips without prior practice or oversight, (iii) the residents used only one method for measurement of nitrate/nitrite (versus use of test strips and the colorimeter in the present study), (iv) the range of concentrations of nitrate present in the tap waters in the Washington study were substantially lower than the range present in Adourékoman, and (v) the goal of the Washington study was to determine agreement between single point in time test strip results run by private residents and results performed on samples collected at approximately the same point by laboratory technicians (versus our goal of seeking variation in nitrate concentration over time).

Despite these differences in methods, these authors note a number of conclusions similar to conclusions drawn from our theoretical and field studies. They note, for example, that the residents were generally able to understand and apply directions for the nitrate/nitrite test strips. As illustrated by high agreement in duplicate results using the test strips, the Washington study also demonstrated that the residents were generally able to reproduce measurements. Finally, these authors argue that test strips may be a valuable tool under conditions for which more sophisticated measures are not possible.

In reviewing the conclusions from the Washington study versus the results presented in this chapter (and in Crane et al., 2009), the differences in the methodology also lead to an interesting, complementary difference in final conclusions from these two studies. Specifically, Nielsen et al. (2008) conclude that test strips should preferentially be limited to use as a screening tool. This conclusion is a result of the single use of the test strips by the residents and the primary measure of utility of the test strip being direct comparison with laboratory measures performed on water samples from the same sources. In contrast, the field studies performed in Bénin imply that simple measures (test strips and colorimetry) can provide high quality estimates of temporal variation in nitrate. While these two conclusions at first appear to be conflicting, the theoretical work provided in Crane et al. (2009) argues that the utility of simple measures is directly related to the parameter to be estimated. Specifically, Crane et al. (2009) demonstrate that simple methods applied to integrated parameters such as average concentration over a large number of samples can provide high-quality estimates. It may be argued that trend analysis as used in our field study is this type of integrated parameter. Crane et al. (2009) continue to argue that single point parameters such as maximum concentration over a sampling period are not well represented by simple measures. It may be argued that one-time agreement of test strip results with laboratory measures is closer to this single point type of parameter. Hence, integration across the three studies (our field work, Crane et al., 2009 and Nielsen et al., 2008) provides insight into conditions under which simple measures may represent valuable tools for collecting high-quality scientific data sets (e.g. integrated parameters) versus conditions for which these same measures are best used as screening instruments (e.g. point parameters).

18.7   CONCLUSIONS

The results obtained in the Bénin case study have demonstrated how involving the local community as critical partners in the process of data collection may provide a critical tool in addressing the challenge of monitoring water quality in rural Africa. Combined with the theoretical efforts reported in Crane and Silliman (2009), these applied results imply that local monitoring may thus offer a new strategy for pursuing water quality monitoring efforts in rural Africa. Details of this case study (i) demonstrated the ability of local monitoring groups to accurately and consistently perform simple analytical methods over an extended period of time, (ii) allowed identification of four likely sources of nitrate in the groundwater of Adourékoman, including identification of trends in the nitrate concentrations in at least one of the wells, and (iii) demonstrated the utility of combining a range of methods, sampling frequencies, and sampling expertise in developing a viable data base for assessment of groundwater quality at a rural field site. Most importantly, this case study provides the opportunity, and the justification, for discourse on rural water quality monitoring in Africa to take an important step towards including local participation in the monitoring process.

REFERENCES

Canter, L.W. (1997). Nitrates in Groundwater, New York: Lewis Publisher, p. 263.

Crane, P.E. (2006). Implementation of a Sustainable Groundwater Quality Monitoring Program in Rural Bénin, West Africa, Master's Thesis, University of Notre Dame.

Crane, P.E. (2007). An Investigation of Data Collection Methods Applicable in Groundwater Research in Rural Regions of Developing Nations, Ph.D. Dissertation, Department of Civil Engineering and Geological Sciences, University of Notre Dame (http://etd.nd. edu/ETD-db/theses/available/etd-12192007-181805).

Crane, P.E. and Silliman, S.E. (2009). "Sampling Strategies for Estimation of Parameters of Ground Water Quality", Ground Water, in press.

Hossain, M.F. (2006). "Arsenic contamination in Bangladesh—An overview", Agriculture, Ecosystems and Environment, 113: 1–16.

L'Hirondel, J. and L'Hinrondel, J.-L. (2002). Nitrate and Man: Toxic, Harmless or Beneficial, New York: CABI Publishing, p. 168.

Knobeloch, L., Salina, B., Hogan, A., Postle, J. and Anderson, H. (2000). "Blue Babies and Nitrate-Contaminated Well Water." Environmental Health Perspectives 108(7), 675–8.

Masoner, J.R. and Mashburn, S.L. (2004). Water Quality and Possible Sources of Nitrate in the Cimarron Terrace Aquifer, Oklahoma, 2003", USGS/SIR, 2004-5221.

Mueller, D.K. and Helsel, D.R. (1996). Nutrients in the Nation's Waters—Too Much of a Good Thing?, U.S. Geological Survey, Circular 1136.

Nielsen, S.S., Mueller, B.A. and Kuehn, C.M. (2008). "An evaluation of semi-quantitative test strips for measurement of nitrate in drinking water in epidemiologic studies", Journal of Exposure Science & Environmental Epidemiology, 18(2), 142–148.

Nolan, B.T., Ruddy, B.C., Hitt, K.J. and Helsel, D.R. (1997). "Risk of Nitrate in Groundwaters of the United States—A National Perspective", Environmental Science and Technology 31: 2229–2236.

Schoenen, D. (2002). "Role of disinfection in suppressing the spread of pathogens with drinking water: possibilities and limitations", Water Research, 36: 3874–3888.

Sepulveda, J., Valdespino, J.L. and Garcia-Garcia, L. (2006). "Cholera in Mexico: The paradoxical benefits of the last pandemic", International Journal of Infectious Diseases 10: 4–13.

Silliman, S.E., Boukari, M., Crane, P.E., Azonsi, F. and Neal, C. (2007). "Observations on Element Concentrations of Groundwater in Central Bénin", Journal of Hydrology, 335(3–4), 374–388, 2007 (dx.doi.org/10.1016/j.jhydrol.2006.12.005).

Somasundaram, M.V., Ravindran, G. and Tellam, J.H. (1993). "Ground-Water Pollution of the Madras Urban Aquifer, India", Ground Water, 31 (1): 4–11.

UNDP. (2006). "World water and sanitation crisis urgently needs a Global Action Plan." United Nations Development Programme, web publication: http://content.undp.org/go/newsroom/november-2006/hdr-water-20061109.en.

Wakida, F.T. and Lerner, D.N. (2005). "Non-agricultural sources of groundwater nitrate: a review and case study", Water Research, 39: 3–16.

WHO. (2004). Uranium in Drinking-water: Background document for development of WHO Guidelines for Drinking-water Quality. World Health Organization, p. 15.

WHO. (2006). Guidelines for Drinking-water Quality: incorporating first addendum, 3 ed. Volume 1: Recommendations. World Health Organization, web publication: (http://www.who.int/water_sanitation_health/dwq/gdwq3rev/en/index.html).

19

Charitable endowments as an institute for sustainable groundwater development and management

Gaathier Mahed
Department of Earth Sciences, University of the Western Cape, South Africa
Council for Geoscience, Pretoria, South Africa

Yongxin Xu
Department of Earth Sciences, University of the Western Cape, South Africa

ABSTRACT: Charitable endowments (Waqf) are a community-based sustainable development initiative. The principle of charitable endowments has its roots entrenched in sustainability. The beauty of it is the fact that the donated asset is never lost but instead the returns stemming from the charitable endowments are utilized. Groundwater management has always posed numerous problems. Everything from financing a water project through to the maintenance of infrastructure incurs cost. Charitable endowments, in conjunction with other Islamic principles such as Shura, still find application in the water resources management field in Middle East countries and make a major contribution to the sustainable management of local groundwater resources through solving the problems relating to the financing of groundwater projects. This requires a multi-faceted approach, in which the clergy, engineers, teachers, social scientists, financial managers as well as media work together in a holistic manner. This chapter examines some of the technical, socio-economic, financial as well as religious issues to illustrate the all-encompassing nature of this subject matter. In the end the principle of Shura (mutual consensus) in conjunction with waqf and Islamic law are the determining factors for the success of any groundwater project.

19.1   INTRODUCTION

Water is fundamental for all life systems and the all living entities are composed from this finite resource, as shown in the Quran. Groundwater is of particuler importance due to the fact that it constitutes a much greater proportion of the total freshwater supply, when compared to surface water. (Miller, 2002)

Local groundwater resources, because of their ubiquitous nature, are the main source for community supply throughout the world and particularly in the semi-arid and arid regions. Figure 19.1 shows the various types of wells or boreholes which were used for groundwater utilization in Libya. Sustainable development principles emphasize the need for local management of such resources. Public participation is a key principle in this regard, but its systematic application has remained a challenge for groundwater management world-wide, particularly where the poorest communities are involved. It is thus important to link up to value systems and approaches of traditional societies that may be supportive of sustainable management at local level. These societies may include various religious organisations. The charitable endowments, still practised in Islamic countries, is an important and well documented example in this regard.

Charitable endowments (waqfs) or pious endowments (Abouseif, 1994), an Islamic based financial tool, has its roots based in charitable donation of assets that have been utilized to fund large scale projects. Waqfs originated in the country of Saudi Arabia around 1400 years ago

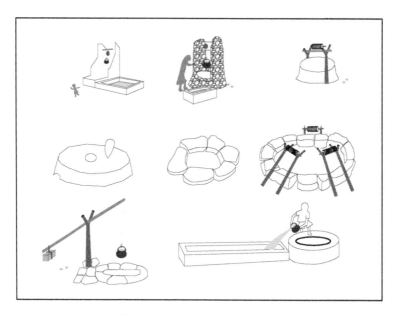

Figure 19.1.   A sketch showing different types of wells used in Libya. (See colour plate section).

during the time of the Prophet Muhammad (Peace be upon him). The first ever call for a waqf was answered with the donation of agricultural land occupied by date palms (Khan, 1971). The underlying principle for the charitable endowments is the fact that the donated asset will never be sold, but the returns stemming from the asset will be used to serve the community. Water waqfs work slightly differently due to the fact that no real financial returns occur. The asset, which is the borehole, is purchased and donated in order for the community to utilize it for their development. An excellent example, among several, is the well of Rumah, which was bought from the Jews. This well was given to the muslims of Madinah by Sayyidinah Uthman at the time of the Prophet Muhammed (Peace Be Upon Him) (Khan, 1971). This shows that the waqf is a selfless act done purely for the sake of uplifting the community.

Sami and Murray (1998) as well as Abdurrazak et al. (2005) called for an effective, community based, sustainable management tool for groundwater. This is where charitable endowments can come into play. For example, it has been shown to act as an institution for the development and management of the Qanats (underground aquaducts) in the Middle East (Faruqui et al., 2001). These underground aquaducts continue to sustainably irrigate farmlands. It is critical that the community is involved in the management of the resource, as it directly affects them. The Islamic term for this process of consultation is known as Shura. Shura is of the utmost importance for the understanding of the requirements of the community and developing an effective management strategy catering to their specefic needs.

The waqf infrastructure in conjunction with the aforementioned principle of Shura perfectly compliment each other. Furthermore, if the financial aspects of the charitable endowments are included we see a powerful resource management tool. The wealth generated by the waqf system was so great that major cities had their infrastructure built and maintained by charitable endowments (Abouseif, 1994).

The well of Rumah, as previously mentioned, was one of the first charitable endowments proving the importance of water for the survival of any community. This community-based approach is the underlying principle behind charitable endowments. Groundwater lends itself to the charitable endowments approach, it is cheap to develop and use and can be brought into supply systems with very little capital investment and wells can be easily drilled whenever the water is needed thereby allowing more freedom to operate and control

(Abdulrazzak et al., 2005). This would mean that the financial strains placed on the charitable endowments would be much less than the funding required for developing surface water impoundments. Unfortunately, in the case of groundwater, the amount of water that can be sustainably pumped out is a major limiting factor.

19.2 PRACTICE IN THE MIDDLE EAST

Countries within the Middle East and North African (MENA) region are the major focus due to the following important points:

- Firstly, most of the countries have an Islamic state in place and thus charitable endowments have played a major role in the development of infrastructure in these countries (Baer, 1984).
- Secondly, the general rate of evapotranspiration in the region by far exceeds the rate of precipitation, and thus, groundwater supplies are playing a pivotal role.
- Lastly, due to an increasing population, particularly in the MENA region, and dwindling freshwater supplies, management of the limited water resources is becoming increasingly important (Faruqui et al., 2001).

These water resources problems have to be tackled together with numerous other socio-political issues within the region. The solutions have to take a large variety of factors into consideration. Therefore the waqf body in the region in relation to other political bodies and NGO's has played a major role in groundwater development. In light of this background, charitable endowments should not just be viewed as a financial tool but instead as an all- encompassing tool for socio economic development (Al Melhem, 2007). Furthermore the management of transboundary water ways in the MENA region, could initiate co-operation and prevent future conflicts as well as create broader benefit sharing (Phillips et al., 2006).

19.3 SOCIO ECONOMIC ISSUES

The Dublin principles of 1992, clearly state that the involvement of women and children in the management of water, is critical for the advancement of societies. From an Islamic viewpoint, women are held in high esteem, due to their important role in nurturing the leaders of tomorrow. Therefore their needs should always be taken into consideration. Thus the mutual consensus (Shura) amongst individuals within a society is important, in order to address their needs as well as for the development of the community as a whole. This process of Shura was practiced by the Prophet S.A.W and should be utilized more broadly Furthermore, it creates transparency and accountability, two issues that are seriously lacking in our societies. Shura also allows the communities to expose issues that affect them, through discussion, and thus allows people to solve their pertinent problems through mutual consultation.

Broadly speaking, groundwater can readily address the basic needs for water, due to its relatively easy accessibility. This allows, inter alia, that a borehole can be sunk closer to a village, and less time can be spent on collecting water. Thus the women and children, who usually collect the water in rural African areas, could, as a result, direct their efforts towards some important issues such as education, and ultimately the upliftment of their community and society at large. It is interesting to note that one third of all charitable endowments, in Turkey in the 16th century were actually donated by females (Baer, 1984). This indicates that women had property rights, unlike in many western cultures of the past, and they would utilize these for the upliftment of the community. It also meant that they had power to effect change, irrespective of the scale. Furthermore Islam still affords those rights to women today.

Education relating to groundwater management on a micro-scale is of critical importance. This would have to involve the schools, and incorporate a basic understanding of groundwater at grassroots levels. Thus the implementation of awareness-raising relating to water management

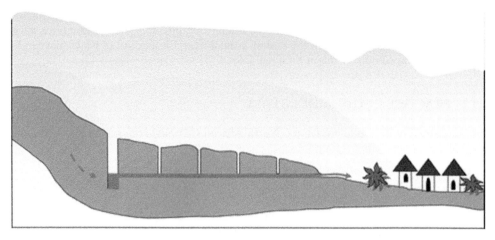

Figure 19.2.   A sketch of cross section of a Qanat or Karez system in Asia. (See colour plate section).

in general would have to be done in school curricula as well as in general media. The clergy could also be brought into play due to their great influence among people (Faruqui et al., 2001).

The construction and management of a Qanat (also known as Karez in China) (Figure 19.2) is a good example of the application of the principle of Shura. Faruqui et al. (2001) clearly explained the entire process. It is initiated with Shura, within the community and the payment of individuals who distribute the water, is then done by means of water. All interested and affected parties would then decide on who should distribute the water These distributors are then able to irrigate their arable lands and grow crops. Everything is done in an effective manner and at a micro-scale. This somehow relieves the main duty of the water ministry as the people would literally be governing themselves. It is estimated that about 75% of all water supplies in Iran stems from Qanats. Similar systems can be found throughout the MENA region (Todd and Mays, 2005). However, it is unfortunate that the construction of a Qanat is extremely laborious and a borehole seems to be a much simpler modern solution.

This is where charitable endowments could come into play to help matters. It would play a role in terms of financing the initial stages of a groundwater supply project as well as the maintenance of other related infastructures. Finally, charitable endowments would also aid in monitoring groundwater quality. Costs associated with groundwater development and operation activities are usually less than respective surface water supplies. Furthermore, irrespective of the source of the supply, costs are fully recoverable as outlined by Islamic law (Faruqui et al., 2001) Thus all funds from charitable endowments projects are effectively utilized.

Some NGO's, such as the Africa Muslim Agency, are also involved with water and sanitation projects funded by waqf. Once they have generated some funds, a certain percentage is then invested and its returns further used for maintenance purposes. This is a self-sustainable approach to resource management, which in turn is the fundamental principle of charitable endowments. These types of projects have a great impact on most afflicted communities and they effectively help in relieving the responsibilities of governments in the SADC region. In many instances NGO's like this are the main controlling bodies for water waqfs in the impoverished regions of the world, like the Indo Pak sub continent.

19.4   TECHNICAL ISSUES

Numerous technical issues are evident when utlising groundwater. These include factors such as recharge, sustainable yield and water quality among others In the MENA region, Qanats have been successfully developed for this purpose and the infrastructure utilizes gravity flow and not pumping (Figure 19.1). The next step would then be to establish a long term sustainable management of the aquifer. It is important to note that maintenance of infrastructure as well as supply of the water should have to be continuously financed.

The supply management of water is extremely important for sustainability issues. Recharge rates determine the amount of water, which can be sustainably extracted from the aquifer over a specific period of time. Groundwater resources need to be strictly monitored because if abstraction exceeds replenishment, then the amount of groundwater available for the subsequent year would dramatically be reduced (Merret, 2005). A unique exemption to this rule is the water stemming from a well in Makkah. Engineers cannot seem to fathom the source of the Zam-Zam, nor can they explain the infinite nature of the resource. The distribution of this water is solely financed by charitable endowments. Furthermore, the mosques in the Kingdom of Saudi Arabia receive Zam-Zam on a regular basis (Shareef, 2007).

Unfortunately the aforementioned principles of the Zam-Zam do not apply to the general management of groundwater. This is due to the finite nature of the resource. Therefore a shift has to be made from supply management to demand management. The prophet Muhammad (S.AW) used a 2/3 litre of water for wudhu (ablution) and about 1.5 litres for bathing. This is a perfect water demand management system whereby the total amount of water utilized by the masses is kept to minimum levels. In this case, the successful implementation of such a strategy relies upon the clergy. This has been proven to be effective, most recently in Jordan, whereby the pulpit has been used in conjunction with mass media with great efficacy (Faruqui et al., 2001).

19.5   CONCLUSION

Overall, it could be said that waqf is a unifying characteristic within the MENA region. This therefore justifies its use as a tool for solving the major problems with regard to water, and specifically groundwater supply and management. Above all, in order for such a system to work, other Islamic water laws also have to be put into place. This is important due to the manner in which the waqf system, Shura and islamic water laws work in conjunction with one another (Faruqui et al., 2001). The implementation of such a strategy would be critical for the development of the region and its economy.

The use of waqf infrastructure, such as mosques and Islamic schools for educating communities with regards to groundwater use and management could be of paramount importance. Furthermore, the use of some monetary aspects of charitable endowments for financing groundwater projects is just as vital as using the water itself. Therefore, charitable endowments provides a holistic solution for the sustainable development of groundwater resources.

Finally, solutions and strategies deduced from the MENA region could be practically applied to great effect in the Southern Africa region because of the similarity in climatic conditions between the two areas. It could also be said that this community-based approach would find good standing on the African soil at large because of its tightly-knit social fabric. It seems that charitable endowments could provide some of the financial, social and political solutions to the problems relating to groundwater in general.

REFERENCES

Abdulrazzak, M.J., Al Weshah, R. and Al Zubari, W. (2005). Prospective on the implications of over development of groundwater resources in Arab countries, Proc Int Workshop on Management and Governance of Groundwater in Arid and Semi-Arid Countries, Cairo Egypt 03–07 April, 2005.

Abouseif, D.B. (1994). Egypts adjustment to Ottoman rule, Institutions, Waqf and Architecture in Cairo (16th and 17th Centuries), E.J. Brill, Leiden, The Netherlands.

Al Melhem, K.A.R. (2007). Kuwait Awqaf Public Foundation Waqf Experience, I In conf proceeding International Waqf Conf of Southern Africa 17–18 August 2007, Cape Town, South Africa.

Baer, G. (1984). Women and Waqf: An analysis of the Istanbul Tahrir of 1546 in Studies in Islamic Society, Warburg and Gilbar (eds), Haifa Universty Press, Haifa University, 3999, Haifa, Israel.

Faruqui, N.I., Biswas, A.K. and Bino, M.J. (eds) (2001). Water Management in Islam, United Nations University Press, ISBN 92-808-1036-7.

Fetter, C.W. (2001). Applied Hydrogeology (Fourth Edition), Prentice Hall Inc, Upper Saddle River, New Jersey ISBN 0-13-122687-8.

Figueres, C.M., Tortojada, C. and Rockstrom, J. (eds) (2003). Rethinking water management-innovative approaches to contemporary issues, Earthscan Publications Ltd, 120 Pentonville Road London ISBN 1-85383-994-9.

Khan, M.M. (1971). The translation of the meanings of Sahih Al Bukhari Arabic-English Vol III, Sethi Straw Board Mills (Conversion) Ltd, Gujranwala, Pakistan.

Merret, S. (2005). The economics of groundwater, Proc Int Workshop on Management and Governance of Groundwater in Arid and Semi-Arid Countries, Cairo Egypt 03–07 April, 2005.

Miller, G.T. (2002). Living in the Environment (12th Edition), Thomson Learning.Wadsworth/Thomson Learning, 10 Davis Drive Belmont CA, ISBN 0-534-37697-5.

Phillips, D., Daoudy, M., McCaffrey, S., Ojendal, J. and Turton, A. (2006). Trans-boundary Water co-operation as a tool for conflict Prevention and Broader Benefit sharing, Global Development studies No. 4, Prepared for the Ministry of Foreign Affairs Sweden.

Sami, K. and Murray, E.C. (1998). Guidelines for the Evaluation of water resources for rural development with an emphasis on groundwater in Water Research Commission (WRC) Report No 677/1/98.

Shareef, A.A. (2007). Awqaf in Makkah Mukarramah: Problems and Prospects, In conf proceeding International Waqf Conf of Southern Africa 17–18 August 2007, Cape Town, South Africa.

Todd, D.K. and Mays, L.W. (2005). Groundwater Hydrology (3rd Edition), John Wiley & Sons, 111 River Street, Hoboken, NJ, ISBN 0-471-05937-4.

Weight, W.D. and Sonderegger, J.L. (2001). Manual of applied Field Hydrogeology, McGraw-Hill, Two Penn Plaza, New York, NY, ISBN No: 0-07-06-9639-X.